A COLOR ATLAS OF PHOTOSYNTHETIC EUGLENOIDS

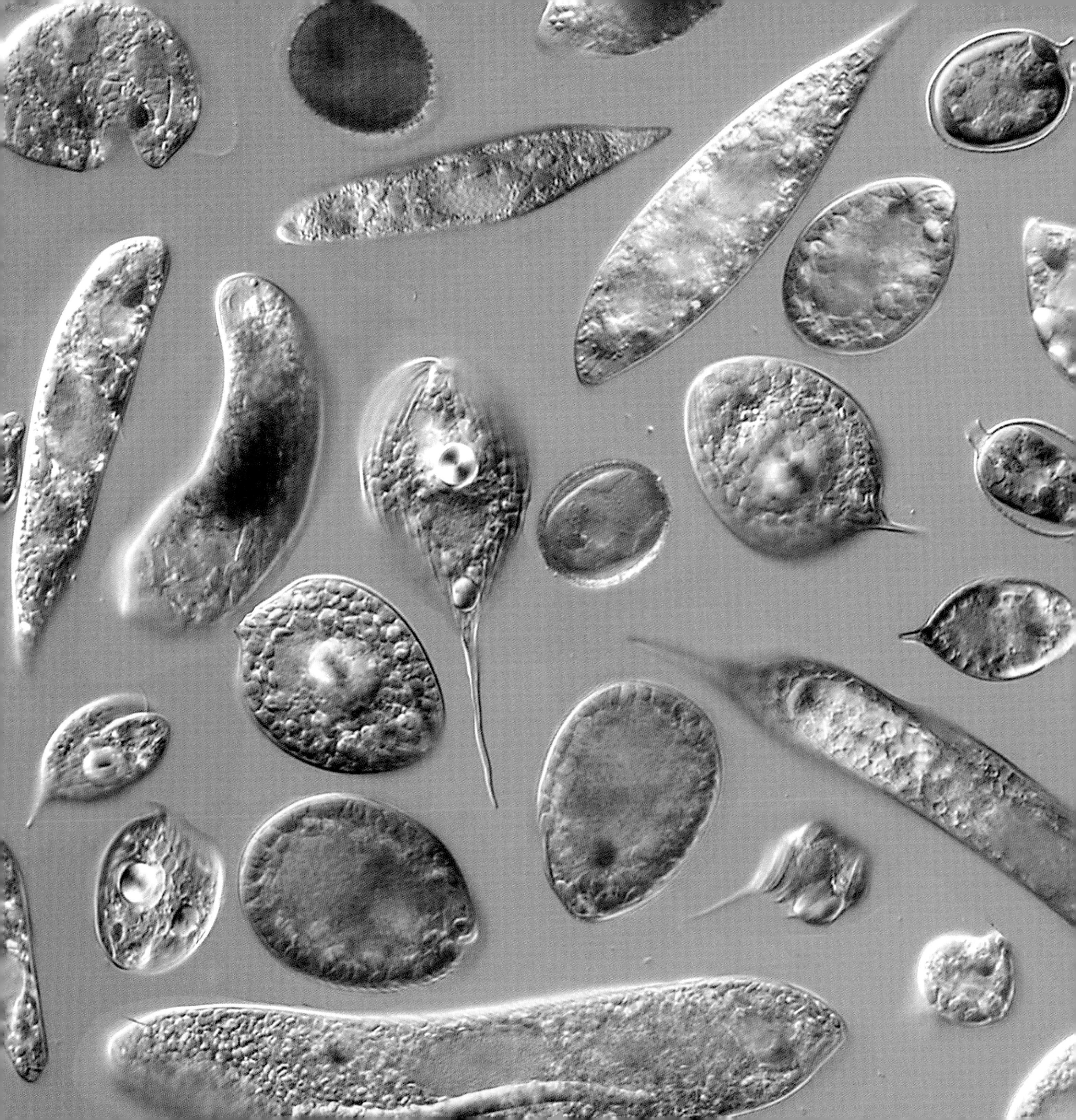

A COLOR ATLAS OF PHOTOSYNTHETIC EUGLENOIDS

Ionel Ciugulea
Richard E. Triemer

MICHIGAN STATE UNIVERSITY PRESS • EAST LANSING

This book was made possible by the generous support of the Phycological Society of America.

MICHIGAN STATE
UNIVERSITY

Publication of this book was supported by an award from the MSU Foundation.

ACKNOWLEDGMENTS

The authors wish to thank all of the former members of the Triemer laboratory, Caroline Becker, Stacy Brosnan, Eric Linton, Caitlin Lowery, Anna Monfils, Maria Alejandra Nudelman, and Woongghi Shin, for providing some of the photos that appear in this book.

We are grateful to Mathew Bennett for maintaining the cultures used during the preparation of the *Atlas* and for his comments, suggestions, and valuable discussions.

We would like to express a special thanks to Dr. Bozena Zakryś of Warsaw University and Dr. Visitacion Conforti of the University of Buenos Aires for the many hours they spent working with us on interpretation, identification, and taxonomy of taxa presented in this book.

We thank the Phycological Society of America for their generous support and sponsorship of this publication. We would especially like to thank the production team of Julie Loehr, Annette Tanner, and Kristine Blakeslee at MSU Press and Charlie Sharp of Sharp Des!gns for all of their help in assembling and refining the *Atlas*.

The authors wish to acknowledge the financial support provided by the National Science Foundation PEET program (Partnership for Enhanced Expertise in Taxonomy, grant no. DEB 4-21348). Many of the images shown in this book were taken as part of research projects funded by the PEET grant.

∞ The paper used in this publication meets the minimum requirements of ANSI/NISO Z39.48-1992 (R 1997) (Permanence of Paper).

Michigan State University Press
East Lansing, Michigan 48823-5245

Printed and bound in China.

16 15 14 13 12 11 10 1 2 3 4 5 6 7 8 9 10

LIBRARY OF CONGRESS CATALOGING-IN-PUBLICATION DATA

Ciugulea, Ionel.
A color atlas of photosynthetic euglenoids / Ionel Ciugulea and Richard E. Triemer.
p. ; cm.
Includes bibliographical references and index.
ISBN 978-0-87013-879-9 (cloth : alk. paper)
1. Euglenoids—Atlases. 2. Photosynthesis—Atlases. I. Triemer, Richard E. II. Title.
[DNLM: 1. Euglenida—Atlases. 2. Photosynthesis—Atlases. QK 569.E93 C581c 2010]
QK569.E93C58 2010
579.8'40222—dc22
2009031986

Cover and book design by Charlie Sharp, Sharp Des!gns, Lansing, Michigan

green press INITIATIVE Michigan State University Press is a member of the Green Press Initiative and is committed to developing and encouraging ecologically responsible publishing practices. For more information about the Green Press Initiative and the use of recycled paper in book publishing, please visit *www.greenpressinitiative.org.*

Visit Michigan State University Press on the World Wide Web at *www.msupress.msu.edu*

DEDICATION

The authors would like to dedicate this publication to those in our personal lives who made this possible.

I would like to dedicate this work to my wife Linda, whose patience, support, and encouragement keep me going. Her enthusiasm for and interest in the "little green beasties" we work with continues to astonish me.

RICHARD E. TRIEMER

I dedicate this book to my sons, Radu and Andrei, as an encouragement to fulfill their aspirations and to my wife, Mioara, for her continuous and energetic support.

IONEL CIUGULEA

CONTENTS

INTRODUCTION

A HISTORICAL PERSPECTIVE ON THE GENERA

In 1674, Antonie van Leeuwenhoek described the very first microscopic protist as an organism "green in the middle and at either end white." Müller later named this creature as *Cercaria viridis* (1786), which subsequently became *Euglena viridis* (Ehrenberg 1830). This led to a series of studies and discoveries such that, by the end of the nineteenth century, most of the major photosynthetic euglenoid genera were described.

Euglenoid systematics began with Ehrenberg describing several of the major photosynthetic taxa: *Euglena* (1830), *Cryptoglena* (1831), *Colacium* (1833), and *Trachelomonas* (1833). The genus *Phacus* was added in 1841 by Dujardin, followed by Perty's description of *Lepocinclis* and *Eutreptia* in 1852 and Mereschkowsky's description of *Monomorphina* in 1877. The remaining photosynthetic genera were established in the twentieth century. Da Cunha added the genus *Eutreptiella* in 1914, and Deflandre created the genus *Strombomonas* in 1930. Finally, Triemer erected the genus *Discoplastis* in 2006, and Karnkowska, Linton, and Kwiatowski established the genus *Euglenaria* in 2010.

Stein (1878) made the first attempt to organize euglenoid systematics and recognized four groups of euglenoids based on the presence or absence of the chloroplast and the nutritional mode. Bütschli (1884) was the first to consider euglenoids equal in rank with the other protists and placed them at the suborder level (Euglenoidina) with protists such as *Mastigamoeba, Trypanosoma, Bicosoeca, Dinobryon, Uroglena* (Monadina), *Bodo, Entosiphon* (Heteromastigoda), *Synura, Chlamydomonas, Haematococcus, Phacotus, Gonium, Pandorina, Eudorina, Volvox,* and *Cryptomonas* (Isomastigoda). Bütschli included also in Euglenoidina genera such as *Chromulina, Gonyostomum,* and *Vacuolaria*, which are no longer recognized as euglenoids. Several other taxonomic schemes were created during the late 1800s and early 1900s (e.g., Klebs 1883, Blochmann 1895, Senn 1900, and Lemmermann 1913) that proposed various groupings of the euglenoid taxa. The taxonomy commonly used today has its basis in that proposed by Hollande (1942). He created a taxonomic system based on the body shape, nutritional mode (phototrophic, osmotrophic, or phagotrophic), flagellar structure, and degree of metaboly (the ability of the cell to change its body shape through peristaltic movements). Leedale (1967) added new physiological and electron microscopic information into Hollande's scheme. Leedale's system (emended in 1978) recognized six separate orders based on the mode of nutrition, flagellar structure, and pellicle (cell surface) construction. Starting in 1987, a major step was made in taxonomy when molecular data (SSU rDNA) was used to construct phylogenies (Sogin and Gunderson 1987). In 1997, Montegut-Felkner and Triemer developed the first molecular phylogeny of euglenoids based on the SSU rDNA sequences from four taxa (one photosynthetic).

Subsequent studies focused largely on SSU rDNA, and as more taxa were added to the dataset, it became clear that several of the recognized genera were paraphyletic or polyphyletic (Linton et al. 1999, 2000). In the beginning of the twenty-first century, new research was done on relationships among the phagotrophic and osmotrophic genera showing that the photosynthetic taxa formed a monophyletic group (Preisfeld et al. 2000, 2001; Müllner et al. 2001; Busse and Preisfeld 2003).

A major taxonomic revision was presented by Marin et al. (2003) in which most of the photosynthetic genera were emended using morphological characters and molecular signatures in taxon diagnosis. They resurrected the genus *Monomorphina* and also dissolved the genus *Strombomonas* back into the genus *Trachelomonas*. A recent phylogenetic study utilizing SSU rDNA and partial LSU rDNA (Triemer et al. 2006) included 84 taxa from 11 genera. The study confirmed the monophyly of the genera *Monomorphina*, *Cryptoglena*, *Colacium*, *Trachelomonas*, and *Strombomonas* (the latter two taxa re-established as independent loricate genera) and created a new genus, *Discoplastis*. This study also established that *Lepocinclis* and *Phacus* were monophyletic. Only the monophyly of *Euglena* remained questionable. More information about euglenoid molecular phylogenetics is available in Triemer and Farmer (2007). They state: "by virtue of their unique cell morphology, gene composition, and ecological diversity, the Euglenozoa remain as one of the most interesting and fascinating groups of protists."

HABITAT

The euglenoids are present worldwide. Photosynthetic euglenoids are encountered especially in fresh water, but are also present in marine or brackish waters and soil. Among the factors that favor the growth of euglenoids are temperature, light, and organic matter. In fresh water, euglenoids are generally not found in springs, fast flowing rivers, or large, deep lakes, and in the latter case they usually occur only near the banks. Generally, photosynthetic euglenoids are rare in clean (oligotrophic) waters. They are common in small bodies of water (ditches, ponds), shallow lakes, and slow flowing rivers. They can tolerate high levels of organic matter and/or metals and are an important algal component in eutrophic lakes and reservoirs that are surrounded by agricultural land (Rosowski 2003). It has been suggested that photosynthetic euglenoids

FIGURE 1. Bloom of *Euglena rubra* from the Gietzel wetland on the campus of Michigan State University. Both the green and red areas contain the same species. The difference in coloration is dependent upon the concentration of red pigment granules found in the cells.

can play a significant role in purification of sewage ponds and oxygenation of polluted waters (Wołowski and Hindák 2005). Some euglenoids can form blooms, resulting in very high numbers of individuals. The blooms can be green (*E. viridis*), brown (*Trachelomonas*), or red (*E. sanguinea, E. rubra*—fig. 1). Usually blooms have a negative impact on the environment, modifying the color, odor, and taste of the water and, most significantly, causing massive deaths in invertebrate, fish, and even bird populations. Fish kills caused by *Euglena sanguinea* are causing important economic losses in fisheries in the United States (Zimba et al. 2004). It has even been suggested that *Euglena sanguinea* could be the species responsible for turning the Nile River red in ancient Egypt, as described in the Biblical passage Exodus 7:20–21 (Hort 1957–58). Some euglenoids can tolerate a wide pH range (*E. mutabilis*, from 1.0 to 8.6; Zakryś and Walne 1994) or high levels of heavy metals, insecticides, or radiation. They have also been found on the surface of acidic hot-mud pools near volcanoes (Sittenfeld et al. 2002). Photosynthetic euglenoids predominantly swim freely, but several can be attached to the substratum (*E. adhaerens*) or can form colonies often attached to invertebrates (*Colacium*).

NUTRITION

There are three primary modes of nutrition within euglenoids: phagotrophy (engulfing living or dead particulate organic matter), osmotrophy (absorbing dissolved nutrients directly from the environment), and phototrophy (utilizing solar energy to synthesize organic compounds through photosynthesis). Phylogenetic studies suggest that phagotrophic euglenoids emerged before phototrophs. At some point in evolution, a phagotrophic ancestor is believed to have engulfed a green alga or its chloroplast, giving rise to the phototrophs via this secondary endosymbiosis (Gibbs 1978, 1981; Cavalier-Smith 1993, 1998, 1999; McFadden 2001). The three chloroplast membranes in euglenoids (as opposed to two membranes in chloroplasts derived via primary endosymbiosis) and the presence of a vestigial feeding apparatus in almost all phototrophic euglenoid genera (e.g., *Colacium, Cryptoglena, Euglena, Eutreptia, Lepocinclis, Phacus, Strombomonas,* and *Trachelomonas;* Shin et al. 2002) strongly support this hypothesis. Photosynthetic euglenoids can supplement photosynthesis by acquiring different compounds, such as vitamins (e.g., cyanocobalamin and thiamine) through osmotrophy (Provasoli and Pintner 1953; Provasoli 1958). Some euglenoids are primarily osmotrophic, while others have become osmotrophic secondarily following the loss of photosynthesis (e.g., *Euglena longa, E. quartana, Lepocinclis acus* var. *hyalina, Phacus ocellatus,* and *Trachelomonas reticulata*).

CELL STRUCTURE

SHAPE

The shape of the euglenoid cell is highly variable within and among genera. Very common is the spindle-like to almost cylindrical shape found in many species of *Euglena, Euglenaria, Discoplastis, Eutreptia, Eutreptiella,* and *Lepocinclis.* Cell bodies can be straight or twisted (*Lepocinclis tripteris, Phacus tortus*). A particular, flattened leaf-shaped cell is typical of many *Phacus* species, while in genus *Monomorphina* the body is pyriform (teardrop-shaped) in lateral view. The cell body can be rigid, as in *Cryptoglena, Monomorphina,* and *Phacus;* it may show slow bending movements (e.g., *Lepocinclis*) or flowing, peristaltic movements (metaboly), as in *Colacium, Discoplastis, Euglenaria, Euglena, Eutreptia, Eutreptiella, Strombomonas,* and *Trachelomonas.* Metaboly generates dynamic changes in the shape of the cell that sometimes make it difficult to determine body shape.

SIZE

Cell size varies greatly among euglenoid genera and, in some cases, even within genera. Some of the smallest *Euglena* species (e.g., *E. minima*) are only 4 µm wide and about 12 µm long, while the larger *Euglena* species can be over 200 µm in length (*E. ehrenbergii*). The majority of species in the genera *Cryptoglena, Monomorphina, Strombomonas,* and *Trachelomonas* vary within a much narrower range and are under 100 µm in length. The smaller rounded species, such as *Trachelomonas volvocina,* may be as small as 6 µm in diameter. The greatest range of sizes is found in the genus *Lepocinclis.* The smaller species (e.g., *L. ovum, L. steinii*) are on the order of 10–20 µm, whereas on the opposite end of the scale, *Lepocinclis helicoideus* can be 500 µm in length.

PELLICLE

The pellicle (fig. 2) is the euglenoid cell surface. It is a complex structure and consists of the plasma membrane, proteinaceous strips lying beneath the membrane, groups of microtubules associated with the strips, and tubular cisternae of endoplasmic reticulum.

The main part of a strip is a frame that is sigmoidal in transverse section (Leander and Farmer 2000a, 2000b, 2001a, 2001b). A TEM cross-sectional view of a cell shows the pellicle organized as a series of ridges and grooves. Strips are arranged in parallel and can have either a longitudinal or

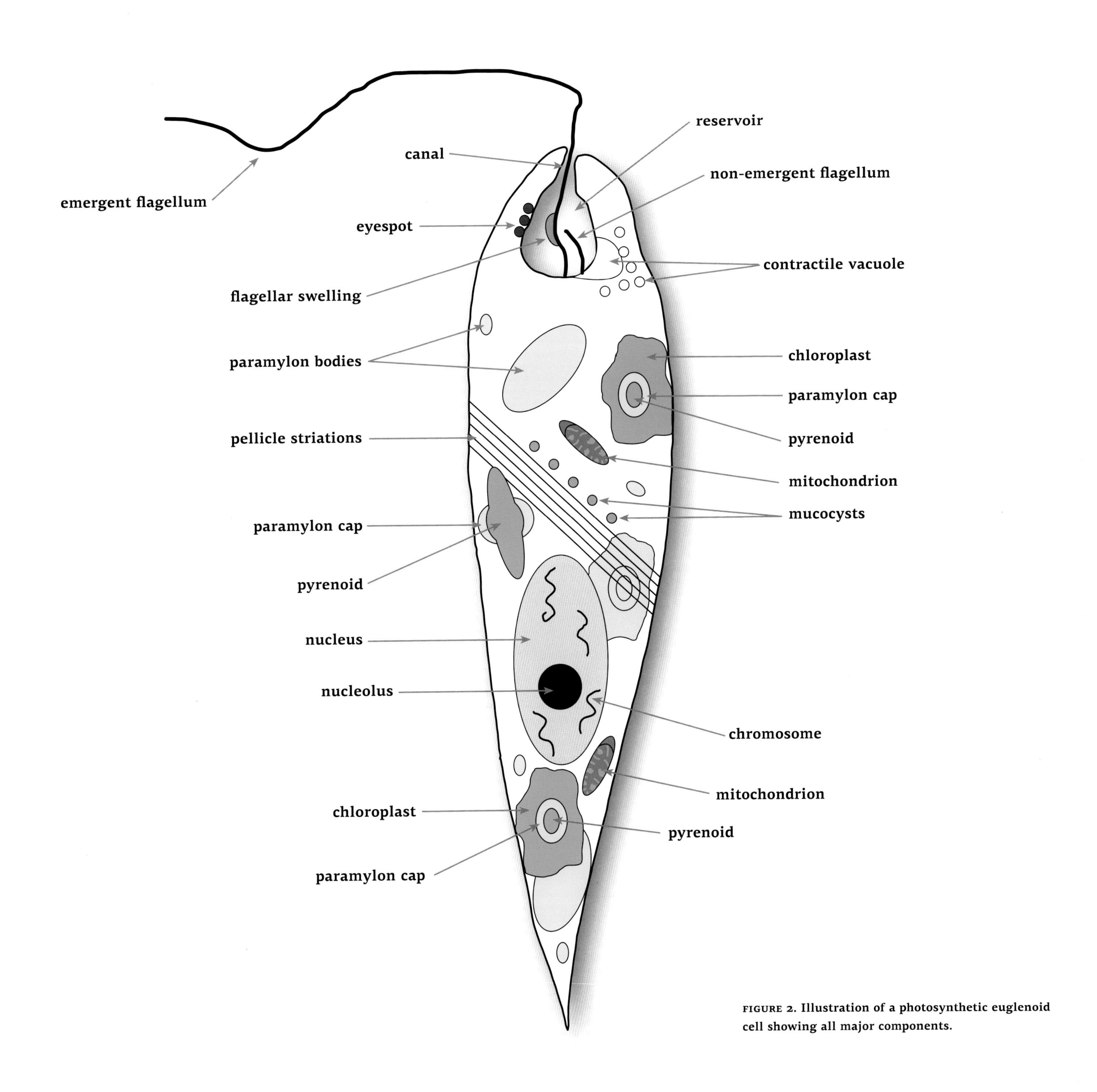

FIGURE 2. Illustration of a photosynthetic euglenoid cell showing all major components.

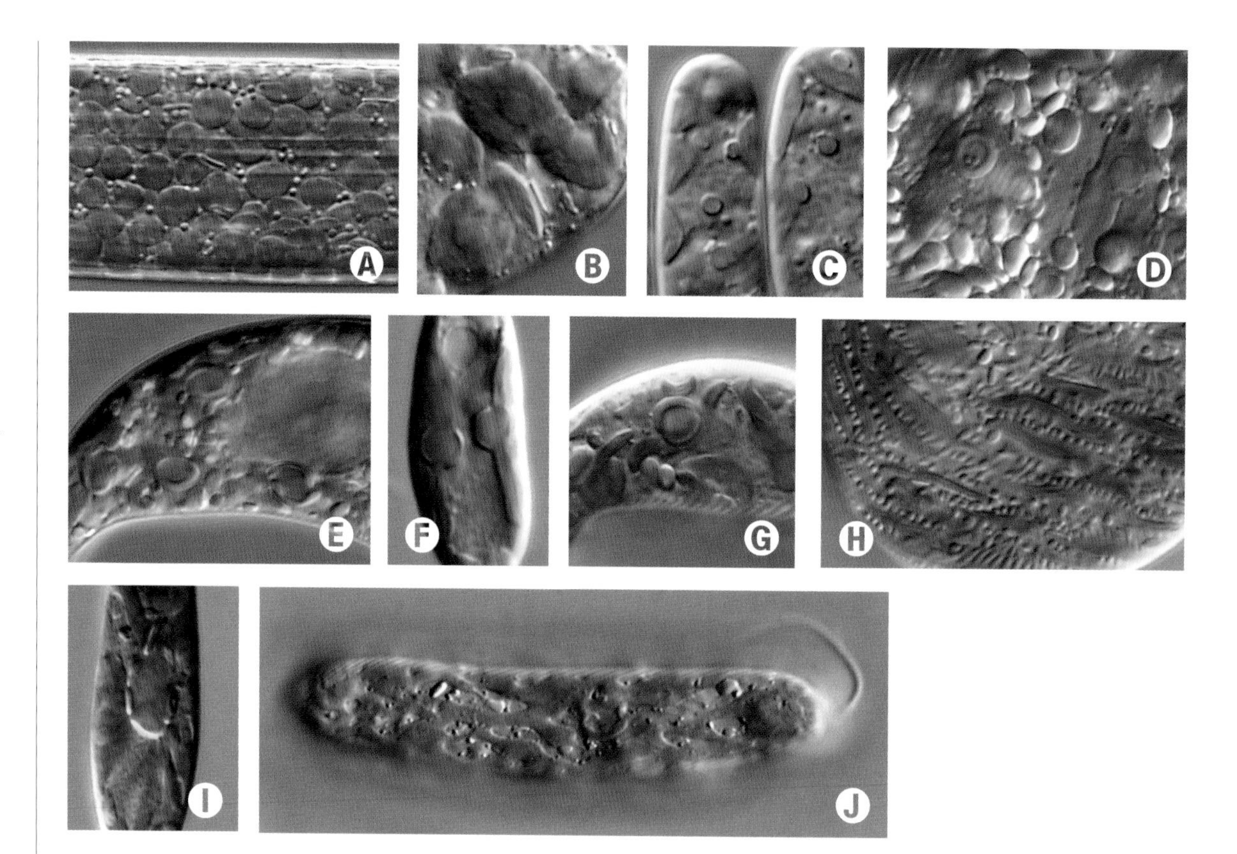

FIGURE 3. **Chloroplast diversity in photosynthetic euglenoids. Small discs with smooth margins lacking pyrenoids (A). Large discs with smooth margins and central, naked pyrenoids (B) or internally projecting haplopyrenoids (C). Large discs with lobed margins and centrally located diplopyrenoids (D, E, G). Plate-like chloroplasts with diplopyrenoids (F). Ribbon-like extensions from the chloroplast as seen near the cell surface (H). Single chloroplast with a large central pyrenoid surrounded by paramylon grains. The ribbon-like lobes of the chloroplast extend from the center outward in a stellate pattern (I). Chloroplast with net-like organization, reticulate (J).**

helical orientation. Some euglenoids show cross-striations between the strips that are visible at the light microscope level (see *Phacus orbicularis*) while other taxa exhibit ornamentations on the strips (see *Lepocinclis fusca* and *L. spirogyroides*).

Euglenoids with fewer, longitudinally oriented strips usually lack metaboly or just exhibit slight bending movements (as in some species of *Lepocinclis*). Taxa possessing many helically oriented strips generally exhibit the wriggling motion known as metaboly.

LORICA

Euglenoid species included in the genera *Trachelomonas* and *Strombomonas* possess an envelope surrounding the cell, known as the lorica. The envelope may be smooth or may exhibit various ornamentations, such as punctae, scrobiculae, warts, and spines. The lorica may have an annular thickening at the anterior end, surrounding the flagellar pore, or it may have a prominent collar. The posterior end of the lorica may taper to form a tail. Loricas can be globular, ellipsoidal, cylindrical, ovoid, trapezoidal, or irregularly shaped. They may be colorless, brown, red, yellow, or black.

The lorica is first secreted as a mucilaginous layer, which is then mineralized. The basic microarchitecture is similar in *Trachelomonas* and *Strombomonas* (Dunlap et al. 1986; Conforti et al. 1994). Iron, silicon, and manganese are the major chemical constituents of the lorica. In *Trachelomonas,* iron is the main mineralizing component of the envelope, with lesser amounts of silicon and manganese, while in *Strombomonas*, the silicon plays the main role (Conforti et al. 1994). Based entirely on the lorica features, Deflandre (1930) created the genus *Strombomonas* by extracting from the genus *Trachelomonas* the species that possessed loricas, which lacked a distinctive collar and lacked ornamentation, but possessed a tailpiece. Many of these taxa also had the ability to accumulate small particles on their surfaces.

In a revision by Marin et al. (2003), the genus *Strombomonas* was dissolved back into *Trachelomonas.* However, later studies (Brosnan et al. 2005; Triemer et al. 2006; Ciugulea et al. 2008) supported the retention of *Strombomonas* as a separate genus. In the study of Brosnan et al. (2005), the authors noted major differences in lorica development between the two genera and used this as evidence for separating the taxa. In *Trachelomonas,* a layer of mucilage was excreted over the entire protoplast in association with the pellicle strips, while in *Strombomonas,* lorica development occurred from the anterior of the cell to the posterior, forming a shroud over the protoplast. Today, as in the past, the taxonomy of species within these genera still relies heavily on the morphology of the lorica.

CHLOROPLAST (CHROMATOPHORE)

The chloroplast (fig. 2) is the organelle in which the photosynthetic process takes place. The photosynthetic pigments typically found in euglenoids are chlorophyll a and b, beta carotene, and the xanthophylls (neoxanthin, diadinoxanthin, diatoxanthin).

Chloroplasts may or may not possess pyrenoids (fig. 3A–J). The pyrenoid is a specialized area of the chloroplast that contains high levels of RuBisCO (ribulose-1,5-bisphosphate carboxylase/oxygenase), the key enzyme in carbon dioxide fixation. Three types of pyrenoids are found in euglenoids.

- NAKED PYRENOIDS, with no paramylon bodies bordering the pyrenoid (fig. 3B).
- HAPLOPYRENOIDS, with a paramylon sheath only on one side of the pyrenoid (fig. 3C).
- DIPLOPYRENOIDS, with a sheath of paramylon on each side of the pyrenoid (also known as a double-sheathed pyrenoid) (fig. 3D–G).

Usually there is only one pyrenoid associated with each chloroplast, but additional pyrenoids are present in some taxa (e.g., two in *Euglena archaeoplastidiata;* Gojdics 1953). All species of *Discoplastis, Lepocinclis,* and *Phacus* lack pyrenoids.

The chloroplast(s) can be located axially, parietally, or between the axis and the plasma membrane. The number of chloroplasts can vary from one (*Euglena stellata*) to many (*Lepocinclis oxyuris*). The morphology of the chloroplasts varies greatly within euglenoids (fig. 3A–J). Euglenoid chloroplasts come in a variety of shapes and sizes. The chloroplasts can be numerous small discs with even margins and without pyrenoids (3A) or larger discs with even margins with naked pyrenoids (3B) or haplopyrenoids (3C). Some chloroplasts have irregular margins, which can be slightly (3D–E) or deeply lobed (3G). Often the chloroplasts can be so deeply incised that each individual lobe gives the impression of being an independent ribbon-like chloroplast. The ribbon-like lobes may radiate from the center of the chloroplast forming a stellate pattern (3I). Alternatively, the elongate ribbon-like lobes may extend toward the cell surface arranged in a spiral pattern (3H). The chloroplasts in some species take the shape of elongated plates (3F). In a few cases, the chloroplast has a reticulate or net-like shape (3J). Using light and confocal microscopic observations, Kosmala et al. (2007) showed that a single, parietal orbicular chloroplast is present in *Monomorphina* species.

Some euglenoids have lost the ability to photosynthesize secondarily and appear colorless (*Euglena longa, E. quartana, Phacus ocellatus,* and *Trachelomonas reticulata*). However, *Euglena longa* still retains a small colorless plastid (Gockel and Hachtel 2000, as *Astasia longa*), and this is likely true for many colorless taxa believed to be derived from closely related photosynthetic taxa.

The shape, number, and location of the chloroplast, as well as the presence and type of pyrenoid, have played a major role in the taxonomy of euglenoids. Now, with large amounts of sequence data becoming available, molecular based phylogenies are clarifying relationships among taxa and helping to determine which morphological characters will be most useful in identifying taxa at the light microscope level.

PARAMYLON

Paramylon (fig. 2), the storage product of euglenoids, is a β-1:3-linked glucan (Leedale 1967). Paramylon, also referred to as paramylum, is a membrane-bound crystal in which fibers traverse it in an overall concentric pattern (Kiss et al. 1987). Paramylon is unique among plants because of its high degree of crystallinity (Marchessault and Deslandes 1979). It also differs from true starch in that it does not show the blue color reaction with iodine (Gojdics 1953).

Paramylon plates may cap the pyrenoids on one or both sides of the chloroplast and/or may be freely scattered through the cell. The free paramylon grains can be very small and numerous or very large and fewer in number. Although large grains are prominent in some taxa, nearly all taxa contain small grains as well. The number of paramylon bodies can vary greatly depending upon the physiological state of the cell. In some instances, the cells are so filled with the storage product that no internal structures can be seen. In species with stellate chloroplasts, there is one paramylon center in the middle of each chloroplast, one (for the single chloroplast) in *Euglena viridis,* two (one for each of the two chloroplasts) in *E. geniculata* (Zakryś and Walne 1998; Shin and Triemer 2004), and three (one for each of the three chloroplasts) in *E. tristella.*

Paramylon is found in many forms among euglenoids and is often used as a diagnostic character. There are various morphological forms and sizes of paramylon bodies.

- SPHERICAL AND ELLIPSOIDAL GRAINS (fig. 4A, B). Numerous small spherical or ellipsoidal paramylon bodies are found in almost all euglenoid cells. In *Colacium, Eutreptia, Strombomonas,* and *Trachelomonas,* free scattered oval granules are the most common paramylon type encountered.
- RODS (fig. 4C, D, O, P). Long rods are found in some *Euglena* and *Lepocinclis* species. In *Euglena ehrenbergii,* rods can be up to 75 µm long and can be straight or bent (fig. 4P). There are two long rods in *Lepocinclis tripteris,* one anterior and one posterior to the nucleus. Short rods are encountered in several taxa, as in *Discoplastis, Euglena agilis,* and *Euglenaria caudata.*
- DISCS (fig. 4E, F, H). Large discs are present in many species of *Phacus,* which usually have one (*Ph. gigas, Ph. orbicularis*) or two (*Ph. alatus, Ph. curvicauda*) discs. The discs are either flat or can be thickened in the middle. When the large discs start to dissolve, they form donut- or ring-shaped bodies, as are often found in *Phacus longicauda.* Small ring-shaped grains are also encountered within euglenoids (*Ph. acuminatus*).
- LINKS (fig. 4I–M). Elongated links can be found in *Lepocinclis oxyuris* and *L. spirogyroides.* Usually there are two, one anterior and the other one posterior to the nucleus (Gojdics 1953). Large,

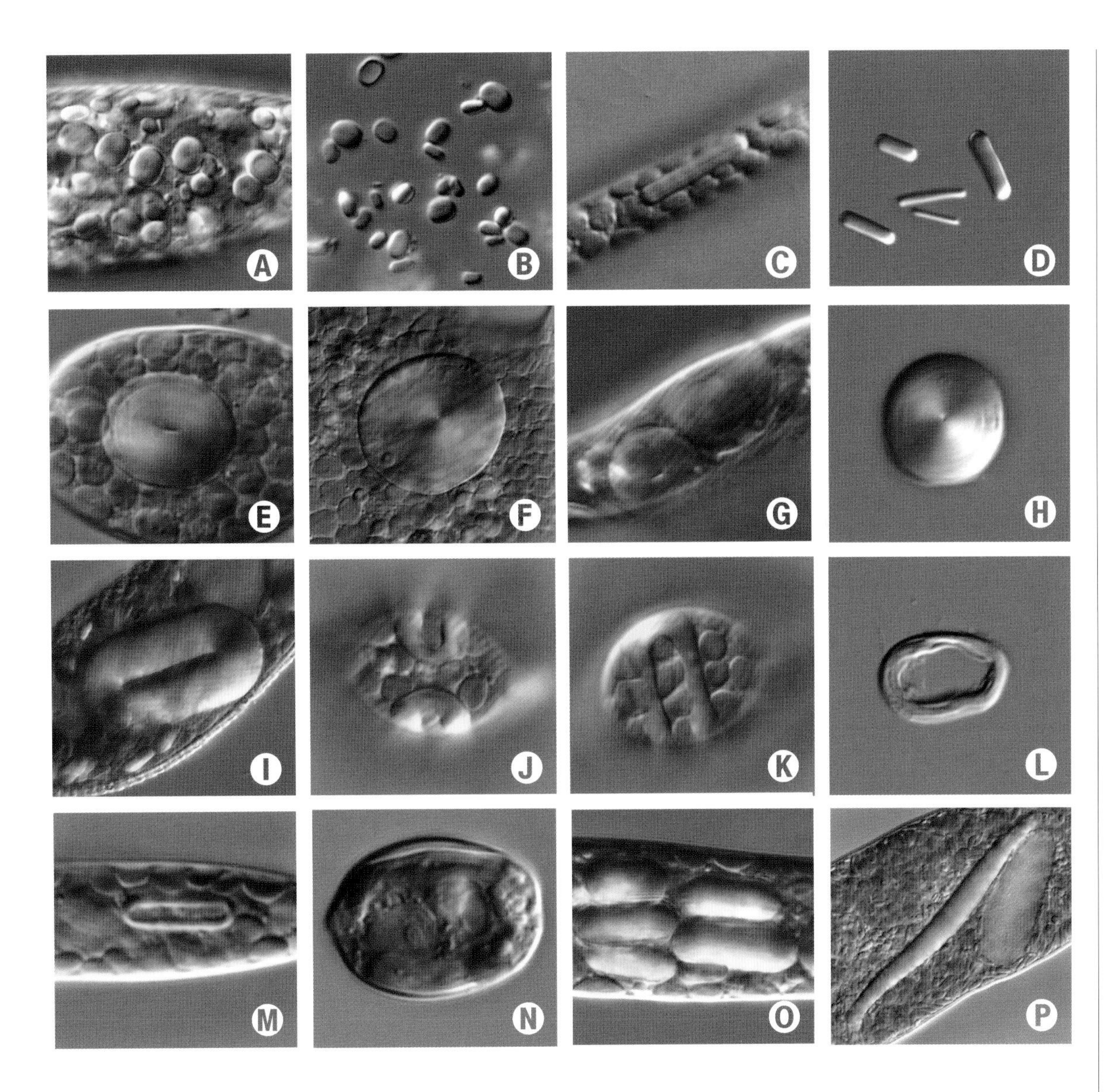

FIGURE 4. Diversity of paramylon types in photosynthetic euglenoids. Spherical to ovoidal bodies within the cell (A) and isolated from the cell (B). Rods of various sizes within the cell (C, O, P) and isolated from the cell (D). Large paramylon discs within the cell (E, F) and isolated from the cell (H). Curved discs or trough-shaped bodies (G). Elongated links within the cell (I, J, K, M) and isolated from the cell (L). Large lateral plate-like bodies (N).

curved, oval links are present in several *Lepocinclis* species (e.g., *L. fusiformis, L. ovum*). Small links are also seen in several taxa (*Phacus longicauda, Lepocinclis fusca*).

- PLATES (fig. 4N). In *Cryptoglena* there are two large lateral plate-like paramylon bodies located between the chloroplast and the pellicle. The paramylon in *Monomorphina* is also plate-like and located between the chloroplast and the pellicle. While two plates is the most common configuration, up to four plate-like paramylon granules have been reported for some species.
- TROUGH-SHAPED (fig. 4G). Found only in *Euglena convoluta*, usually there are seven to nine large curved plates, up to 18 µm long and lying parallel to the long axis of the cell (Gojdics 1953).

MITOCHONDRION

The mitochondrion (fig. 2) is an organelle that plays a major role in providing energy to the cell by generating ATP (adenosine triphosphate). As in other eukaryotes, the euglenoid mitochondrion is surrounded by two membranes. The inner membrane folds into the matrix and forms paddle-shaped cristae. When viewed from the surface (en face) the cristae are disc-shaped, but appear as flattened sacs when viewed from the side. The short "handle" of the paddle connects the disc with the inner membrane of the mitochondrion.

RESERVOIR

The euglenoid cell has an opening at the anterior end, the flagellar pore, which leads to a tubular passage known as the canal (fig. 2). The length and the shape of the canal are dependent upon the shape of the cell. Proceeding posteriorly, the canal widens into a flask-like reservoir (fig. 2). The reservoir exhibits a smooth wall and was considered to be the main vacuole by early writers (Gojdics 1953). The flagella are inserted into the base of the reservoir. The emergent flagellum projects into the reservoir, extends through the canal, and emerges from the flagellar pore to the exterior of the cell. The contractile vacuole is positioned adjacent to the reservoir and empties its contents into the reservoir, where they are expelled to the exterior of the cell.

CONTRACTILE VACUOLE

The contractile vacuole (fig. 2), which functions in water regulation, is located at the anterior end of the cell, near the base of the reservoir and on the opposite side of the flagellar insertion. This structure arises by progressive merging of smaller vacuoles (Chadefaud 1937; Huber-Pestalozzi 1955) and eliminates the water when in excess. The main factor that regulates the "pulsation" of the vacuole is temperature. As an example, in *Euglena deses* the contractile vacuole pulsates every 22 seconds at room temperature (Gojdics 1934) and every 30 seconds at 42°C and 18°C (Dangeard 1901).

EYESPOT AND FLAGELLAR APPARATUS

The flagellum (fig. 2) is the primary locomotory organelle of euglenoids. The most common number of flagella for photosynthetic euglenoids is two, but four flagella are present in *Eutreptiella pomquetensis.* However, in many species the second flagellum is very short and does not emerge from the reservoir, thereby making the cells appear to have only a single flagellum. Near the base of the emergent flagellum is an enlargement called a paraxial swelling (= flagellar swelling = paraxial body) (fig. 2). The swelling, which contains a rhodopsin-like protein (Evangelista et al. 2003) and a photoactivated adenylyl cyclase (Iseki et al. 2002), is considered to be the active photoreceptor (Walne et al. 1998). The eyespot (= stigma) (fig. 2) is located at the anterior end of the cell, adjacent to the paraxial swelling. It is always independent of the chloroplast and ranges in coloration from yellowish-red to bright crimson (Gojdics 1953). The eyespot is built up of a matrix of red granules bearing a carotenoid pigment (Gojdics 1953). The size of the eyespot varies from 2–3 µm to 7–10 µm. The eyespot functions in photoreception by shielding the paraxial swelling and, in this way, changing the light intensity that reaches the swelling, which is the actual photoreceptor. Some euglenoids, especially those that have lost the ability to photosynthesize, lack an eyespot (e.g., *Euglena longa*). Other colorless species retain the eyespot although photosynthesis has been lost (*Euglena quartana, Trachelomonas reticulata, Phacus ocellatus*).

MUCOCYSTS

Mucocysts (or muciferous bodies) (fig. 2) are single membrane-bound organelles. They are located beneath the pellicle and are considered to be a part of the vacuolar system. Mucocysts exhibit two major morphologies: spherical and spindle-shaped. In healthy cells, they are spirally arranged parallel to the pellicle strips, but may be fewer in number and appear randomly dispersed in older cells or in cells growing in unfavorable conditions. Mucocyst content is discharged through minute pores in the pellicle. They can easily be seen by staining with neutral red.

NUCLEUS

Euglenoids are eukaryotic organisms; thus, they possess a true nucleus, surrounded by an envelope that protects the genetic material (fig. 2). The outer membrane of the envelope is not associated with the chloroplast, as is common in many of the chlorophyll a and c containing algae. The nucleus is usually spherical to oval and located near the center or toward the posterior of the cell. It may contain a single prominent nucleolus (referred to also as an endosome in euglenoids) or may have several nucleoli.

CELL DIVISION

Within euglenoids, cell division (bipartition) is the only known mode of replication. The parent cell divides to yield two daughter cells. Before cell division is initiated, the cell doubles its organelles (nucleus, flagella, chloroplast, stigma, etc.). The cleavage furrow begins to develop at the anterior of the cell and progresses posteriorly until the cells are connected only by a thin tube of cytoplasm. This will break as the cells actively writhe about and/or pull apart. The two daughter cells are clones of the parent cell (for reviews of cell division, see Triemer and Farmer 1991a, 1991b).

At interphase, unlike most other eukaryotes, the chromosomes are condensed and easy to see with the light microscope. In most eukaryotes, the beginning of mitosis is signaled by the disappearance of the nucleolus and the breakdown of the nuclear envelope. In euglenoids, the nucleolus and the nuclear envelope remain intact throughout division. In the light microscope, euglenoid mitosis is signaled by the elongation of the nucleolus. This elongation of the nucleolus and concomitant elongation of the nucleus may continue until the nucleus is stretched across the cell. The nuclear envelope then begins to constrict around the center of the elongate nucleus and gives the telophase nucleus a dumbbell shape. In some euglenoids, continued constriction of the nuclear envelope pinches off the nucleus and generates the two daughter nuclei. In other euglenoids, the central portion of the dumbbell is severed as the developing cleavage furrow moves toward the posterior end of the cell.

Euglenoids are not known to undergo sexual reproduction. A previous report of sexuality in *Scytomonas* discussed by Leedale (1967) has not been confirmed.

KEY TO THE PHOTOSYNTHETIC GENERA

1A. Cells with one emergent flagellum 3
1B. Cells with two or more emergent flagella 2

2A. Cells with two equal emergent flagella .... *Eutreptia*
2B. Cells with two unequal emergent flagella (rarely two pair of unequal flagella) .... *Eutreptiella*

3A. Cells enclosed in an envelope (lorica) 4
3B. Cells without an envelope (lorica) 5

4A. Lorica with a definite collar; walls usually ornamented with pores, warts, or spines . . . . *Trachelomonas*
4B. Lorica with a long collar and usually with a conical end; walls smooth with no ornamentation . . . . *Strombomonas*

5A. Cells usually attached to substratum by mucilaginous branched stalks forming dendroidal colonies . . . . *Colacium*
5B. Cells free-swimming 6

6A. Cells rigid or with slight bending movements 7
6B. Cells showing wriggling motion (metaboly) 10

7A. Cells generally flattened 8
7B. Cells ovoid or cylindrical; paramylon grains large and rod-, disc- or ring-shaped . . . . . . *Lepocinclis*

8A. Cells compressed, most species flat and leaf-shaped; paramylon bodies usually 1–2 large discs . . . . *Phacus*
8B. Cells pyriform or coffee bean–shaped 9

9A. Cells pyriform with posterior spine; pellicle with helically arranged ribs . . . . *Monomorphina*
9B. Cells coffee bean–shaped with longitudinal furrow; chloroplast an open cylinder (C- or U-shaped) . . . . *Cryptoglena*

10A. Cells generally spindle-shaped; pellicle finely to visibly striated; various chloroplast shapes with or without pyrenoids . . . . *Euglena/Euglenaria*
10B. Cells spindle-shaped; numerous small disc-shaped chloroplasts without pyrenoids . . . . *Discoplastis*

(NOTE: Euglena *and* Euglenaria *share common morphological characteristics, but can be distinguished at the molecular level by distinct molecular signatures—see* Linton et al. 2010*)*

A COLOR ATLAS OF
PHOTOSYNTHETIC EUGLENOIDS

COLACIUM

EHRENBERG 1833

Cells obovoid to obpyriform, ellipsoidal to cylindrical, sometimes spherical *Euglena*-like cells, usually attached to a substratum by production of mucilaginous stalks at the anterior end; one emergent flagellum present in free-swimming cells, up to twice the length of the body; canal opening subapical; pellicle fine; chloroplasts parietal, discoid, with or without a pyrenoid; paramylon bodies small and numerous; eyespot at the anterior end; longitudinal cell division leads to dendroidal colonies; metaboly present; cysts unknown; common in eutrophic fresh waters on algae or planktonic animals (copepods, rotifers).

A NOTE ON THE LEGENDS

The figure legends include many of the diagnostic features necessary to identify the taxon. However, features such as cell size and flagellar length are highly variable in the literature. For example, Hortobagyi (1943) reports that the body length of *Euglena ehrenbergii* ranges from 129 to 139 µm, while Johnson (1944) reports that it ranges from 190 to 400 µm, with most authors reporting an intermediate value. Similarly, Skuja (1948) reports that the flagellar length of *Lepocinclis tripteris* (as *Euglena tripteris*) is one-eighth to one-sixth body length, while Hübner (1886) states that the flagellum is not quite body length. Moreover, flagella are labile structures that are easily lost or broken, making it difficult to accurately determine flagellar length. Therefore, the values presented in the legends should be taken as approximations based upon our personal experience with the taxa.

Colacium calvum

Stein 1878

SIZE: 42–48 µm long × 15–29 µm wide.

Motile cell showing numerous chloroplasts (A, C) with haplopyrenoids (arrows) (A, B). Surface view showing finely striated pellicle; flagellum (usually body length) partially visible; red eyespot and small paramylon bodies present (C). The clear area at the anterior end of the cell is filled with mucus-containing vesicles (A, C). Cell shape changes dramatically during metaboly (compare A and B).

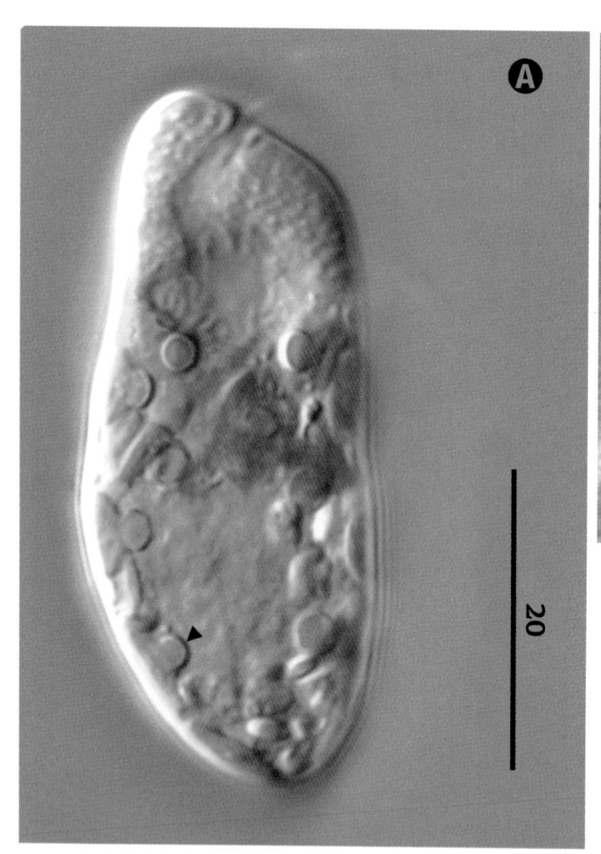

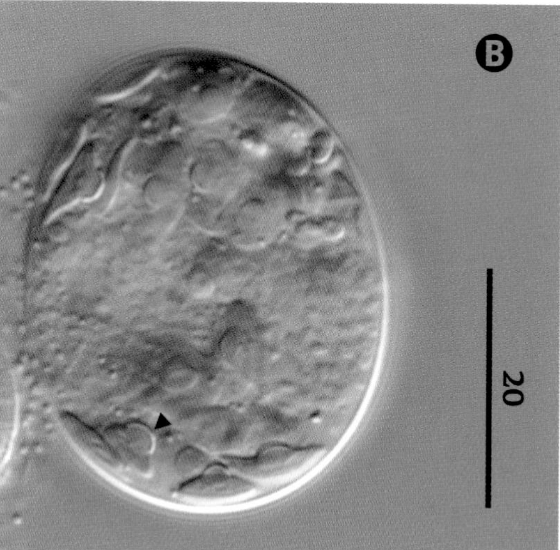

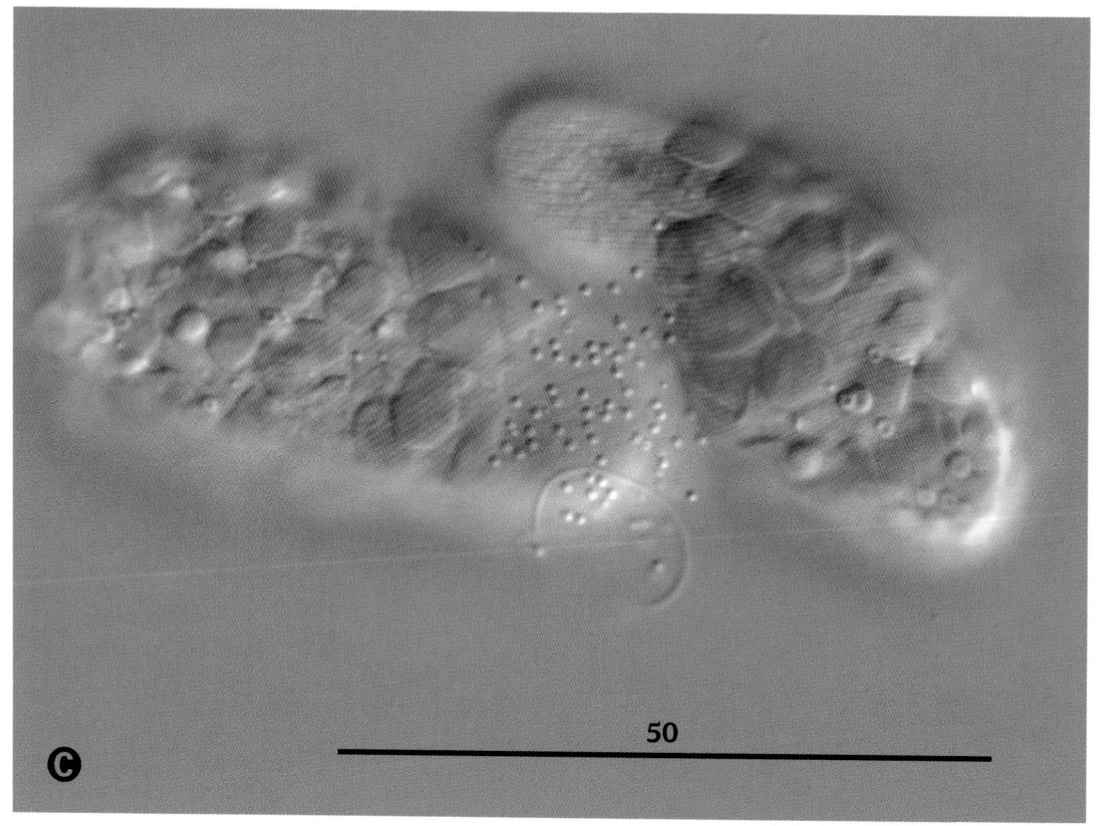

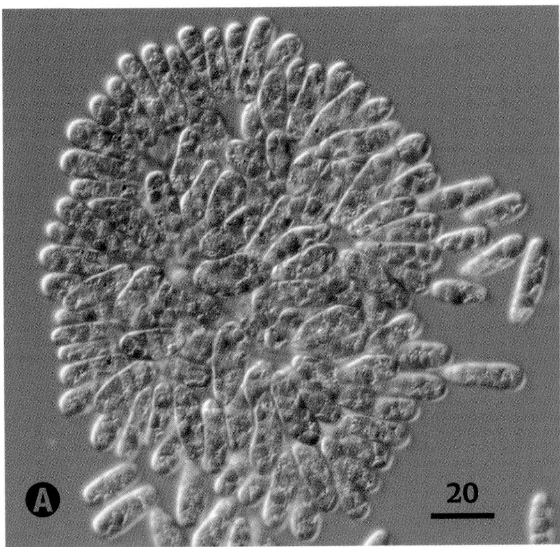

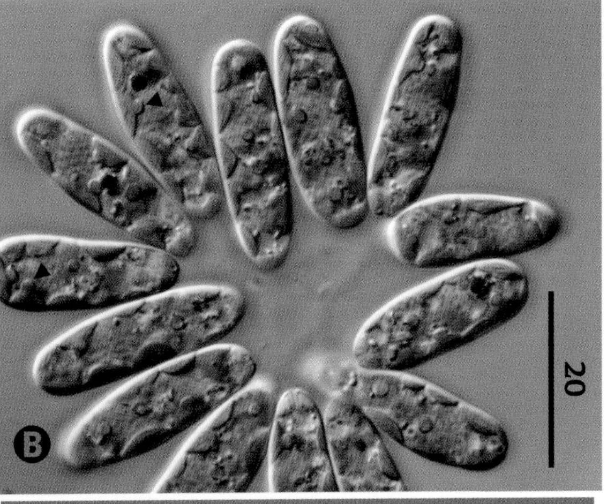

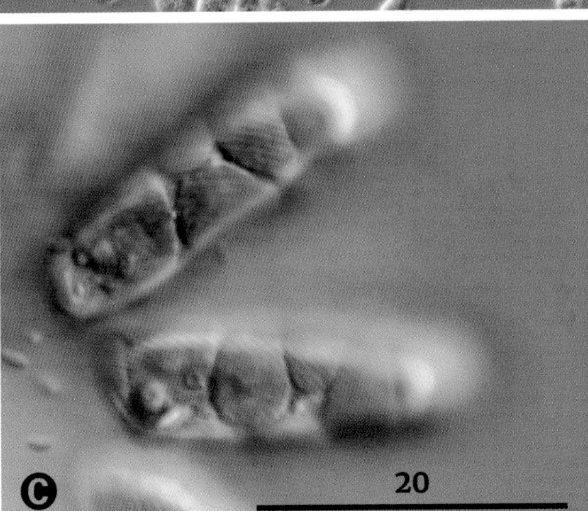

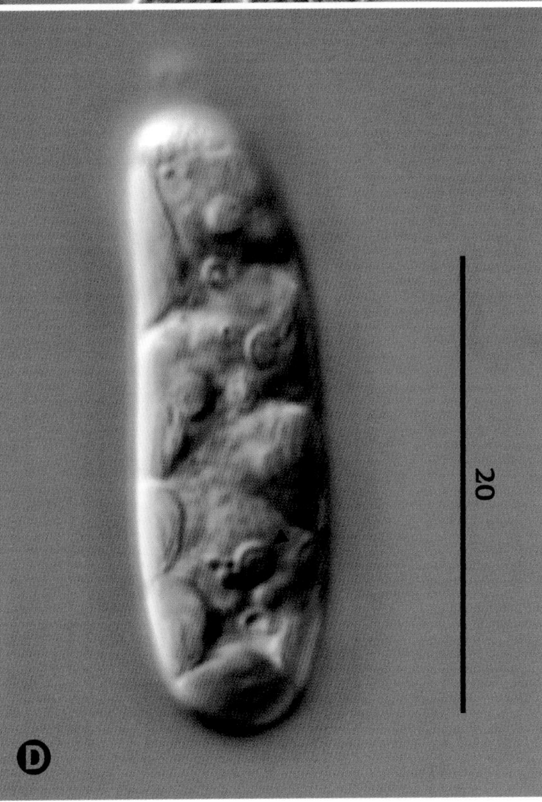

Colacium elongatum

Playfair 1921

SIZE: 12–28 µm long × 4–11 µm wide.

Elongate cylindrical cells, each forming a mucilaginous stalk, group into colonies (Ⓐ, Ⓑ). Cells have several disc-shaped chloroplasts, each bearing a haplopyrenoid with a U-shaped paramylon cap, as seen in lateral view (arrows) (Ⓑ), and a circular shape in frontal view (arrow) (Ⓓ). Fine pellicular striations are visible at the cell surface (Ⓒ).

Colacium mucronatum

Bourrelly et Chadefaud 1948

SIZE: 28–35 µm long × 12–16 µm wide.

Cells grouped into colonies with individuals connected through dendroidally branched stalks (**A**, **B**). Ovoid cell with several disc-shaped chloroplasts, each with haplopyrenoid (arrow) (**C**). The finely striated pellicle is visible at the cell surface (**D**).

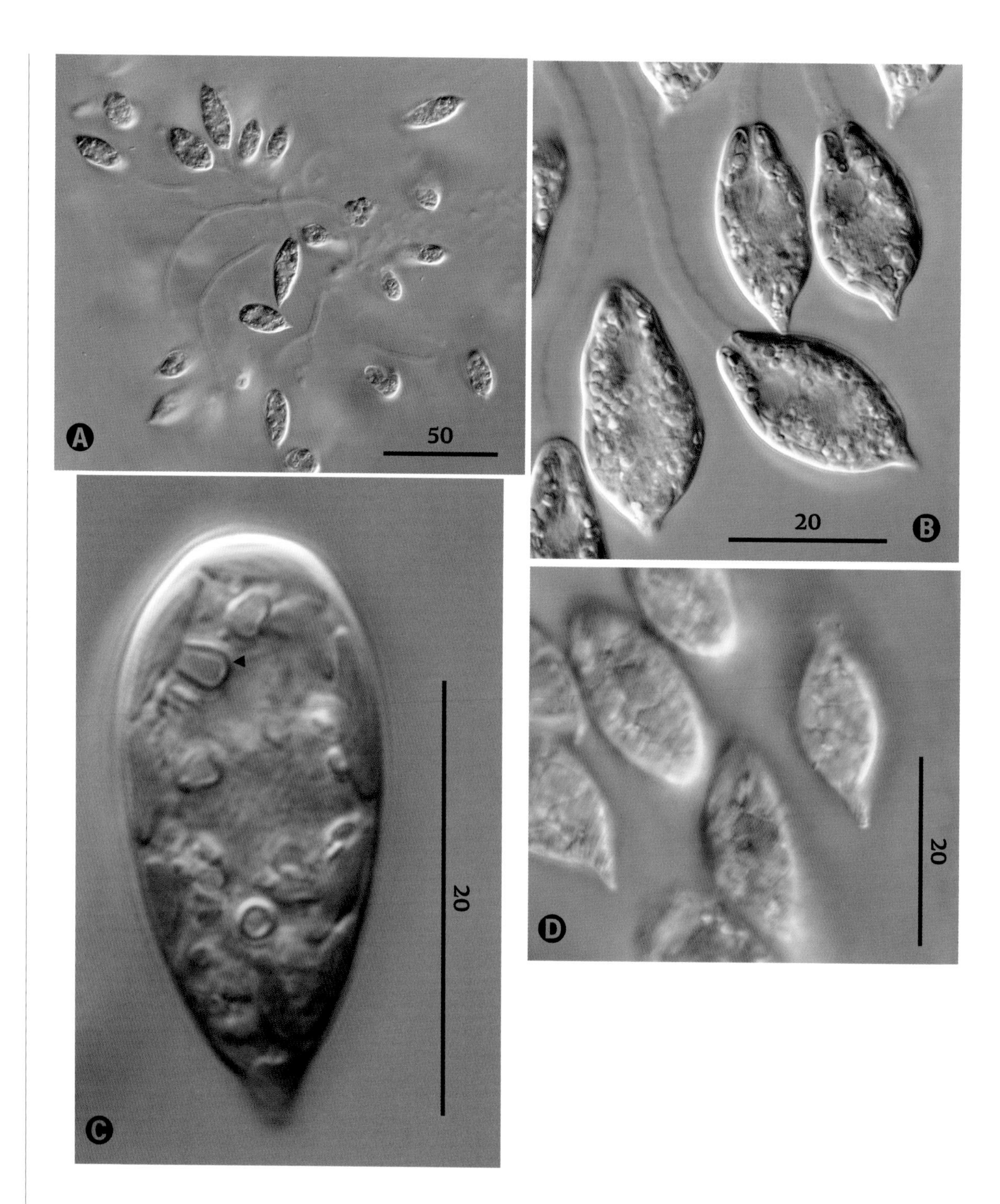

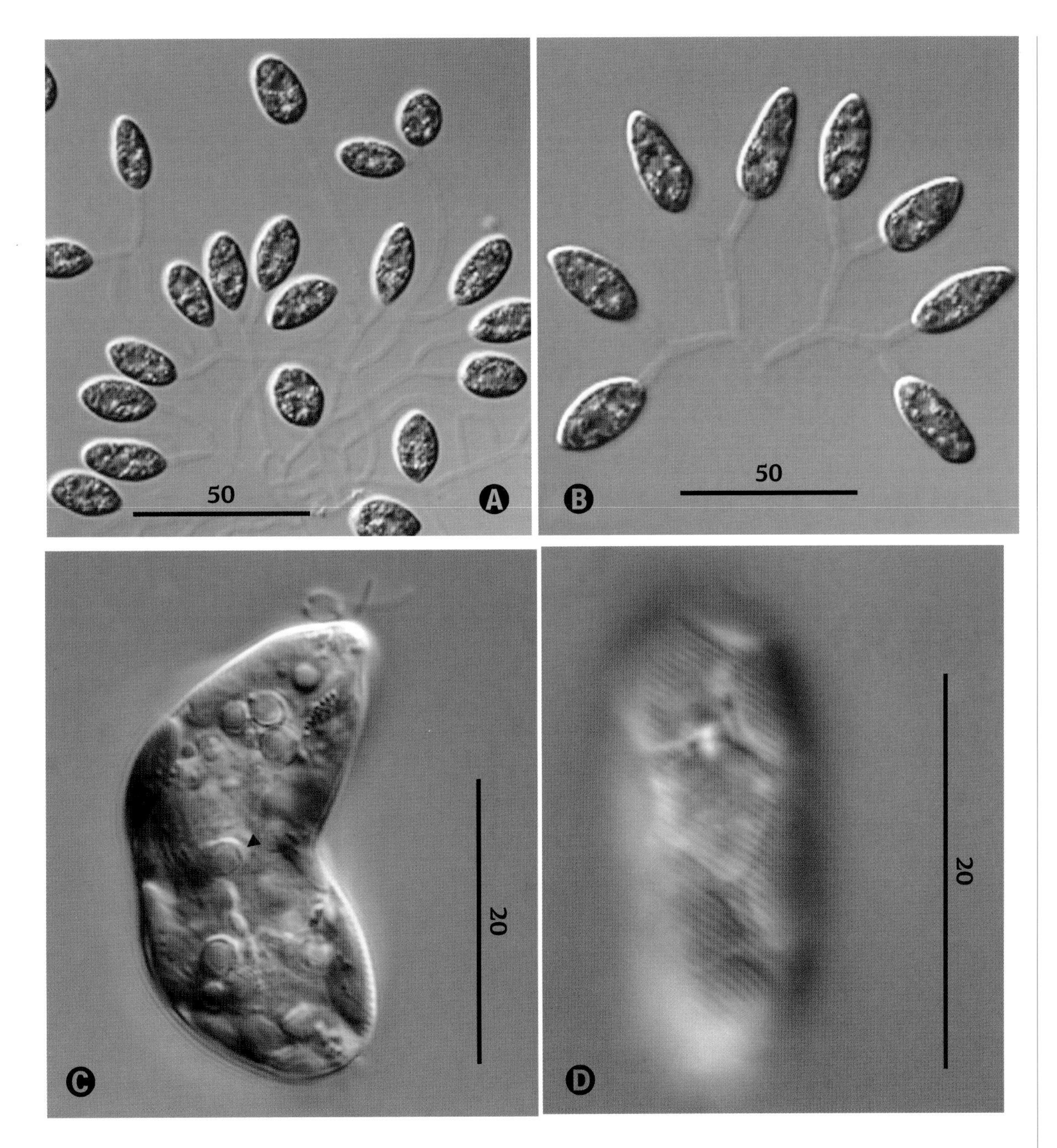

Colacium vesiculosum

Ehrenberg 1838

SIZE: 18–29 µm long × 9–17 µm wide.

Colonies with cells connected through dendroidally branched mucilaginous stalks (A, B). Several disc-shaped chloroplasts are present (D), each bearing a haplopyrenoid with a U-shaped paramylon cap (arrow) (C); flagellum (up to twice body length) is visible (C). The pellicular striations are fine (D). Dividing cell; red eyespot visible at anterior end (C).

CRYPTOGLENA

EHRENBERG EMEND. KOSMALA AND ZAKRYŚ 2007

Cells oval, short, slightly acute posteriorly, coffee bean–shaped, compressed, with a ventral longitudinal furrow; a single emergent flagellum; pellicle thick, as in *Phacus;* one large parietal chloroplast, as an open cylinder in form of letter C or U with two lateral paramylon plates; eyespot at the anterior end; metaboly not present; cells free-swimming; freshwater.

Cryptoglena pigra

Ehrenberg 1831

SIZE: 11–18 µm long × 6–10 µm wide.

Obovoid cell; U-shaped chloroplast with two large plate-like paramylum bodies and red eyespot (A). Groove runs longitudinally along the ventral face of the cell (B). Additional small paramylon bodies and a portion of the groove are visible (C). Flagellum is body length.

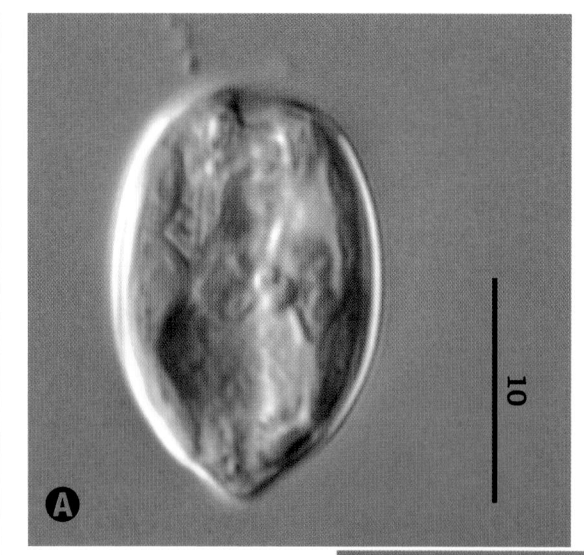

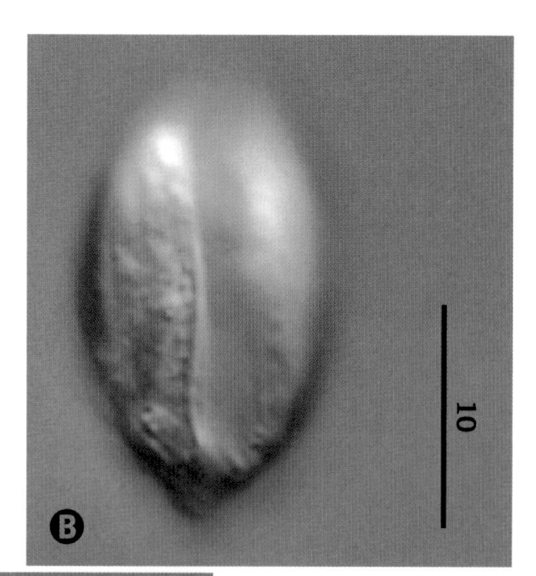

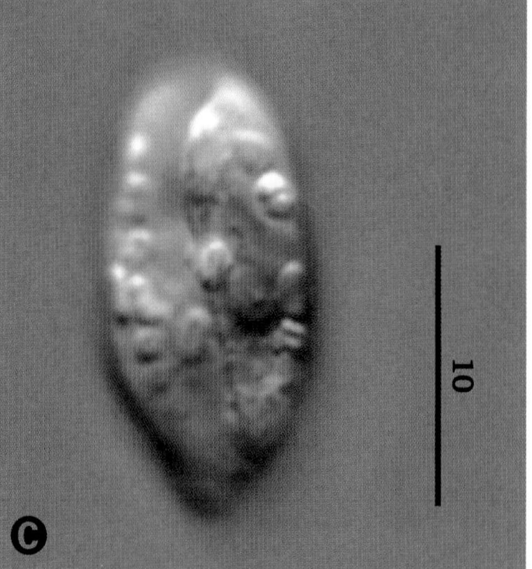

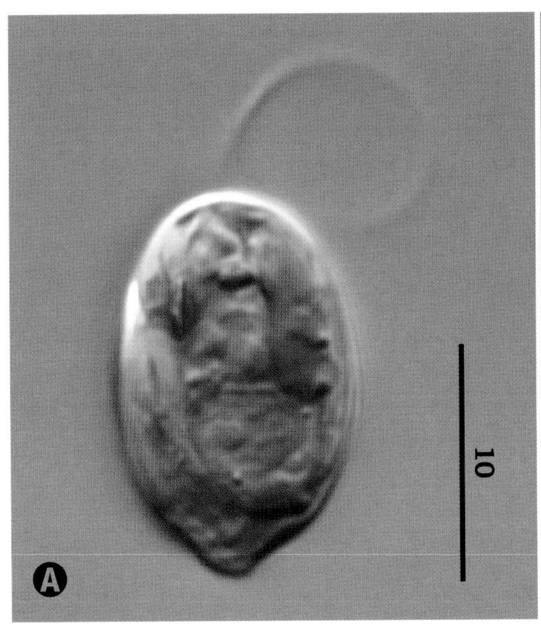

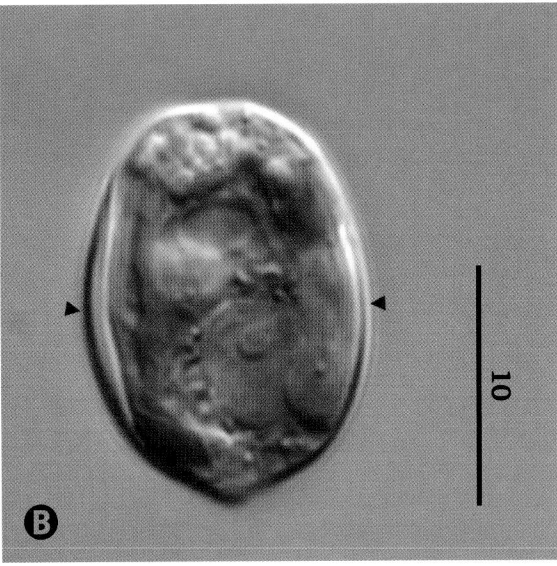

Cryptoglena skujai

Marin et Melkonian 2003

SIZE: 11–18 µm long × 8–13 µm wide.

Coffee bean–shaped cells showing flagellum (body length) (A), red eyespot, nucleus with nucleolus, and one U-shaped chloroplast (A, B) with two large plate-like paramylon bodies (arrows) (B).

DISCOPLASTIS

TRIEMER ET AL. 2006

Cells spindle-shaped, middle somewhat expanded, anterior end obliquely truncated, posterior end attenuated to a long, straight or curved, pointed tailpiece; a single emergent flagellum; metaboly present, spirally striated; numerous chloroplasts (20–40), discoid, parietal; no pyrenoids; paramylon granules small, more or less abundant, short rods or ellipsoids; eyespot at the anterior end; slow metabolic movements; freshwater.

Discoplastis adunca

(Schiller) Triemer 2006

SIZE: 35–54 μm long × 4–10 μm wide.

Spindle-shaped cells showing flagellum (Ⓐ, Ⓑ), eyespot (Ⓐ, Ⓑ, Ⓒ), paramylon grains (arrows) (Ⓑ, Ⓒ), and numerous disc-shaped chloroplasts lacking pyrenoids (Ⓐ–Ⓓ). Pellicular striations spiral down the length of the cell (Ⓓ).

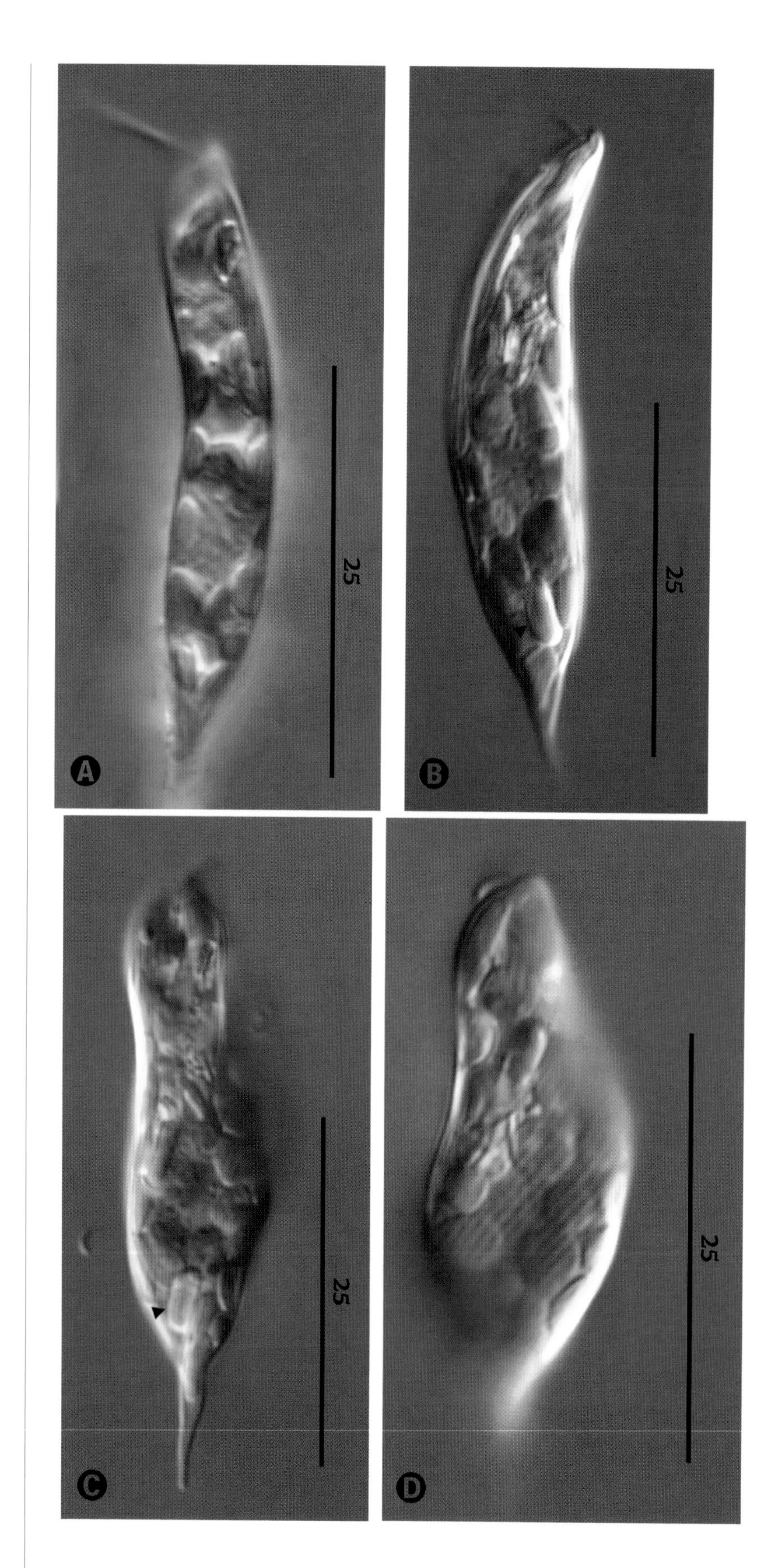

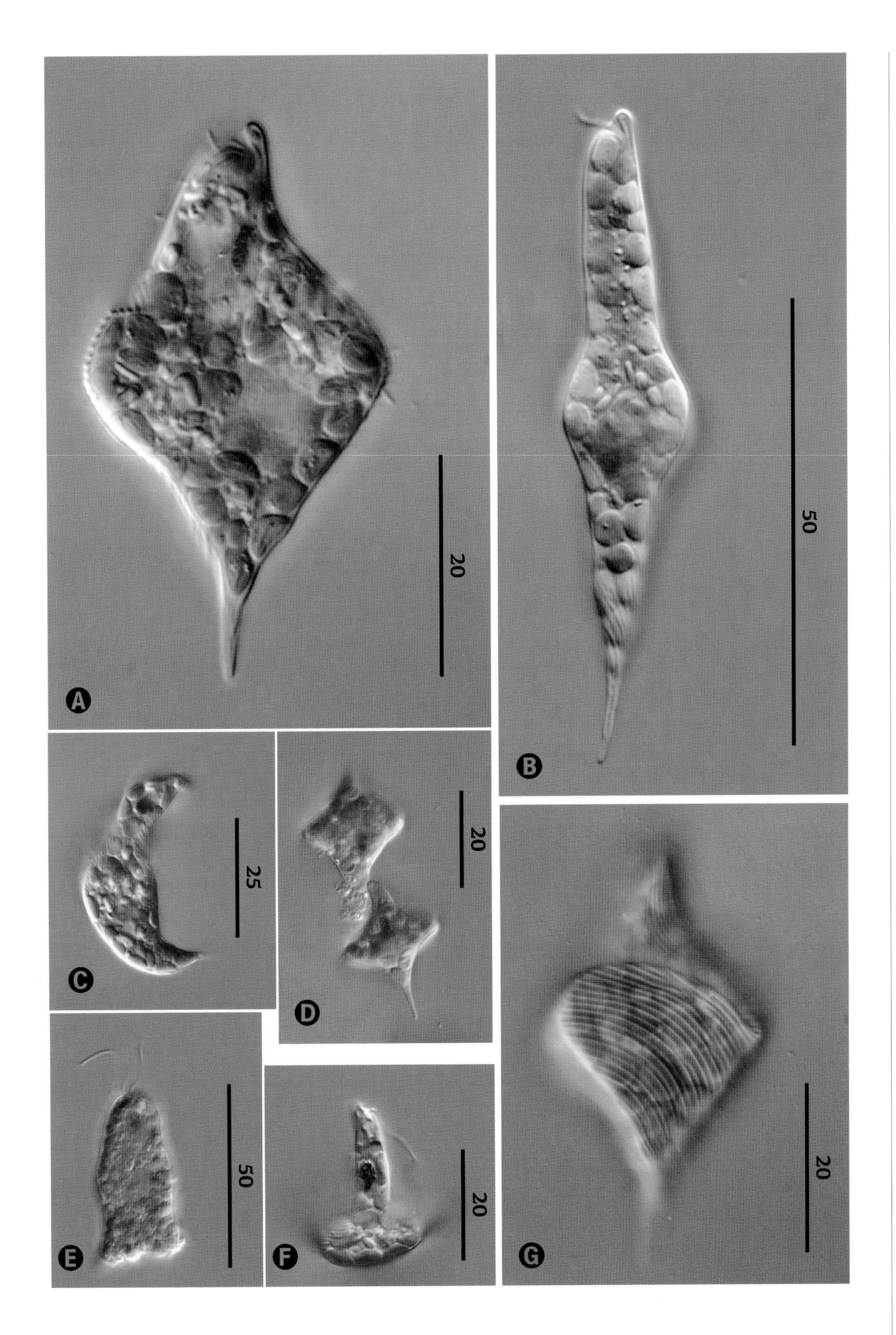

Discoplastis spathirhyncha

(Skuja) Triemer 2006

SIZE: 60–85 µm long × 12–20 µm wide.

Cells showing various shapes assumed during metaboly (A–G). A typical shape of the cell during swimming (B). Spirally arranged pellicular striations (G). Cells contain numerous disc-shaped chloroplasts without pyrenoids (A, B, G); small paramylon bodies present (B, C). A dividing cell with two emergent flagella visible (E); flagellum (B, E, F) usually one-half to three-fourths body length.

EUGLENA

EHRENBERG 1830, 1838

Cells vary greatly in shape with spindle shape being the most common; cell body is asymmetrical and may be somewhat flattened; a single emergent flagellum, but in some species the flagellum is so short that it is not detectable with the light microscope; pellicle finely to visibly striated; chloroplasts may be discoid, shield-shaped, or ribbon-shaped, with smooth or incised margins, with or without pyrenoids; pyrenoids may be naked or sheathed with two paramylon caps (diplopyrenoids); each chloroplast type is characteristic of a particular group of species; some species have lost the ability to photosynthesize and are colorless (e.g., *E. longa, E. quartana*); eyespot present at the anterior end in photosynthetic and some colorless species; cells are never completely rigid but show some level of euglenoid movement (metaboly) ranging from slight to extreme; cysts may be present; found in fresh, brackish, or marine water; there tends to be much variation in shape, metaboly, chloroplasts and pyrenoids, and flagellar length among the species, making identification difficult and many descriptions suspect. *(Note: The genus was first named in 1830 but was not diagnosed until 1838.)*

Euglena agilis

Carter 1856

SIZE: 26–36 µm long × 4–7 µm wide.

Cylindrical cell exhibiting flagellum (one to two times body length), eyespot, and two plate-like chloroplasts, each with a diplopyrenoid (arrows) (**A**, **B**). Encysted cells (**C**).

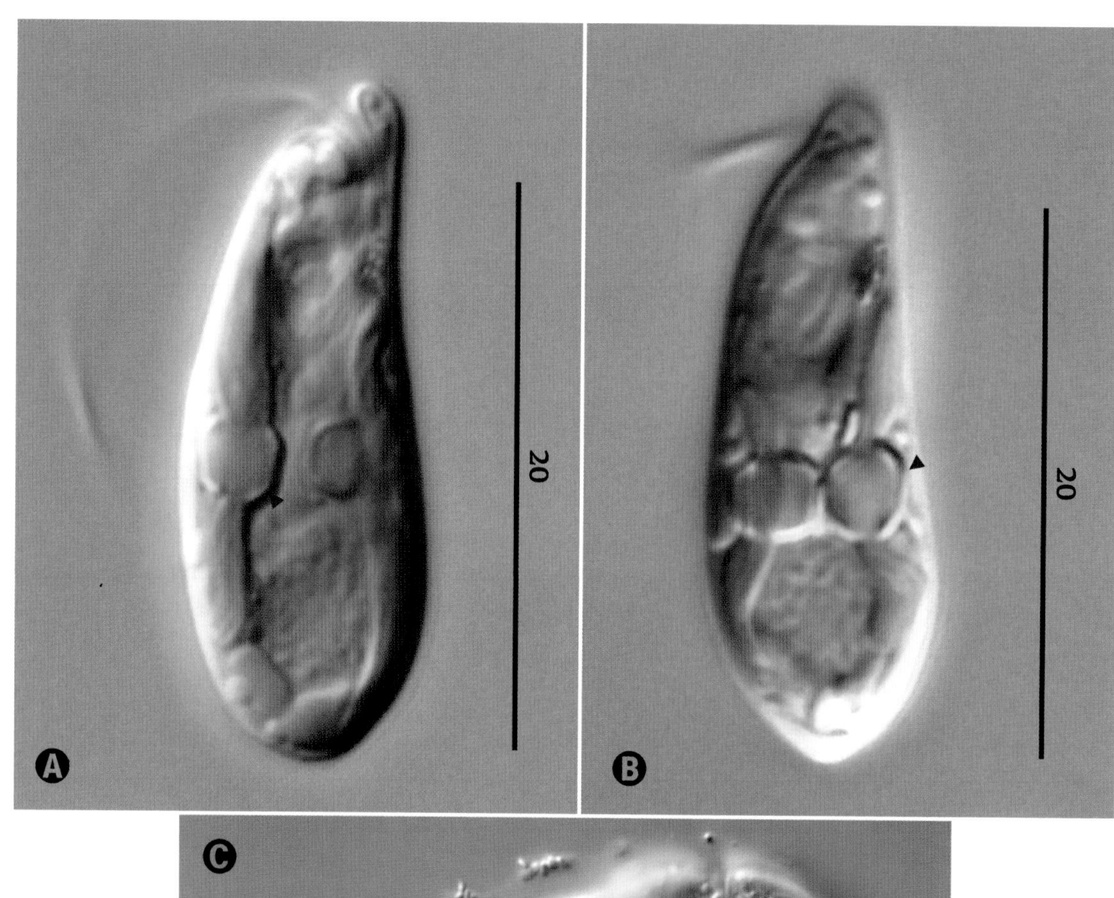

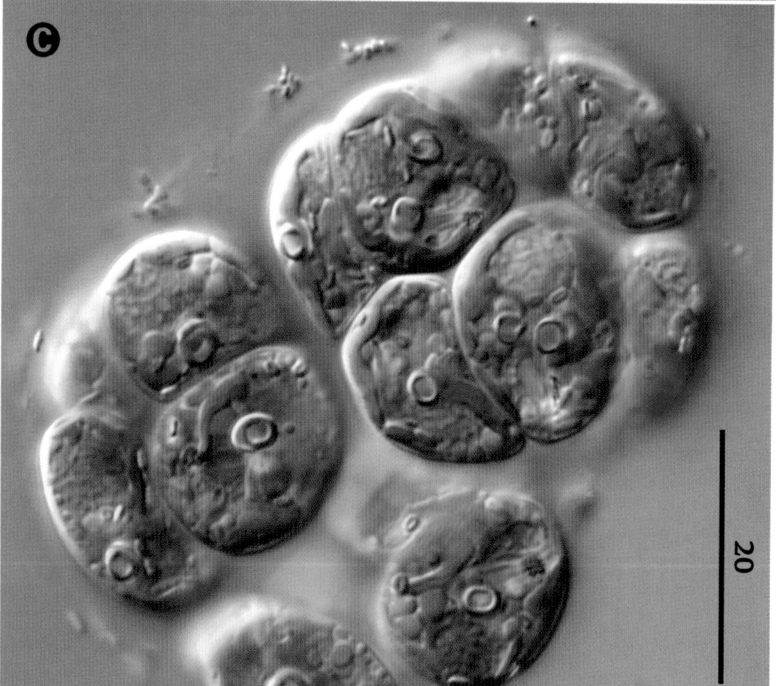

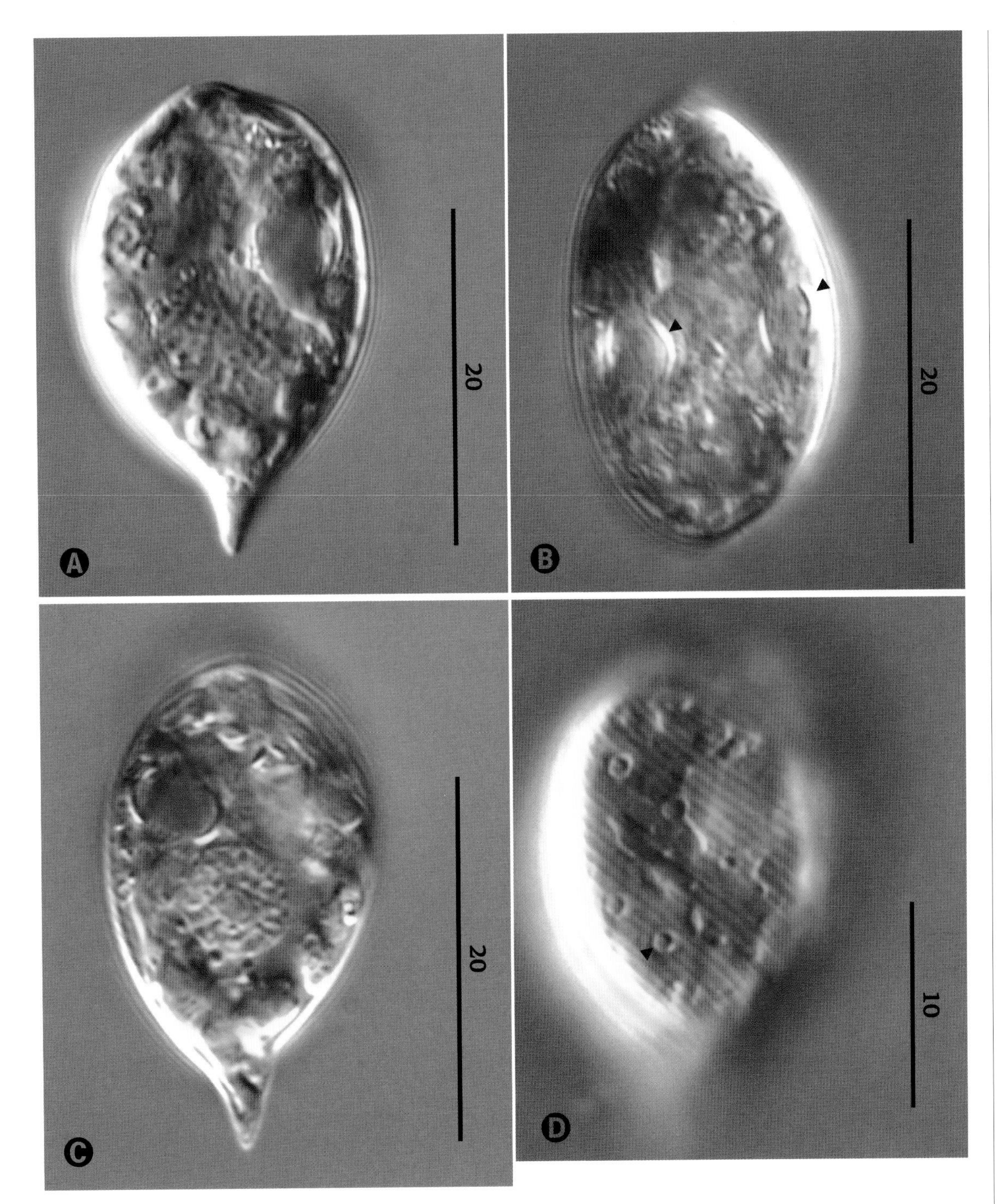

Euglena archaeoplastidiata

Chadefaud 1937

SIZE: 25–50 µm long × 13–22 µm wide.

Ovate swimming cell showing a diplopyrenoid (A). Immobile cell with a large parietal plate-like chloroplast, open in front, and two diplopyrenoids (arrows) (B). Cell exhibiting one diplopyrenoid, central nucleus with nucleolus, and several small paramylon bodies (C). Detail of pellicle with fine spiral strips and spherical subpellicular mucocysts (arrow) (D). Flagellum is about body length.

Euglena archaeoviridis

Zakryś et Walne 1994

SIZE: 33–39 µm long × 7–9 µm wide.

Spindle-shaped cell, almost ovate through metaboly showing flagellum (about body length), canal, eyespot, one axial deeply lobed chloroplast with a large central diplopyrenoid, and small paramylon grains (**A**). Detail of elongated chloroplast lobes extending toward the cell extremities in a stellate pattern (**B**). Pellicle is densely and spirally striated (**C**).

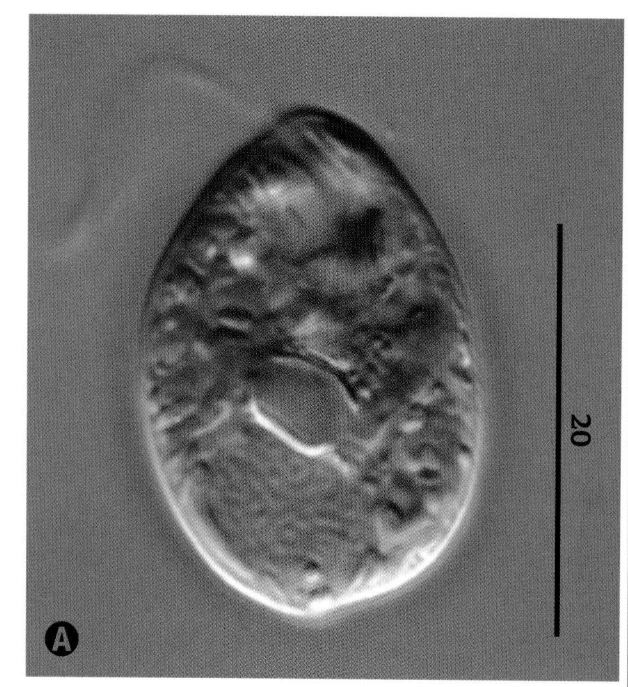

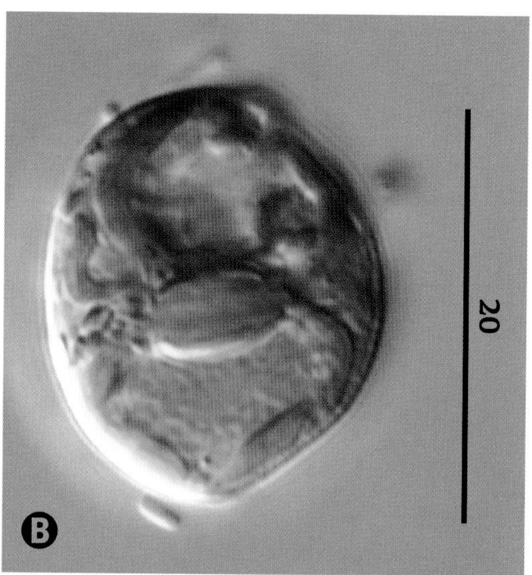

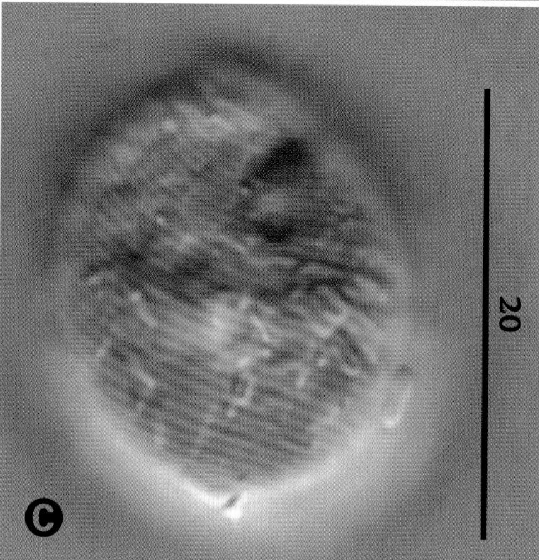

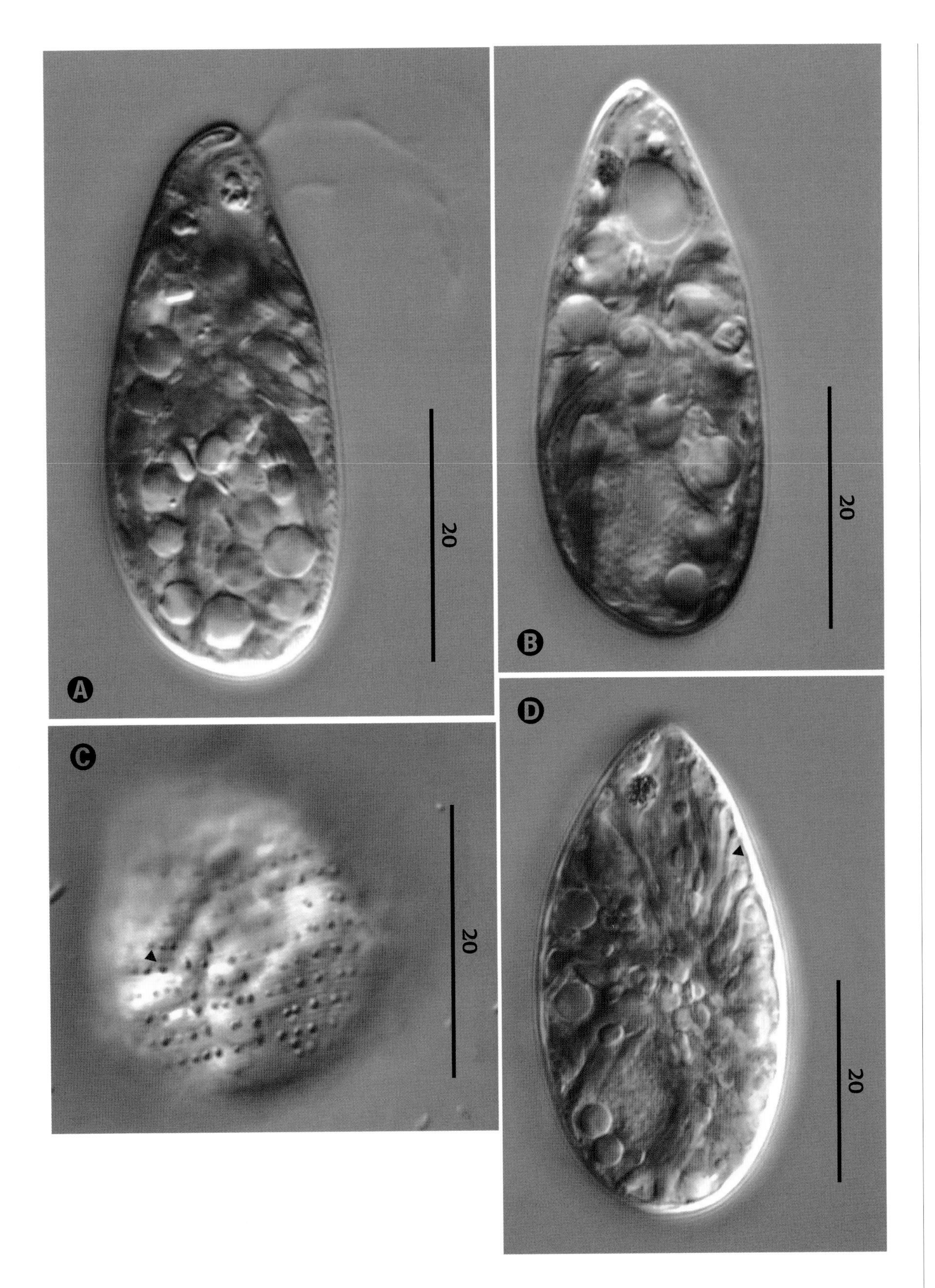

Euglena cantabrica

Pringsheim 1956

SIZE: 54–62 µm long × 20–25 µm wide.

Cell (usually spindle-shaped) showing flagellum (about twice body length), canal, chloroplast, and paramylon grains (**A**). Cell exhibiting eyespot, canal, reservoir, paramylon bodies, and deeply lobed chloroplast (**B**). Metabolic cell; rows of spherical mucocysts (arrow), spirally disposed along fine pellicular striations, are clearly seen in cells stained with neutral red (**C**). Cell with eyespot, ribbon-like lobes of the chloroplast (arrow), and paramylon grains (**D**).

Euglena chlorophoenicea
Schmarda 1846

SIZE: 104–145 µm long × 17–21 µm wide.

Cylindrical cell showing red eyespot, central nucleus, haematochrome granules, and several small disc-shaped chloroplasts toward the posterior (Ⓐ). Cell with chloroplasts, haematochrome granules, and pellicle strips (Ⓑ). Detail of ovoid paramylon bodies (Ⓒ). Several cells (Ⓓ). Cell rounded through metaboly showing pellicle striations and haematochrome granules (Ⓔ). Anterior end of a cell; portion of the flagellum, red eyespot, chloroplasts, and haematochrome granules are visible (Ⓕ).

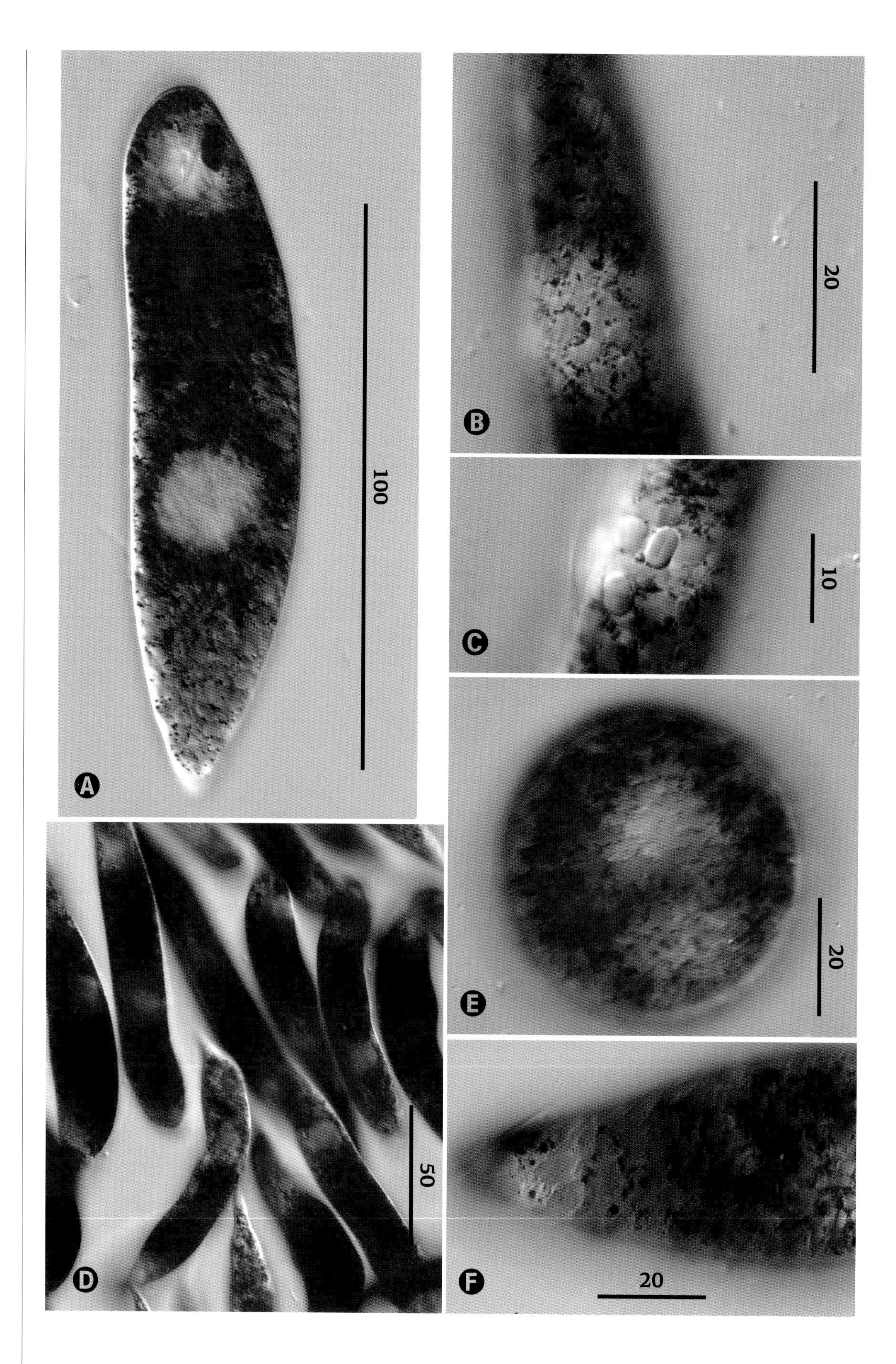

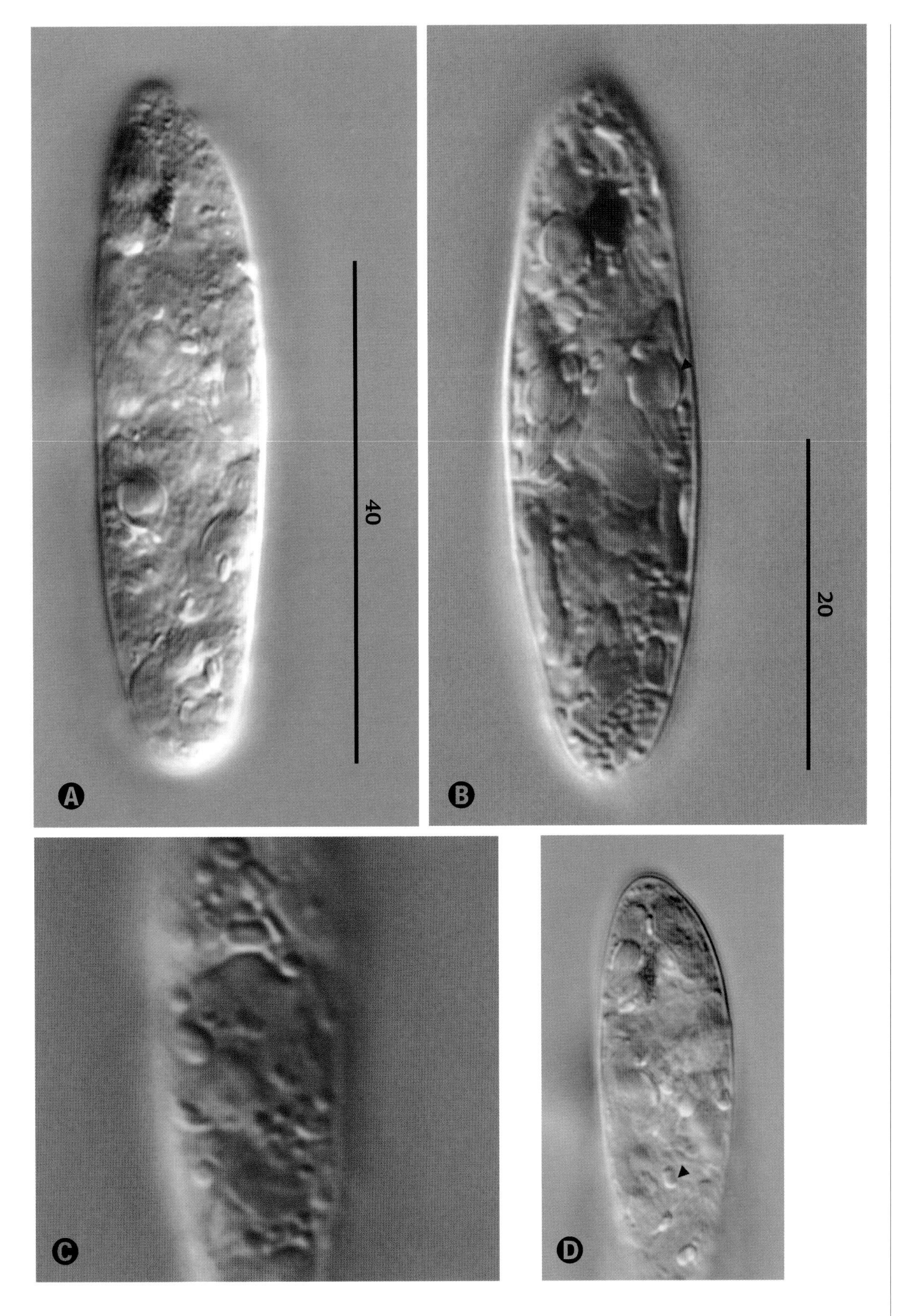

Euglena clara

Skuja 1948

SIZE: 35–78 µm long × 15–25 µm wide.

Two elongate-elliptical cells showing eyespot, disc-shaped chloroplasts, each with a diplopyrenoid (arrow) (Ⓑ), nucleus, and small paramylon grains (Ⓐ, Ⓑ). Detail of chloroplast with uneven margins (Ⓒ). Cell with eyespot and small paramylon bodies (arrow) (Ⓓ). Flagellum is body length.

Euglena convoluta

Korshikov 1941

SIZE: 120–145 μm long × 10–12 μm wide.

Spindle-shaped cell, anterior end truncate with flagellum (about one-sixth body length), red eyespot, chloroplasts, and large paramylon bodies (A). Pellicular striations clearly visible as well as large paramylon bodies (B, F). Cell showing flagellum and eyespot (C). Numerous disc-shaped chloroplasts present; large trough-like paramylon bodies visible (D). Cell exhibiting typical twisted cell body (E).

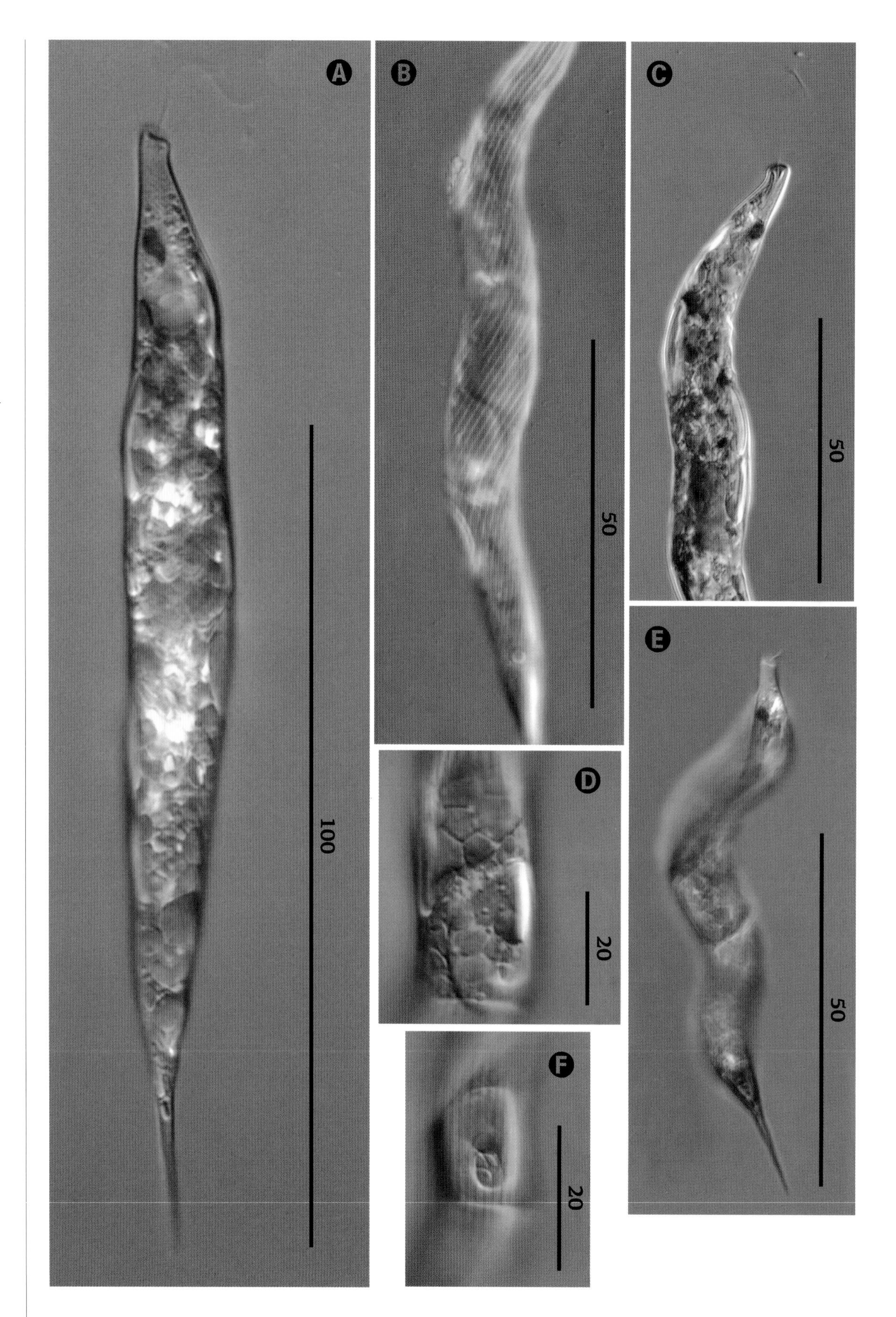

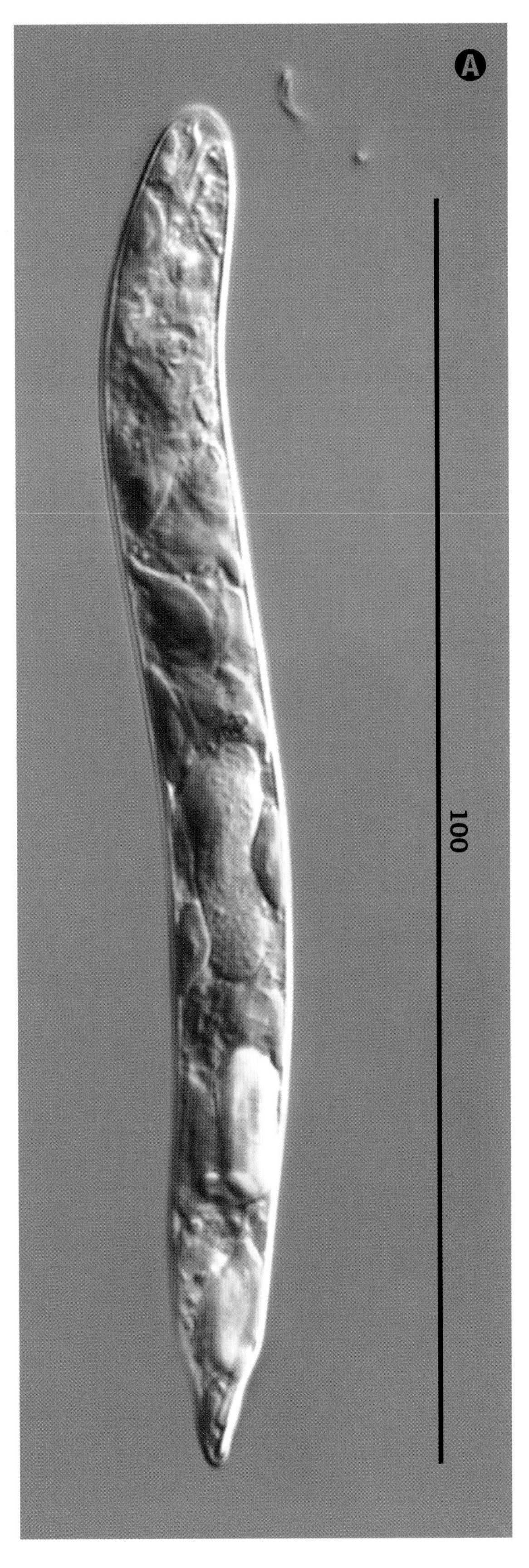

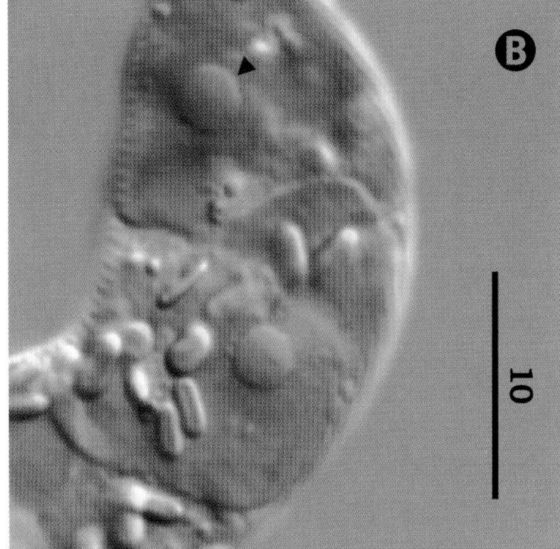

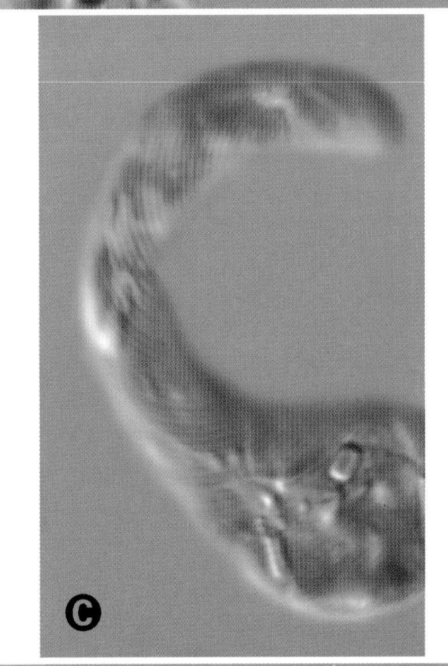

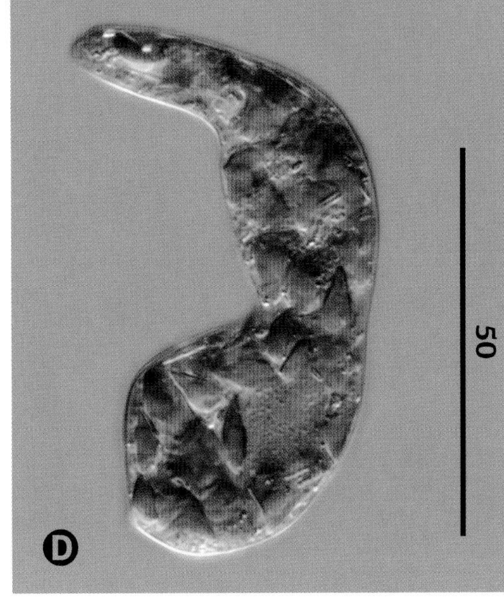

Euglena deses

Ehrenberg 1833

SIZE: 68–206 µm long × 6–23 µm wide.

Typical cylindrical cell with flagellum (one-sixth to one-half body length), eyespot, nucleus, disc-shaped chloroplasts, and posterior paramylon body (**A**). Chloroplast with central naked pyrenoid (arrow) and several small paramylon grains (**B**). Pellicle with fine spiral striations (**C**). Cell in metaboly with eyespot and numerous chloroplasts (**D**).

Euglena ehrenbergii

Klebs 1883

SIZE: 165–373 µm long × 16–50 µm wide.

Cylindrical cell showing flagellum, eyespot, nucleus, and small disc-shaped chloroplasts without pyrenoids (Ⓐ). Pellicular striations spirally disposed (Ⓑ). Cells in metaboly with very long rod-like paramylon bodies (Ⓒ, Ⓓ).

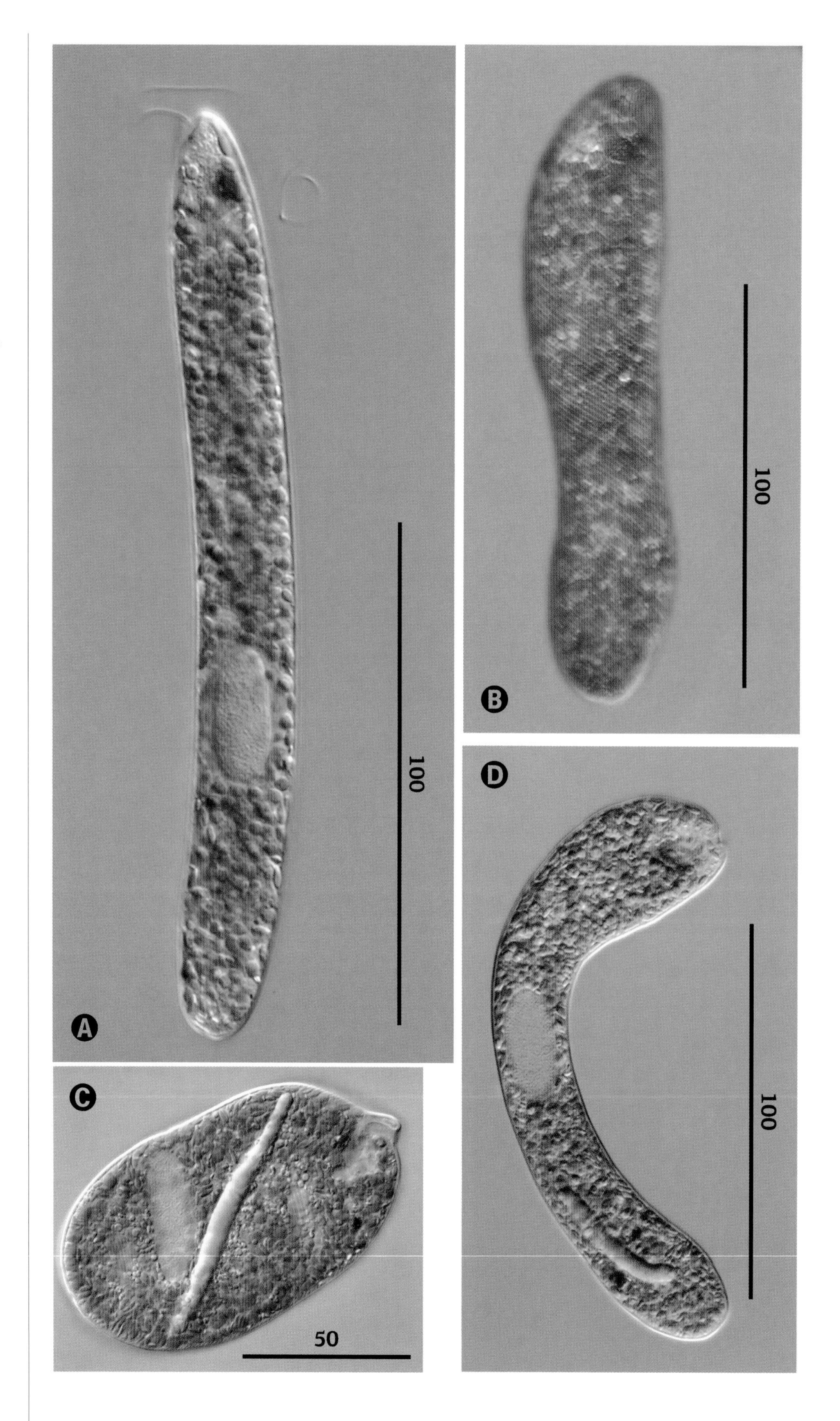

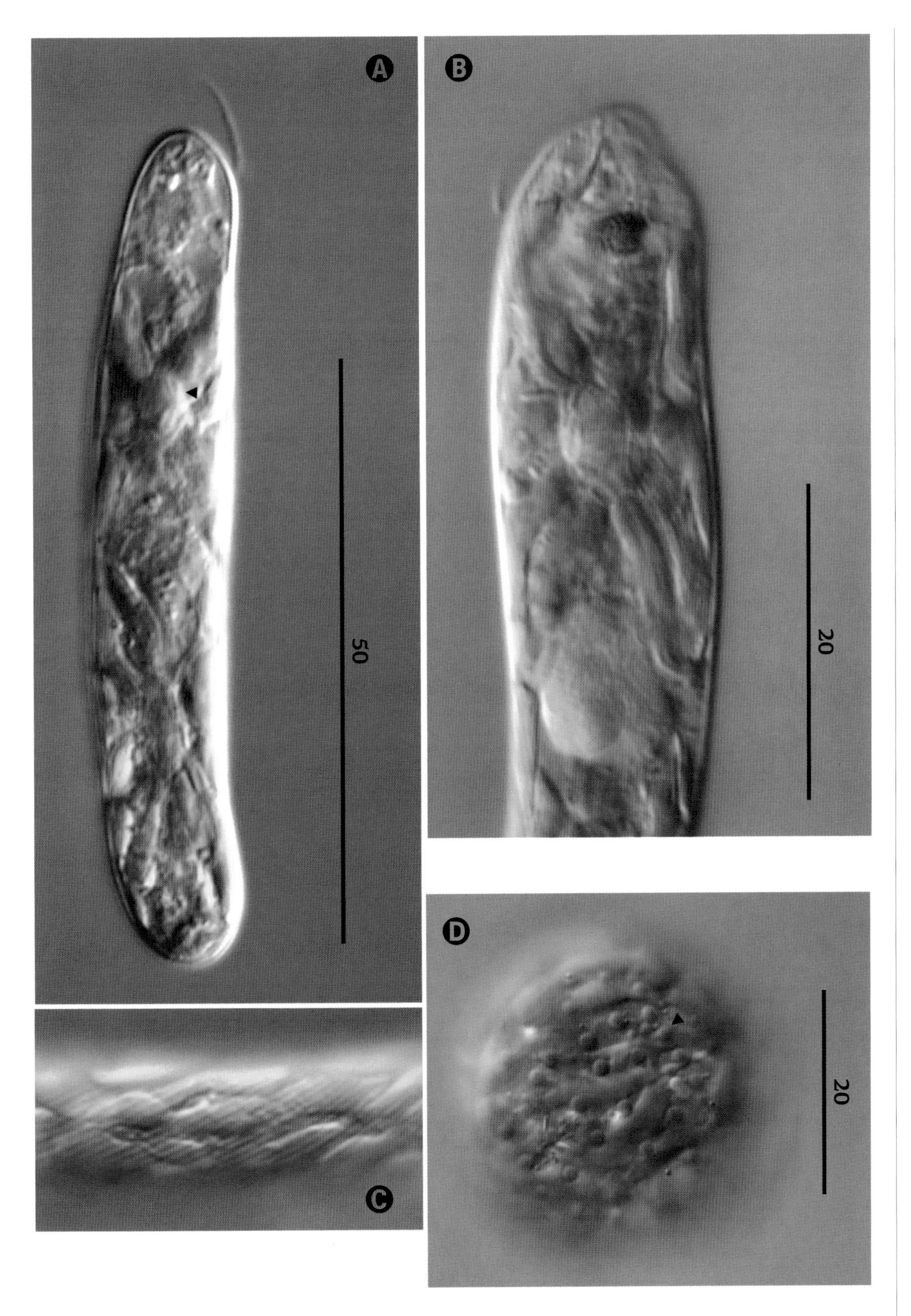

Euglena geniculata

Dujardin 1841

SIZE: 61–88 µm long × 9–18 µm wide.

Cylindrical cell showing flagellum (about body length) and two stellate chloroplasts, each with a paramylon center (arrow) (A). Eyespot, one stellate chloroplast, and nucleus (B). Pellicle is finely striated (C). Spherical mucocysts (arrow) stained with neutral red; cell rounded through metaboly (D).

Euglena gracilis

Klebs 1883

SIZE: 31–75 μm long × 5–12 μm wide.

Cylindrical cell with flagellum (one-half to full body length) and nucleus (Ⓐ). Cell in metaboly with flagellum, eyespot, small paramylon bodies, and diplopyrenoids (arrow) (Ⓑ). Disc-shaped chloroplast with irregularly lobed margin (Ⓒ). Fine spirally arranged pellicular striations (Ⓓ).

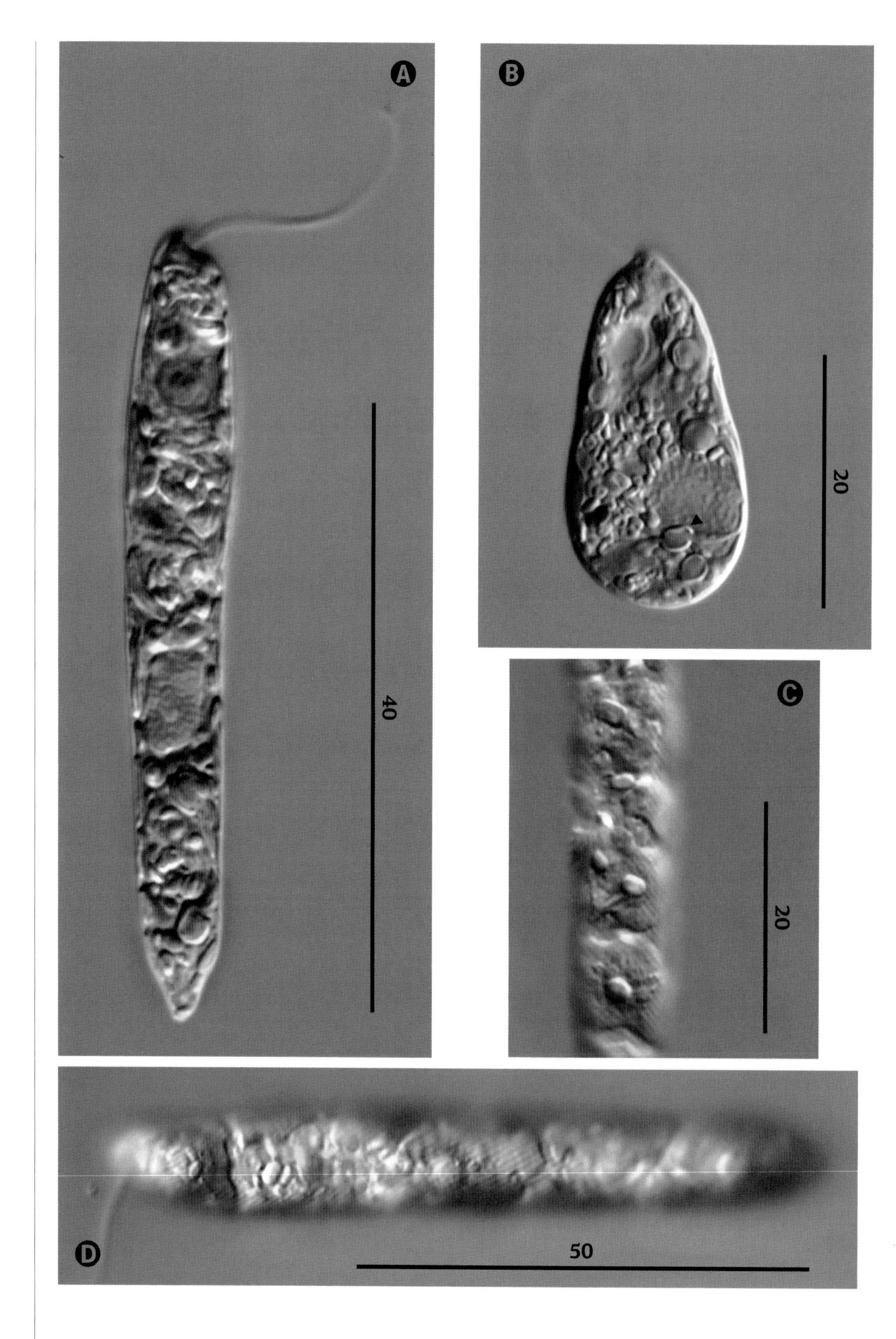

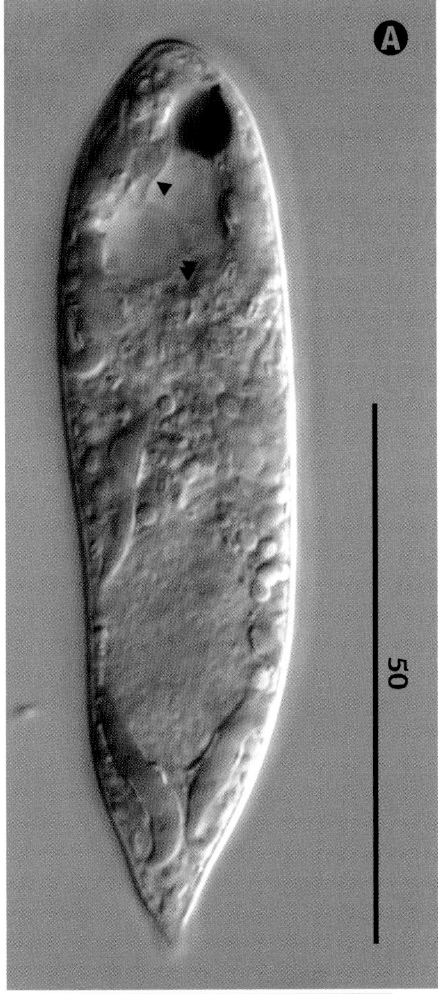

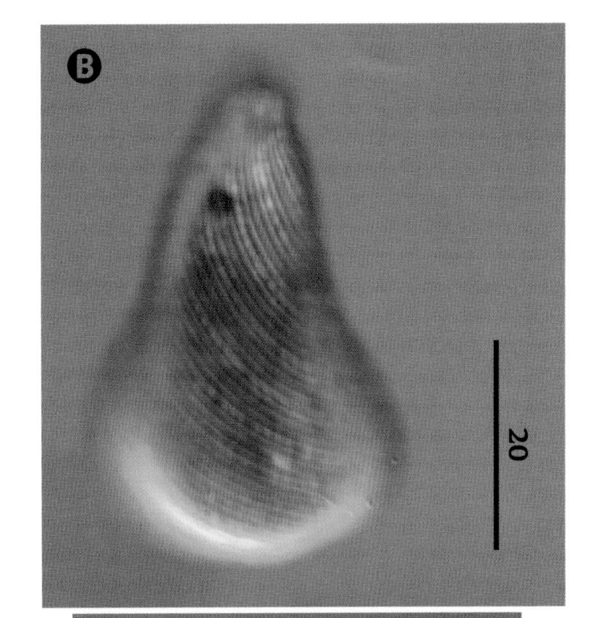

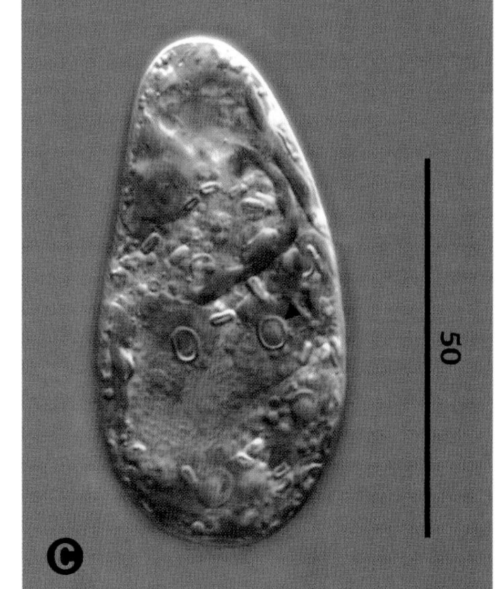

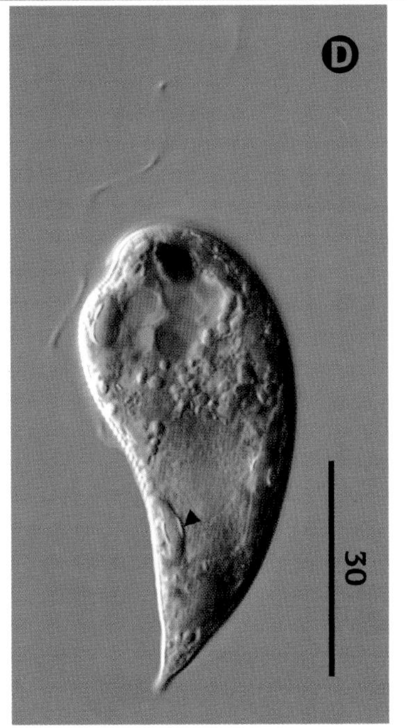

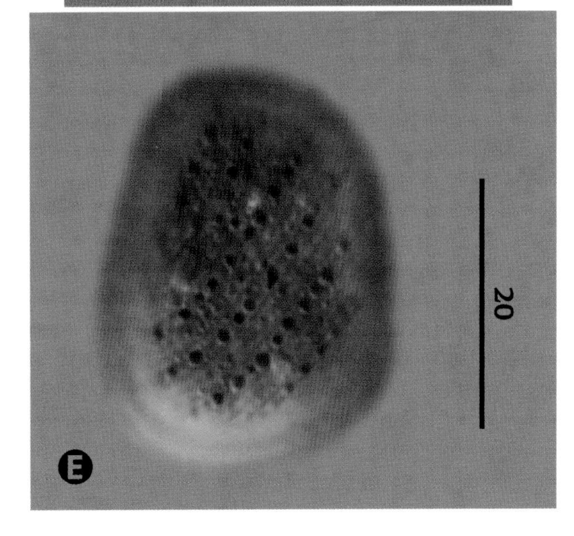

Euglena granulata

(Klebs) Schmitz 1884

SIZE: 66–100 μm long × 15–20 μm wide.

Spindle-shaped cell showing eyespot, basal part of emergent flagellum (arrow), contractile vacuole (double arrow), and disc-shaped chloroplasts with diplopyrenoids (A). Spirally oriented pellicular striations (B). Chloroplasts with lobed margins, diplopyrenoids in face view (arrow), and small paramylon bodies (C). Cell with flagellum (three-fourths to more than body length), eyespot, and chloroplasts with diplopyrenoids (arrow) (D). Mucocysts are spherical and arranged in rows running parallel to the pellicular striations (E, stained with neutral red). Cells are highly metabolic (B–E).

Euglena güntheri

(Günther) Gojdics 1953

SIZE: 67–99 µm long × 5–7 µm wide.

Elongated cell showing red eyespot, hyaline anterior end, disc-shaped chloroplasts, and small paramylon bodies (Ⓐ). Metabolic cell with flagellum, red eyespot, chloroplasts, and paramylon grains (Ⓑ). Detail of chloroplasts, each bearing central naked pyrenoid (arrow) and small paramylon grains (Ⓒ). Pellicle striations are fine (Ⓓ).

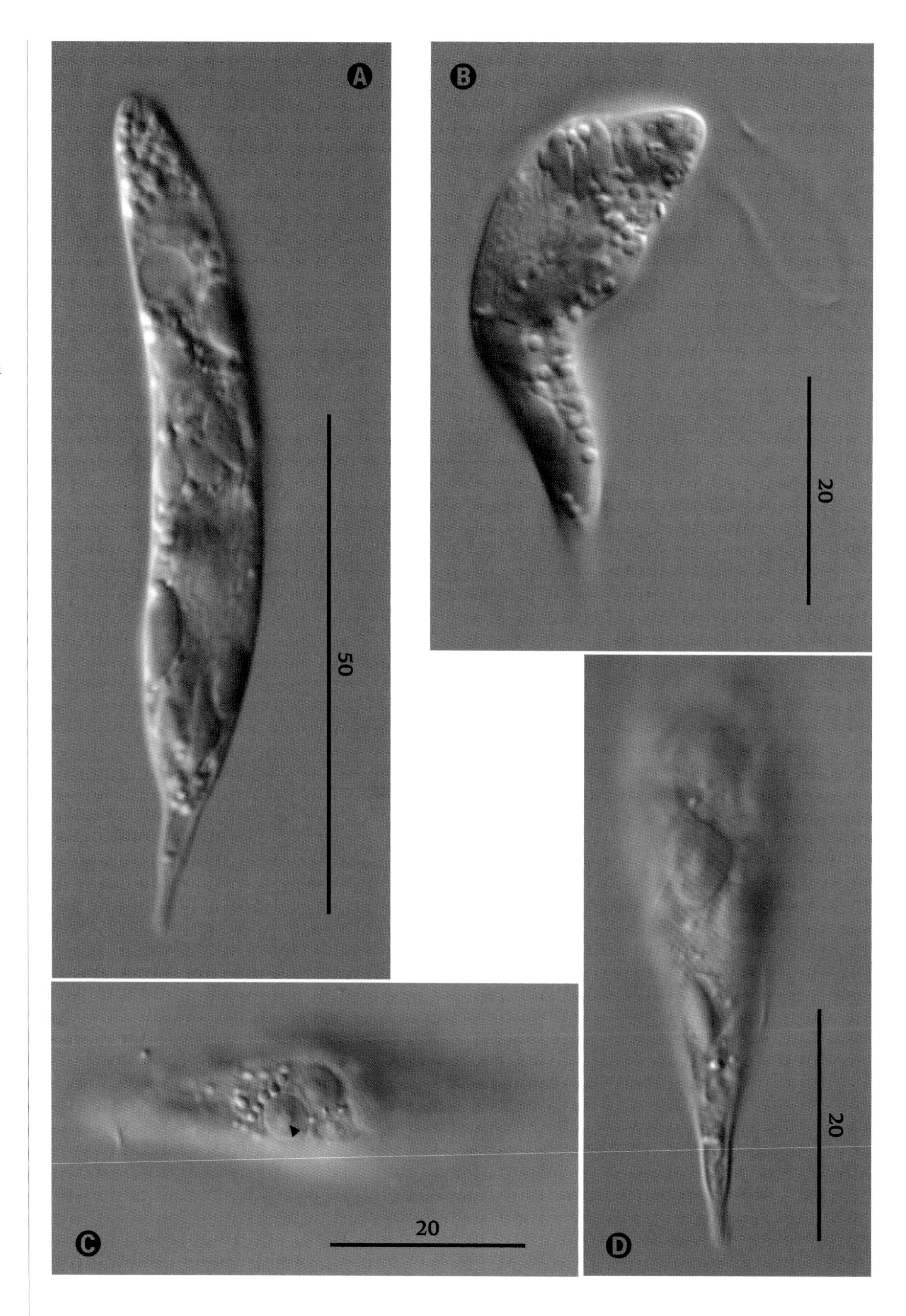

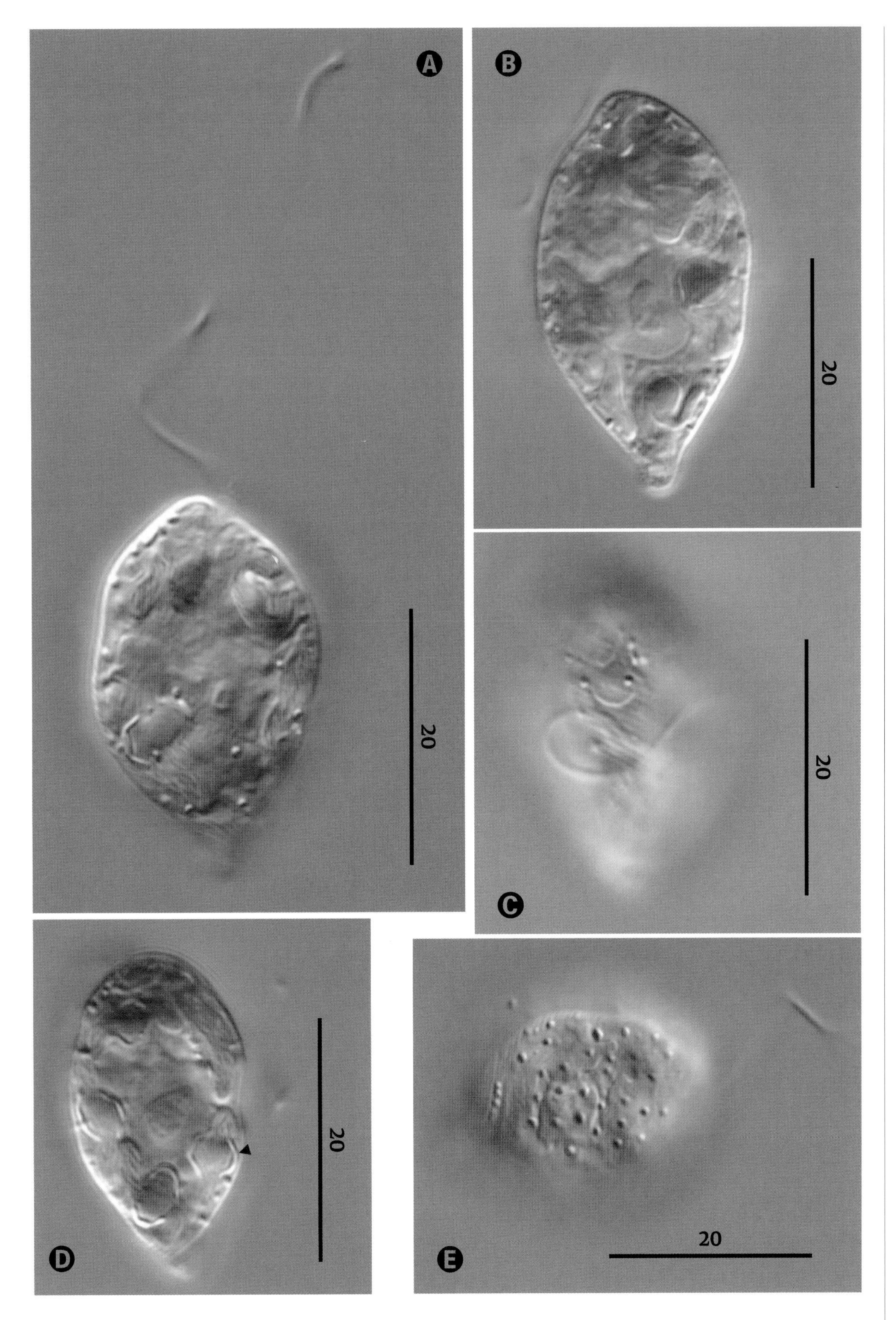

Euglena gymnodinioides

Zakryś 1986

SIZE: 27–36 µm long × 15–21 µm wide.

Cell with long flagellum (one-and-a-half to two times body length), red eyespot, and a few diplopyrenoids (A). Broadly spindle-shaped cell with flagellum, red eyespot, disc-shaped chloroplasts with diplopyrenoids, and central nucleus (B). Detail of spirally fine striated pellicle, with flagellum also visible (C). Multiple diplopyrenoids (arrow) (D). Chloroplasts with uneven margins, spherical mucocysts, and flagellum (E).

Euglena hiemalis

Matvienko 1938

SIZE: 46–67 µm long × 7–15 µm wide.

Cylindrical cell showing plate-like chloroplasts with diplopyrenoids (arrow) and small ovoid paramylon bodies (Ⓐ). Chloroplasts (arrow) and paramylon bodies (Ⓑ). Metabolic cell showing flagellum (body length or slightly shorter), red eyespot, reservoir (arrow), chloroplasts, and paramylon bodies (Ⓒ). Spiral pellicle striations are faint (Ⓓ).

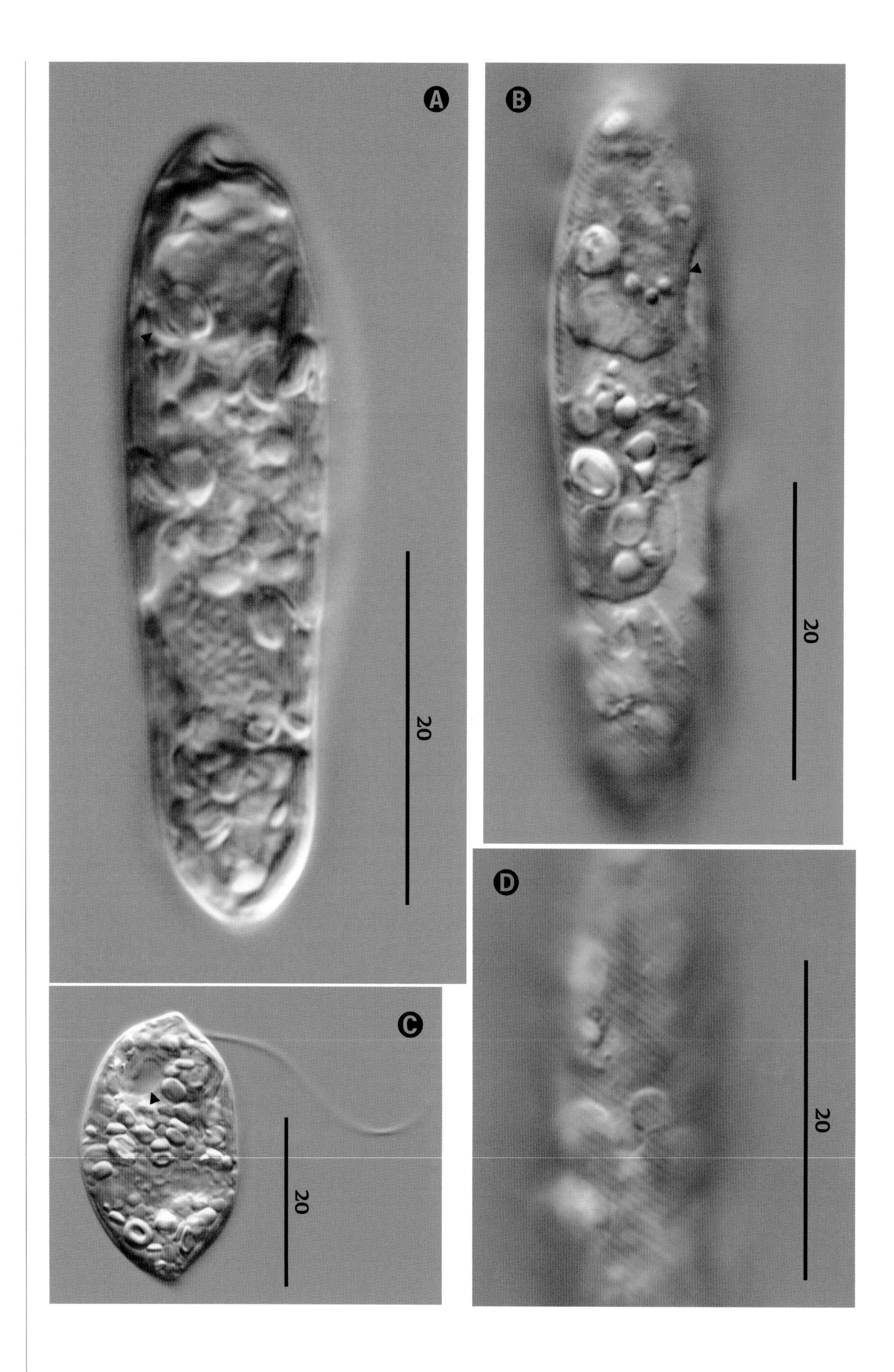

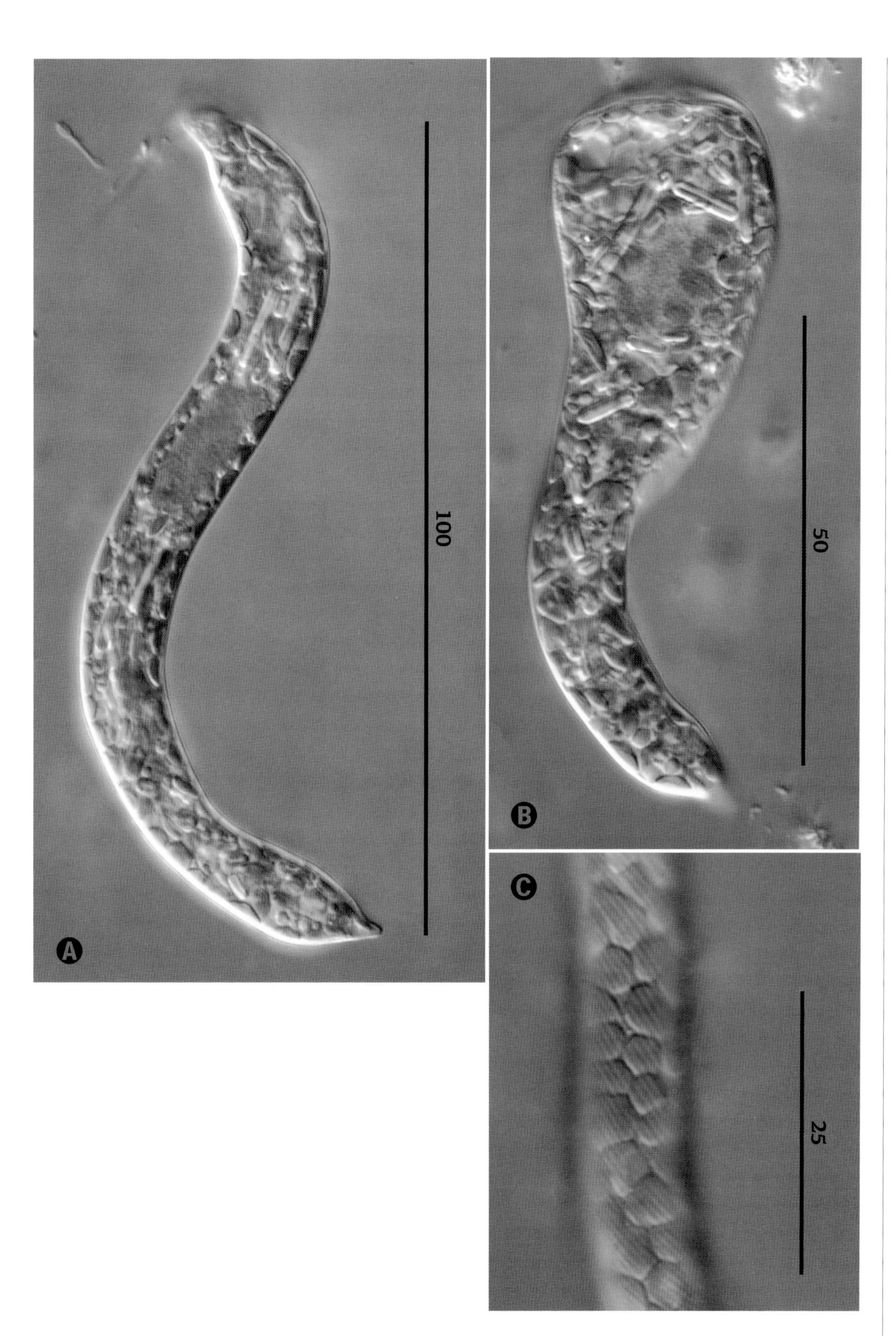

Euglena intermedia

(Klebs) Schmitz 1884

SIZE: 85–165 µm long × 8–20 µm wide.

Cylindrical cell with flagellum (about one-sixth body length), eyespot, and rod-like paramylon bodies (A). Cell in metaboly (B). Pellicular striations and numerous disc-shaped chloroplasts lacking pyrenoids (C).

Euglena klebsii

(Lemmermann) Mainx 1927

SIZE: 76–100 µm long × 6–9 µm wide.

Cylindrical cell showing red eyespot, large plate-like chloroplasts, rod-like paramylon bodies, and nucleus toward the posterior (A). Pellicle is finely and spirally striated (B).

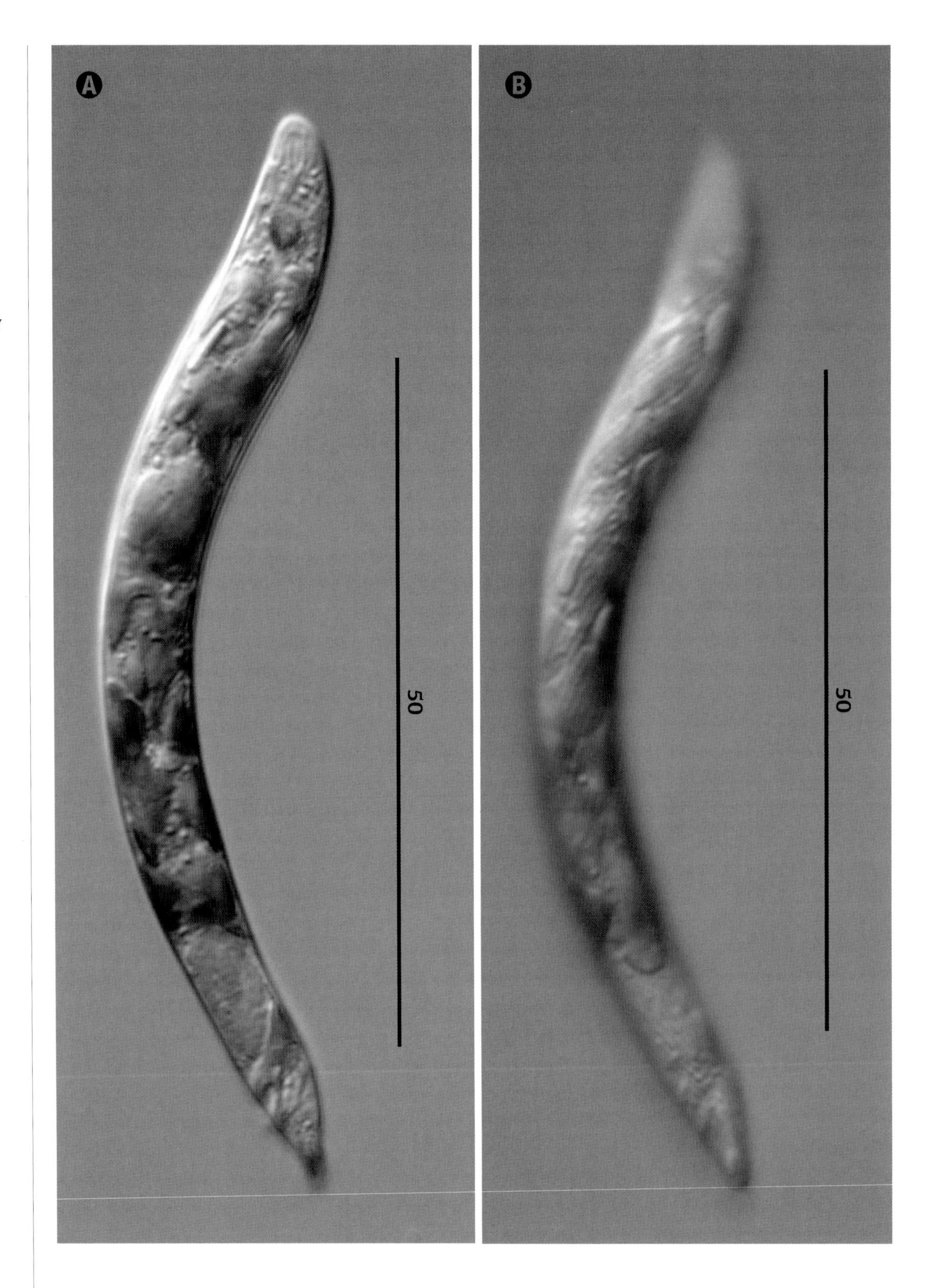

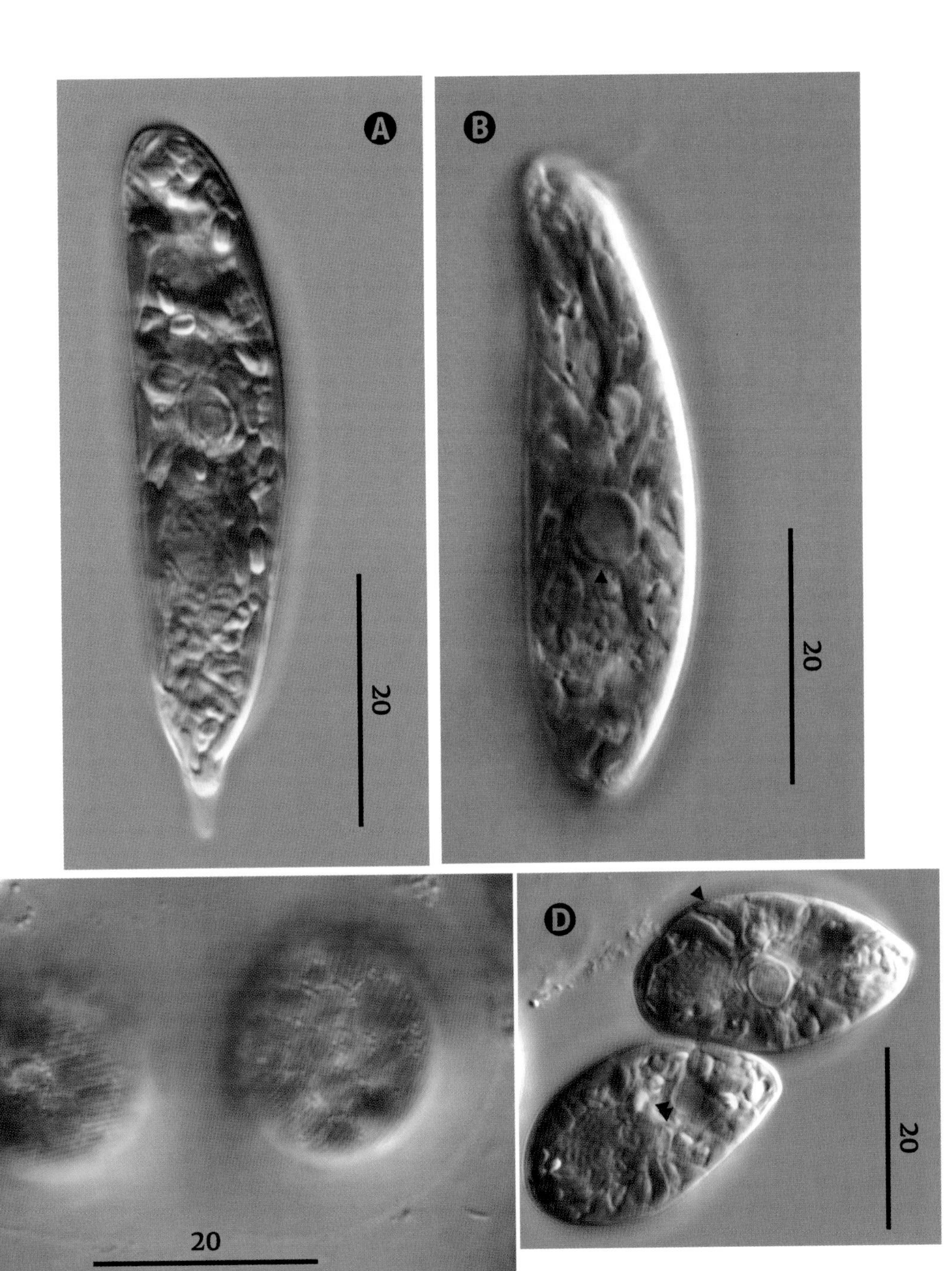

Euglena laciniata

Pringsheim 1956

SIZE: 65–72 µm long × 17–20 µm wide.

Fusiform cell showing eyespot, chloroplast with diplopyrenoid, and small paramylon grains (A). Cell with flagellum (body length or longer) and band-like extensions of the fimbriate chloroplast radiating from a large central diplopyrenoid (arrow) (B). Two encysted cells with spiral striations (C). Two cells showing lobes of the chloroplast (arrow), contractile vacuole (double arrow), and eyespot (D). Cell stained with neutral red showing spindle-shaped mucocysts (arrow) (E).

Euglena longa

(Pringsheim) Marin et Melkonian 2003

SIZE: 50–60 µm long × 5–8 µm wide.

Elongate cylindrical, colorless cells with flagellum (up to one-half body length) and numerous paramylon bodies (Ⓐ, Ⓑ). Cell showing both flagellar bases (arrows) and numerous paramylon grains (Ⓒ). Metabolic cell with flagellum, paramylon bodies, and nucleus with nucleolus (arrow) (Ⓓ).

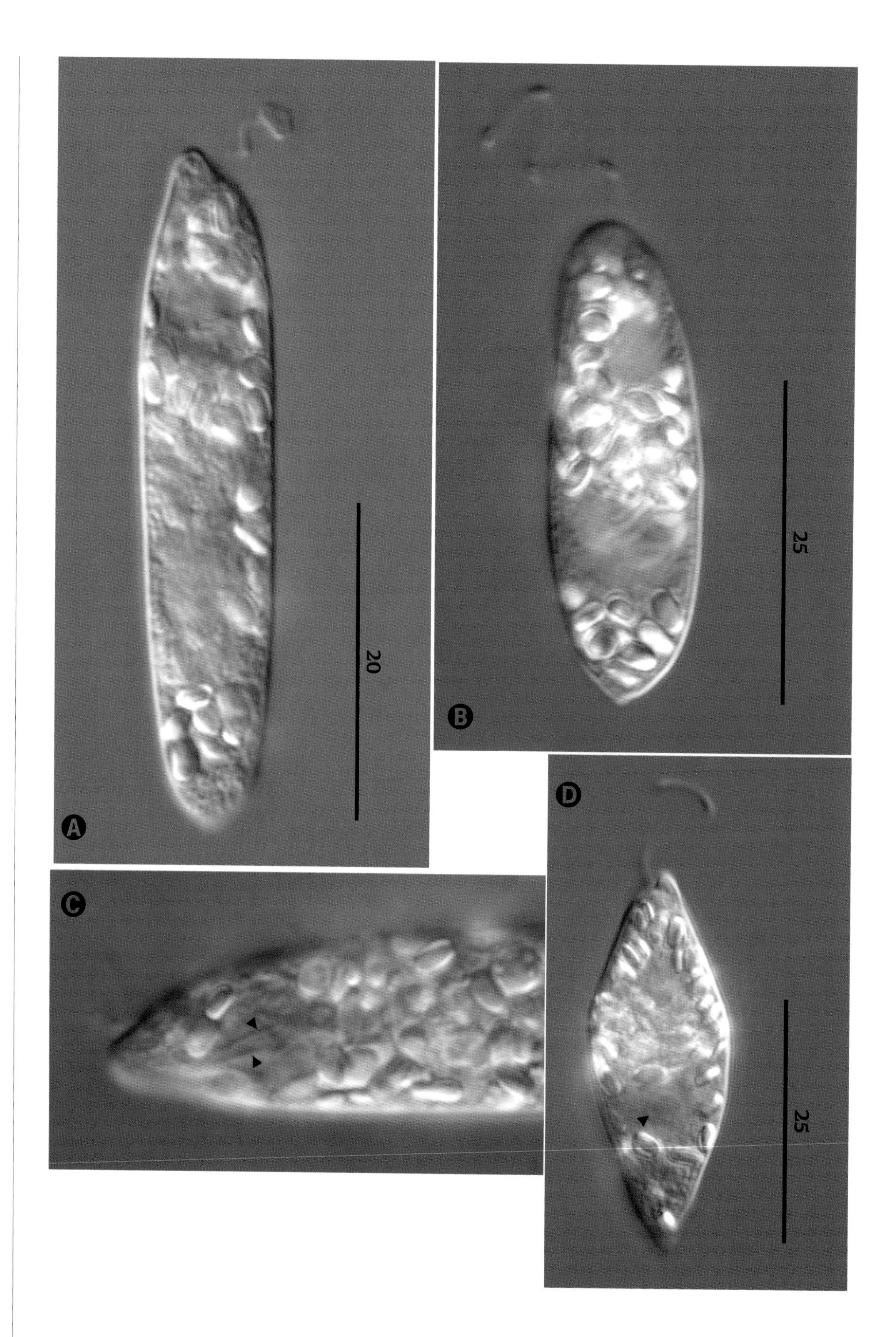

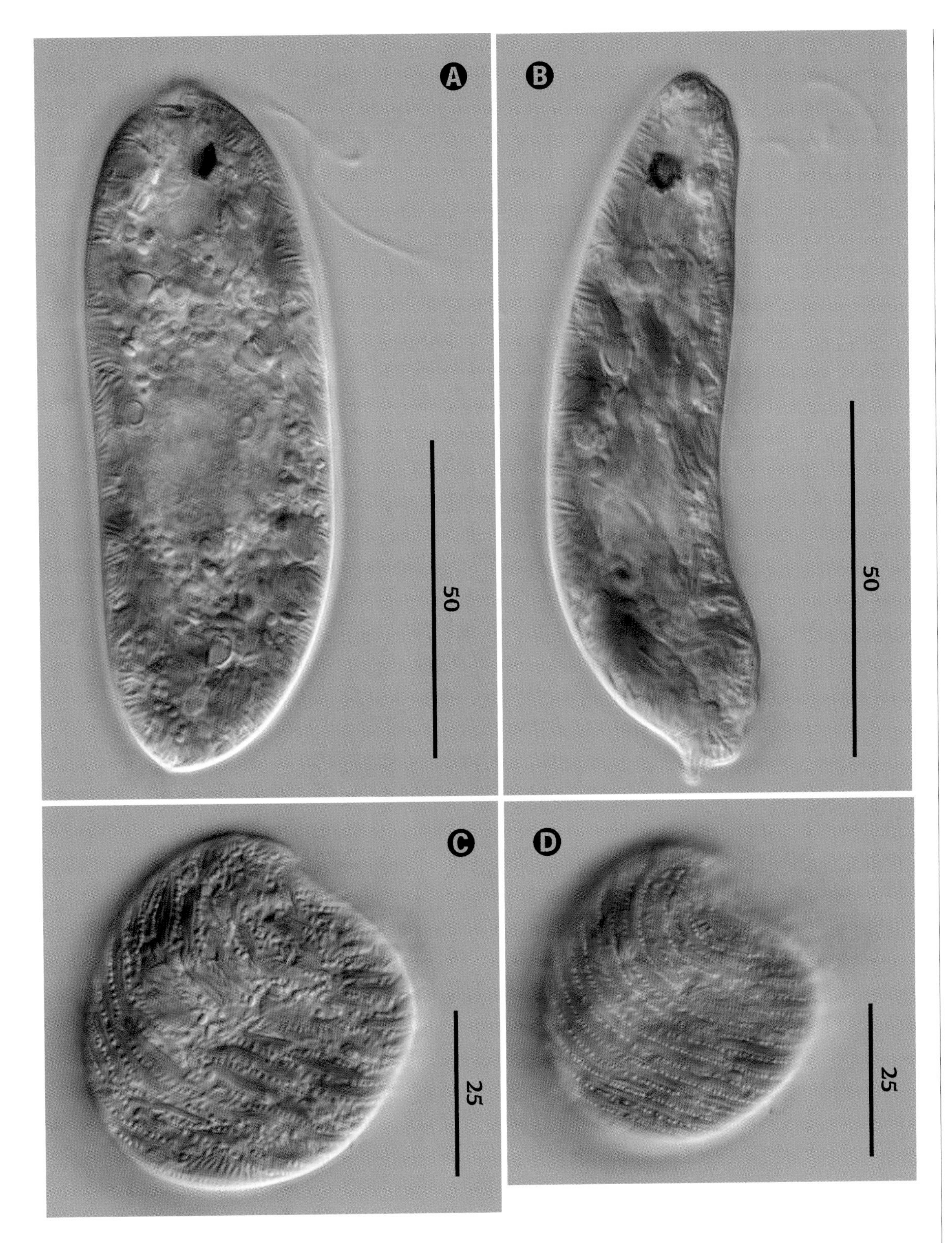

Euglena magnifica

Pringsheim 1956

SIZE: 90–120 µm long × 25–35 µm wide.

Cylindrical cells with long flagellum, red eyespot, chloroplasts with diplopyrenoids, small paramylon grains, and spindle-shaped mucocysts visible along the lateral sides of the cell (A, B). Metabolic cell showing chloroplasts, small paramylon bodies, and spindle-shaped mucocysts (C). Detail of finely striated pellicle with rows of spindle-shaped mucocysts following pellicle strips; chloroplasts with deeply incised margins; ribbon-like chloroplast lobes extend toward the cell surface and are spirally arranged; mucocysts appear spherical when viewed end on (D).

Euglena mutabilis

Schmitz 1884

SIZE: 70–122 µm long × 4–12 µm wide.

Elongate cylindrical cell with red eyespot, plate-like chloroplasts, small paramylon grains, and central nucleus; naked pyrenoids are visible (arrow) (A). Finely striated pellicle and spherical mucocysts scattered randomly (B). Chloroplast with naked pyrenoid (arrow) (C). Metabolic cell showing chloroplasts with naked pyrenoids (D). Emergent flagellum not observed.

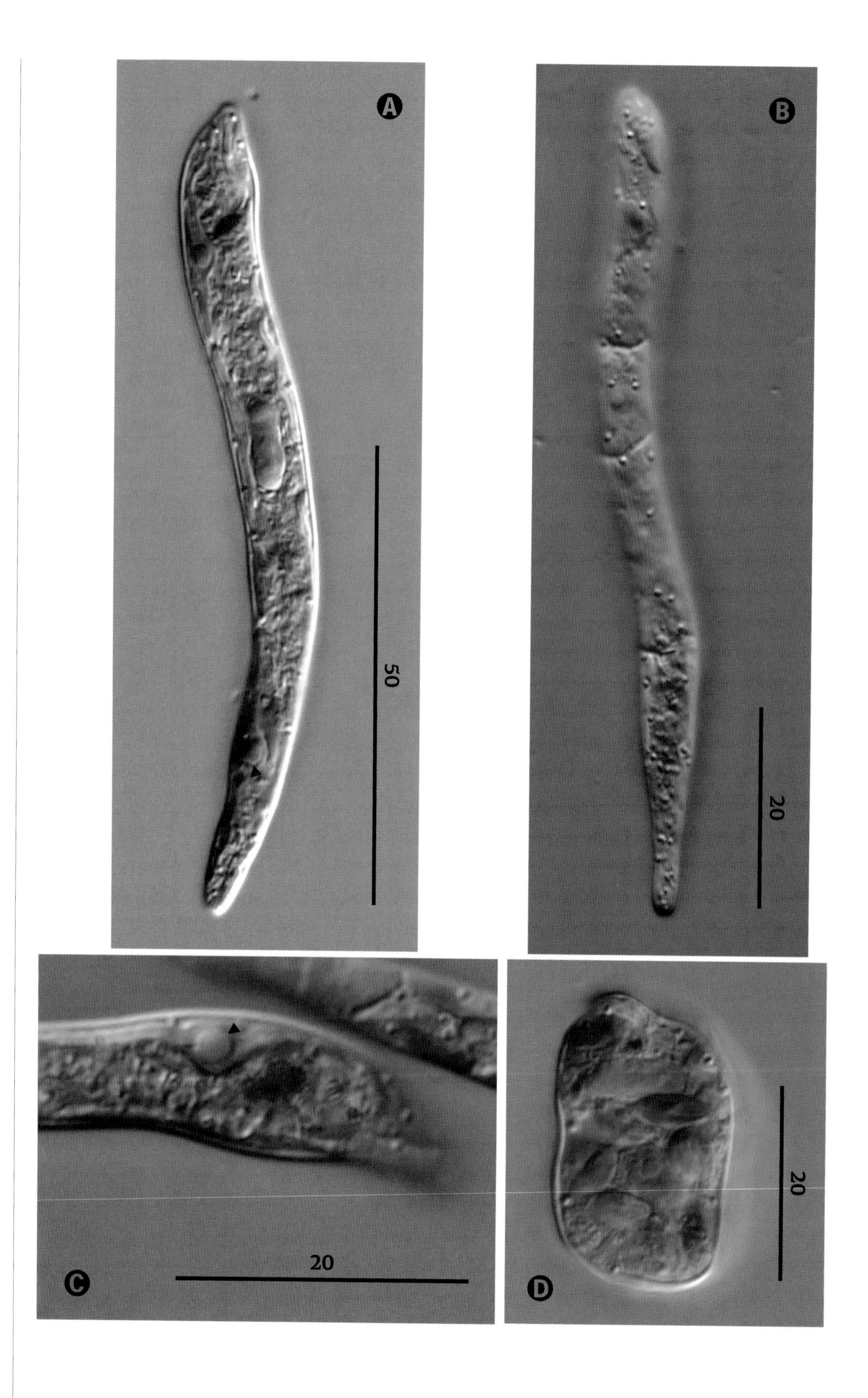

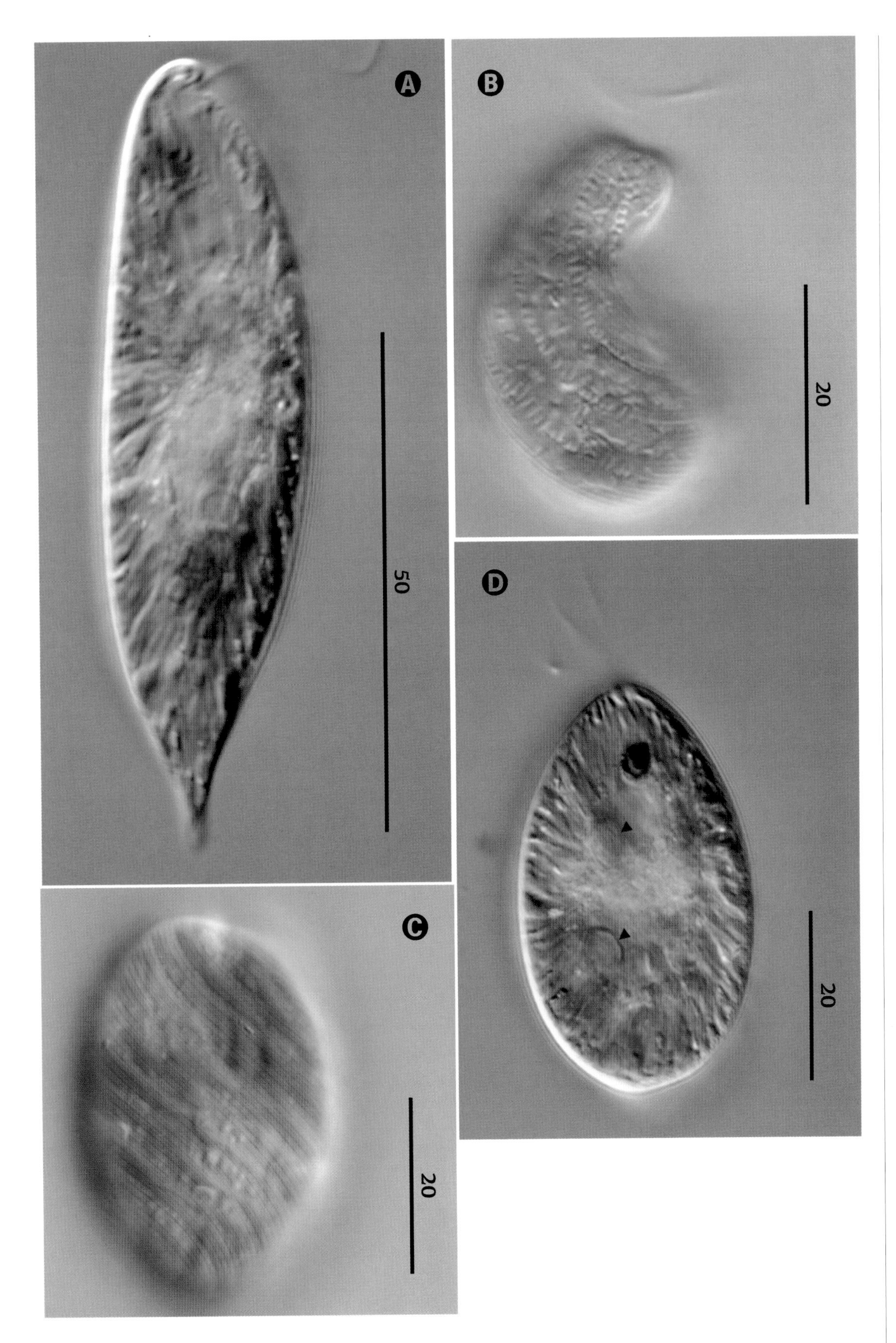

Euglena oblonga

Schmitz 1884

SIZE: 36–76 µm long × 17–36 µm wide.

Spindle-shaped cell exhibiting flagellum (three-fourths to full body length), eyespot, and chloroplasts (**A**). Cell in metaboly with flagellum, fine pellicular striations, and rows of spindle-shaped mucocysts along striations (**B**). Cell showing fine, spirally striated pellicle and few small paramylon bodies (**C**). Metabolic cell with flagellum, red eyespot, and deeply incised chloroplasts with long ribbon-shaped lobes; two diplopyrenoids are visible (arrows) (**D**).

Euglena proxima

Dangeard 1901

SIZE: 60–90 µm long × 11–25 µm wide.

Spindle-shaped cell with eyespot, chloroplasts lacking pyrenoids, and several paramylon grains (Ⓐ). Same cell in a different focal plane showing flagellum (one-third to one-and-a-half times body length), fine pellicular striations, and disc-shaped chloroplasts (Ⓑ). One cell rounded in metaboly next to a cell showing pellicular striations (Ⓒ).

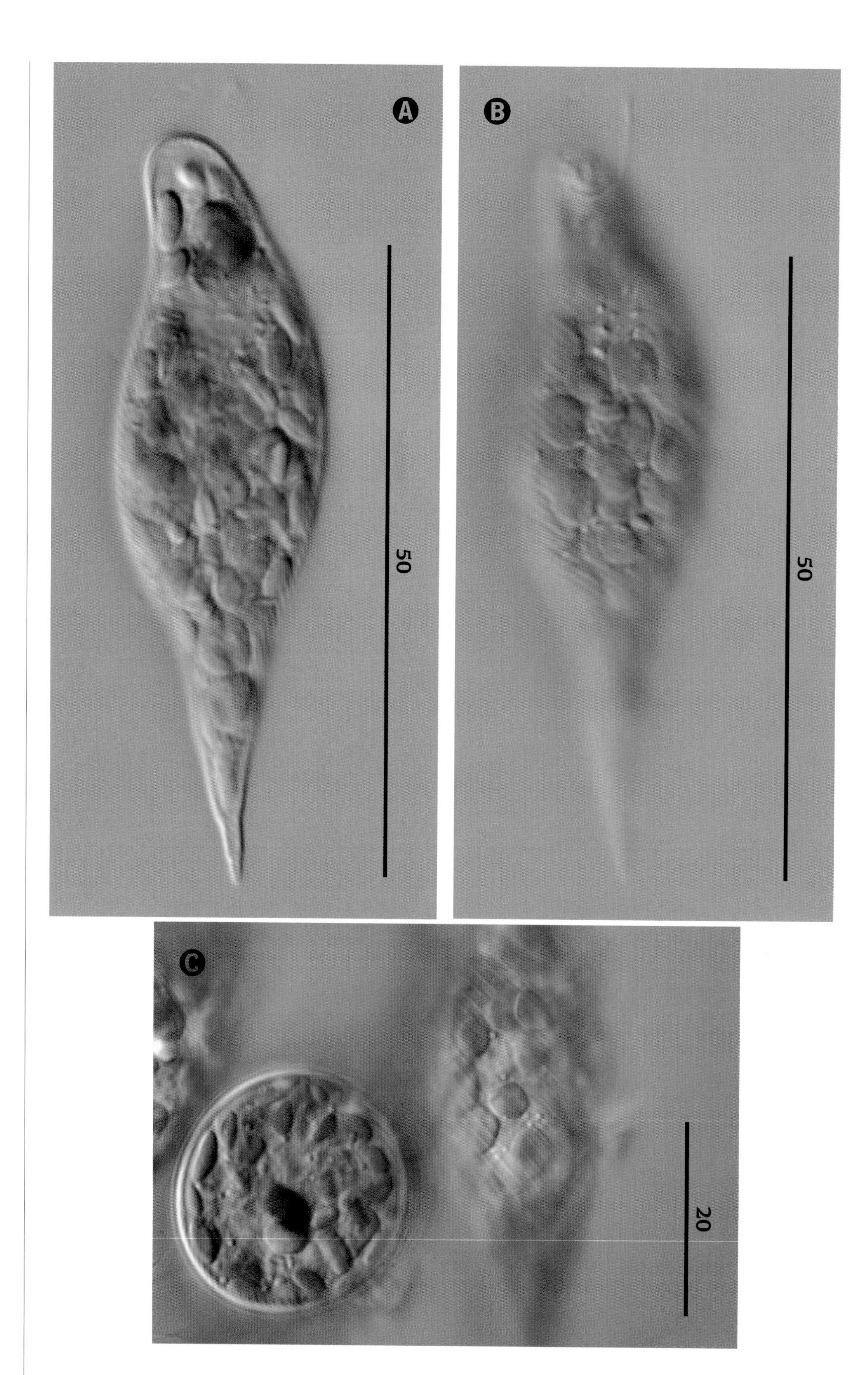

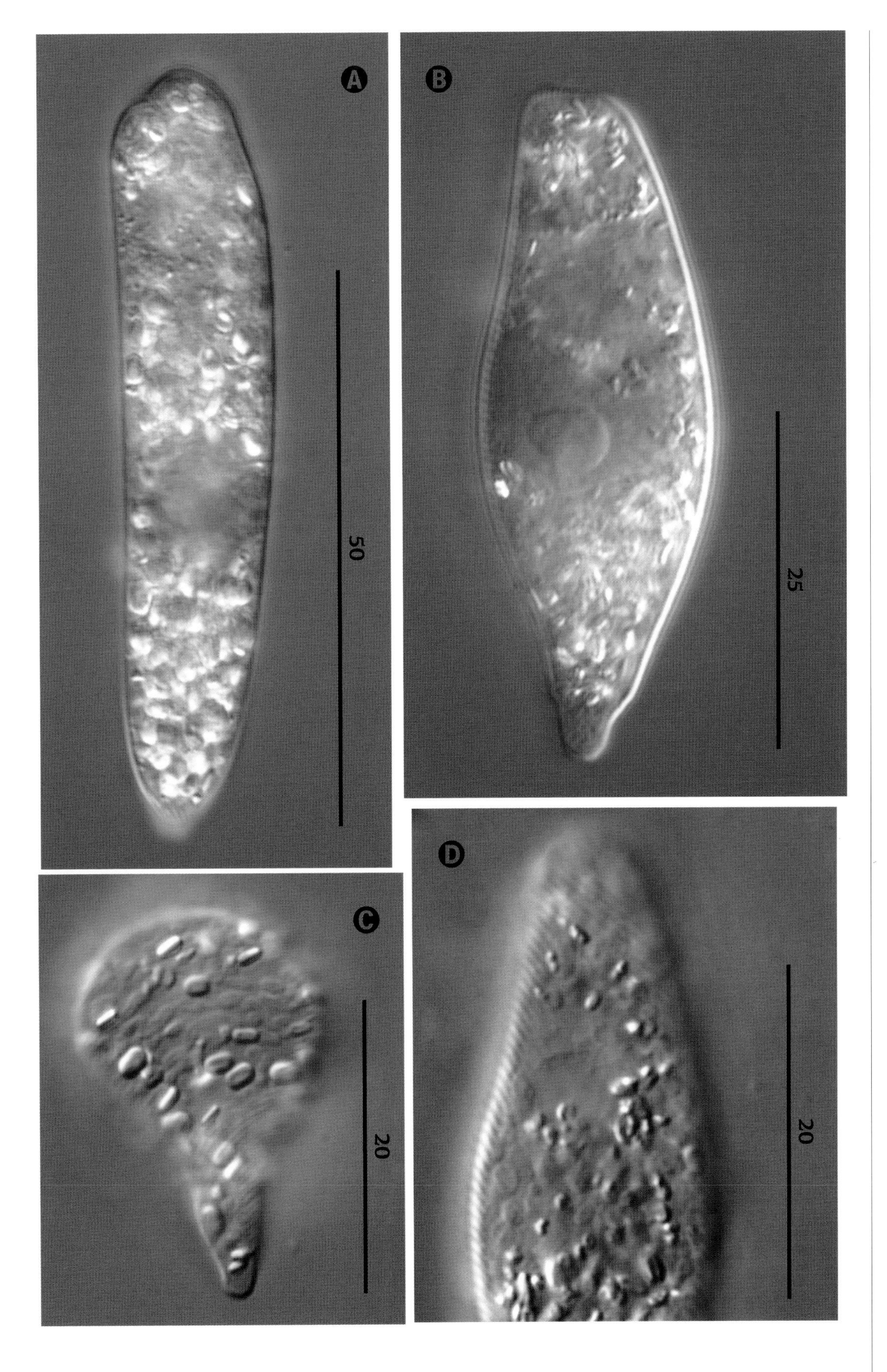

Euglena quartana

Moroff 1904

SIZE: 50–65 µm long × 14–15 µm wide.

Colorless, spindle-shaped cells, orange eyespot, central nucleus with nucleolus (B), and numerous paramylon bodies (A, B). Cell changes shape through metaboly; ellipsoid paramylon grains visible (C). Fine spirally arranged pellicle striations (D). Flagellum is one-and-a-half times body length.

Euglena rostrifera

Johnson 1944

SIZE: 90–145 µm long × 14–48 µm wide.

Spindle-shaped cell with anterior rostrum; red eyespot, central nucleus with nucleolus, chloroplasts with diplopyrenoids, and small paramylon grains (A). Cell with disc-shaped chloroplasts, each with diplopyrenoid (B). Pellicle striations are spirally arranged (C). Cell showing chloroplasts with diplopyrenoids and small paramylon bodies (D). Flagellum is body length.

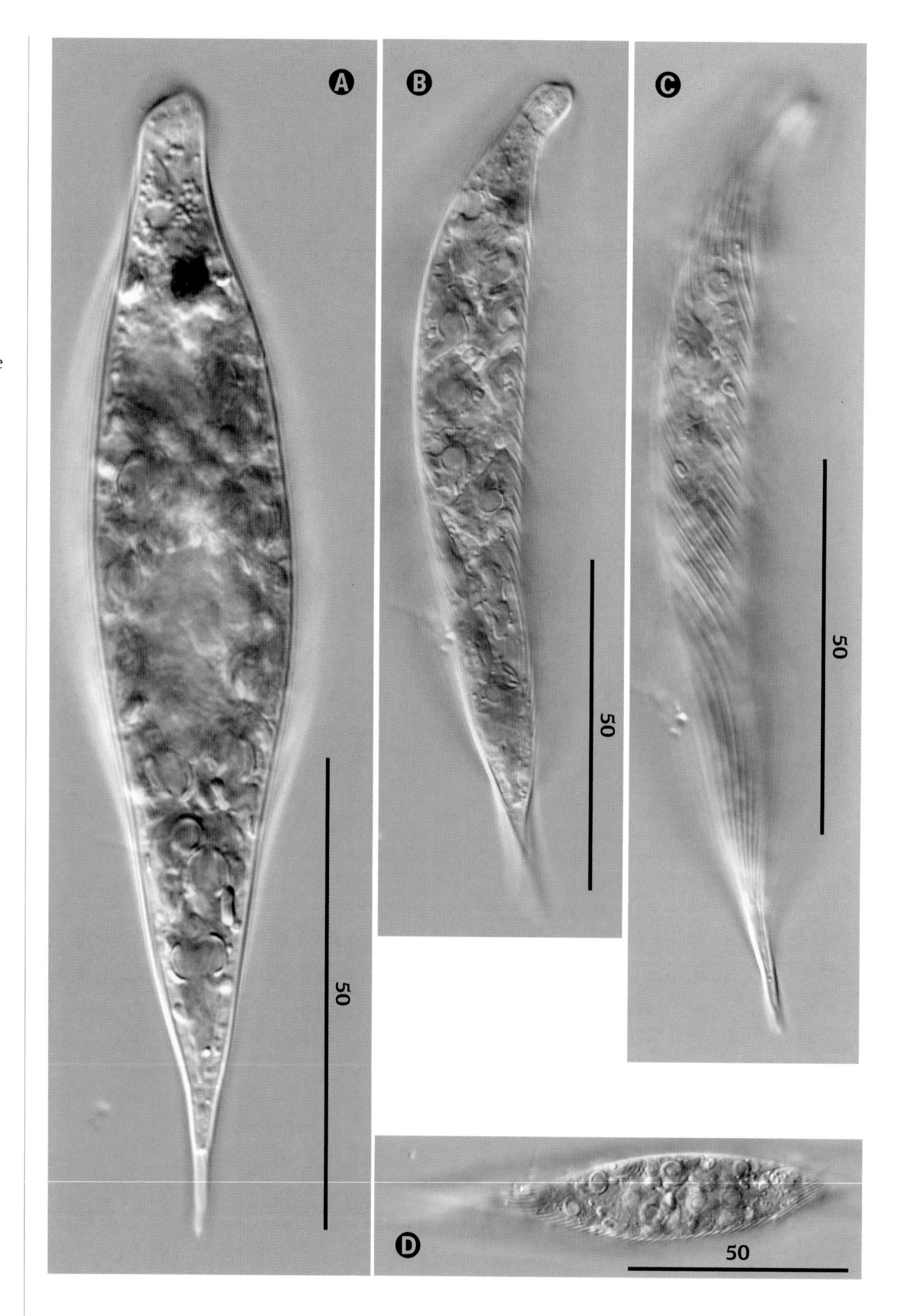

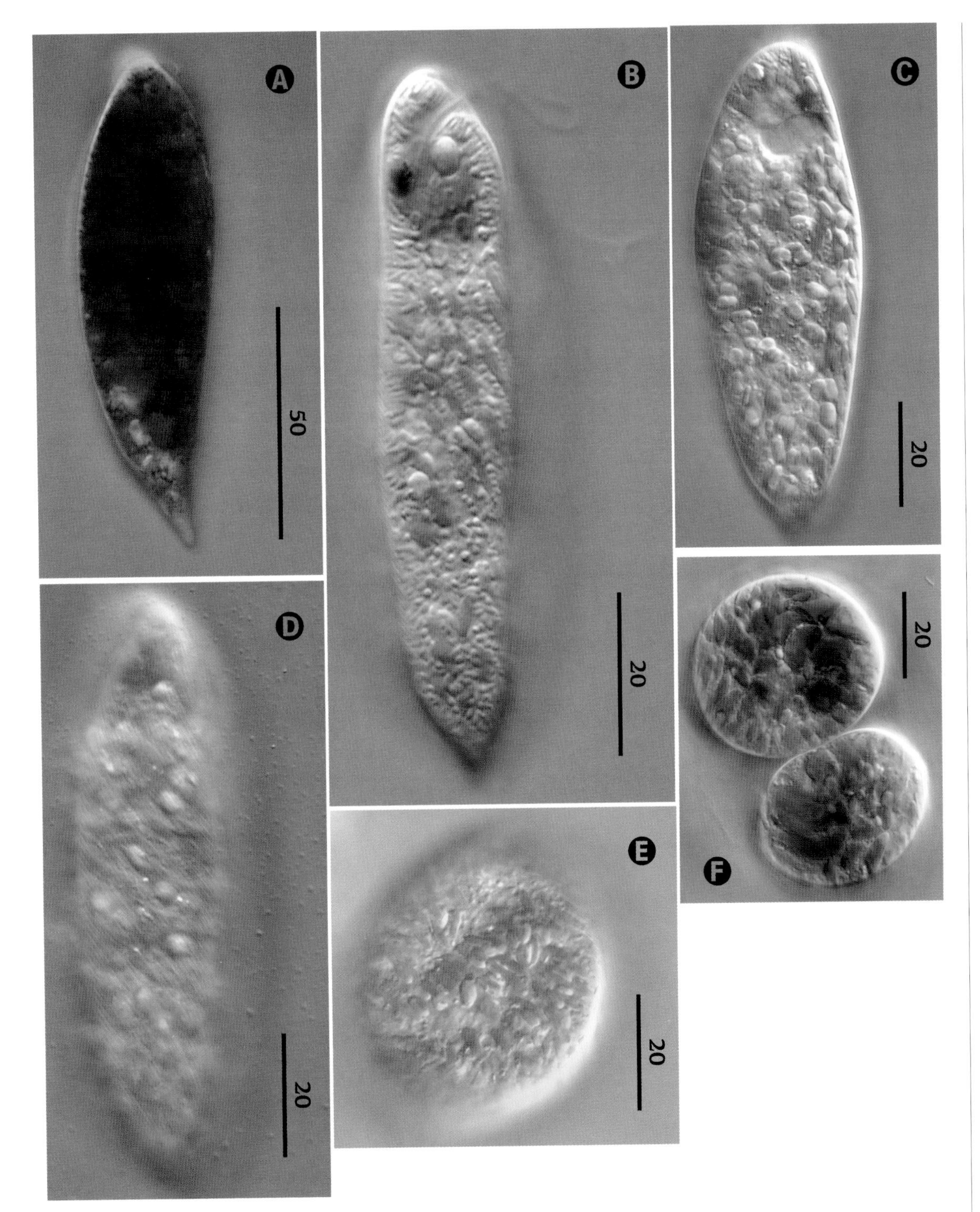

Euglena rubra

Hardy 1911

SIZE: 79–154 µm long × 16–25 µm wide.

Cell filled with red haematochrome granules and showing several paramylon bodies at the posterior (A). Cell with flagellum (one-half to full body length), red eyespot, numerous small disc-shaped chloroplasts seen in lateral view, spindle-shaped mucocysts, and a paramylon grain adjacent to the eyespot (B). Cell with numerous paramylon bodies (C). Fine spirally striated pellicle (D). Cells in metaboly showing chloroplasts lacking pyrenoids (E, F). *Note:* Red pigment is only produced under certain environmental conditions.

Euglena sanguinea

Ehrenberg 1830

SIZE: 70–140 µm long × 20–31 µm wide.

Cell with flagellum (one to one-and-a-half times body length), red eyespot, and an ovoid paramylon grain at the anterior end; the cell is filled with red haematochrome granules (A). Cells showing deeply lobed chloroplasts with diplopyrenoids (arrow) (B), rows of spindle-shaped mucocysts, and paramylon bodies (B, C). The chloroplasts' margins are so deeply incised that they appear ribbon-like near the cell surface (C). Pellicle is finely and spirally striated (D). Cells are metabolic (C, D). *Note:* Red pigment is only produced under certain environmental conditions.

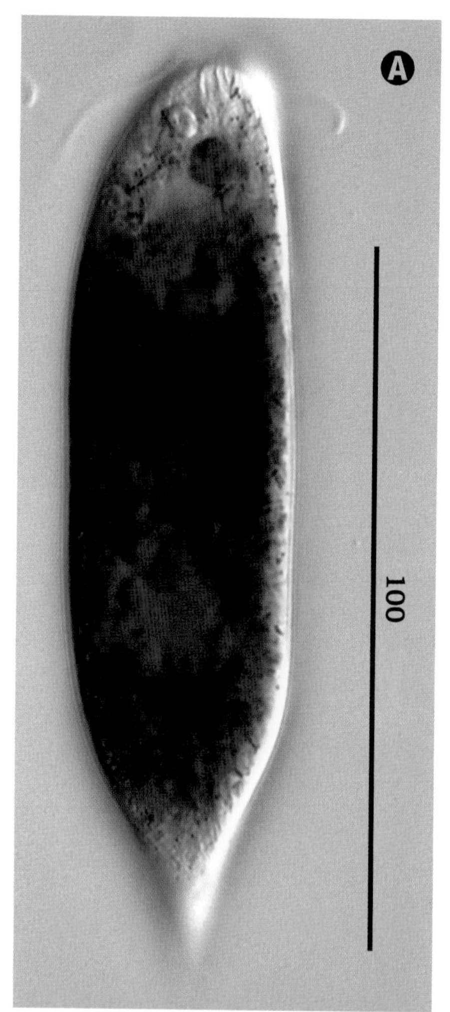

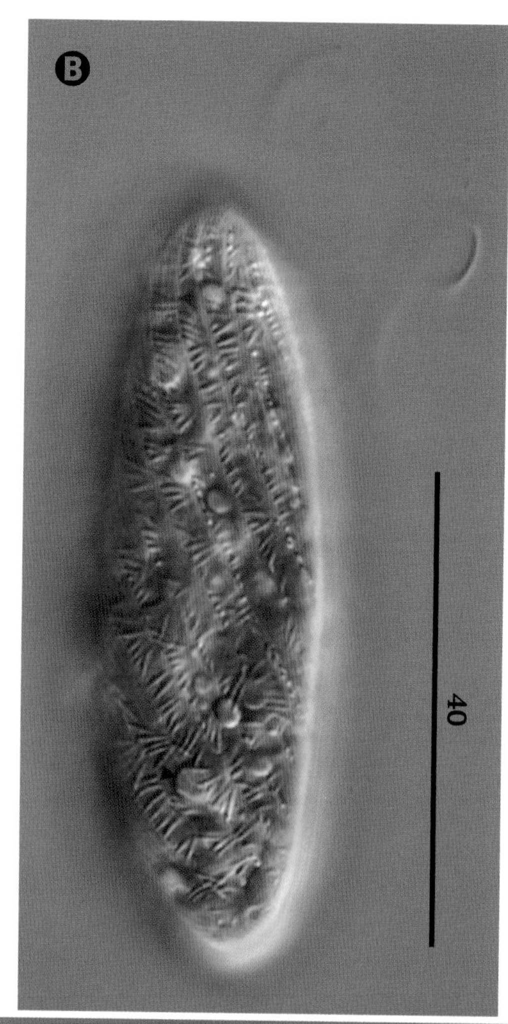

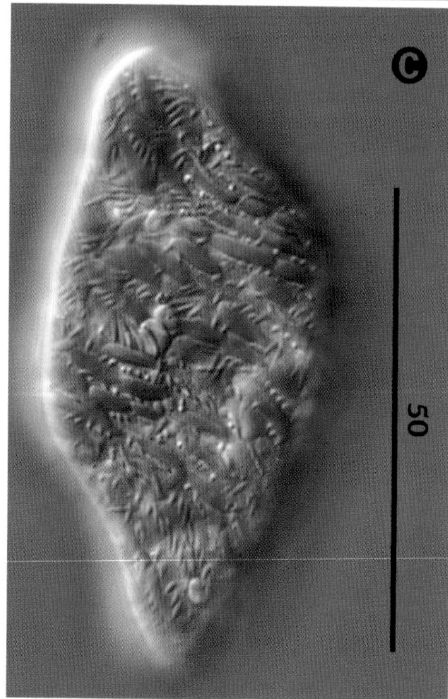

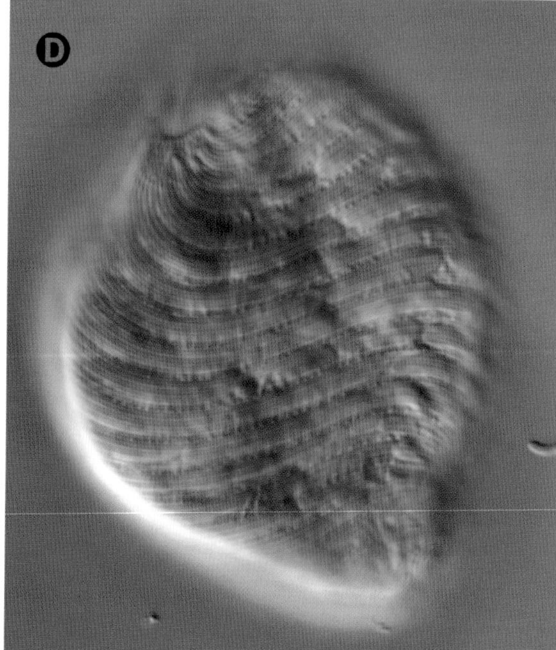

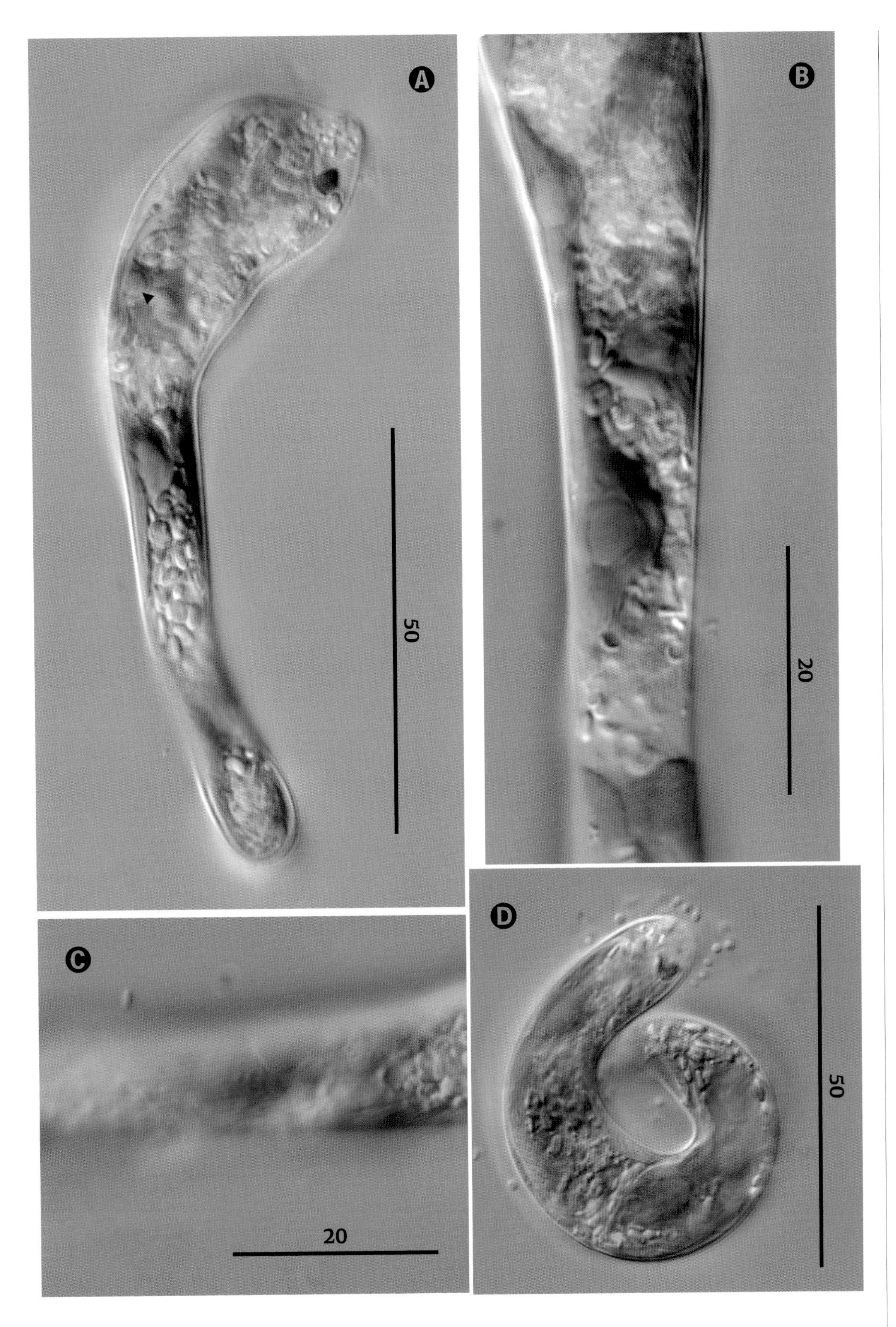

Euglena satelles

Braslavska-Spektorova 1937

SIZE: 87–148 µm long × 7–21 µm wide.

Cylindrical cell with red eyespot, elongated chloroplasts with naked pyrenoids (arrow), and numerous small paramylon grains (A). Chloroplasts with naked pyrenoids (B). Pellicle striations delicate and difficult to resolve (C). Cell showing metaboly (D).

Euglena sociabilis

Dangeard 1901

SIZE: 68–96 µm long × 17–28 µm wide.

Broadly spindle-shaped cell with red eyespot, contractile vacuole below the eyespot, and chloroplasts with diplopyrenoids (A). Two cells in metaboly; red eyespot, chloroplasts with deeply incised margins, and a diplopyrenoid (arrow) are visible (B). Chloroplasts as seen near the cell surface; the long lobes of the chloroplasts spiral around the cell in a parallel array (C). Cell with many paramylon grains; contractile vacuole lies adjacent to the red eyespot (D). Pellicle with numerous, delicate striations (E). Flagellum about body length.

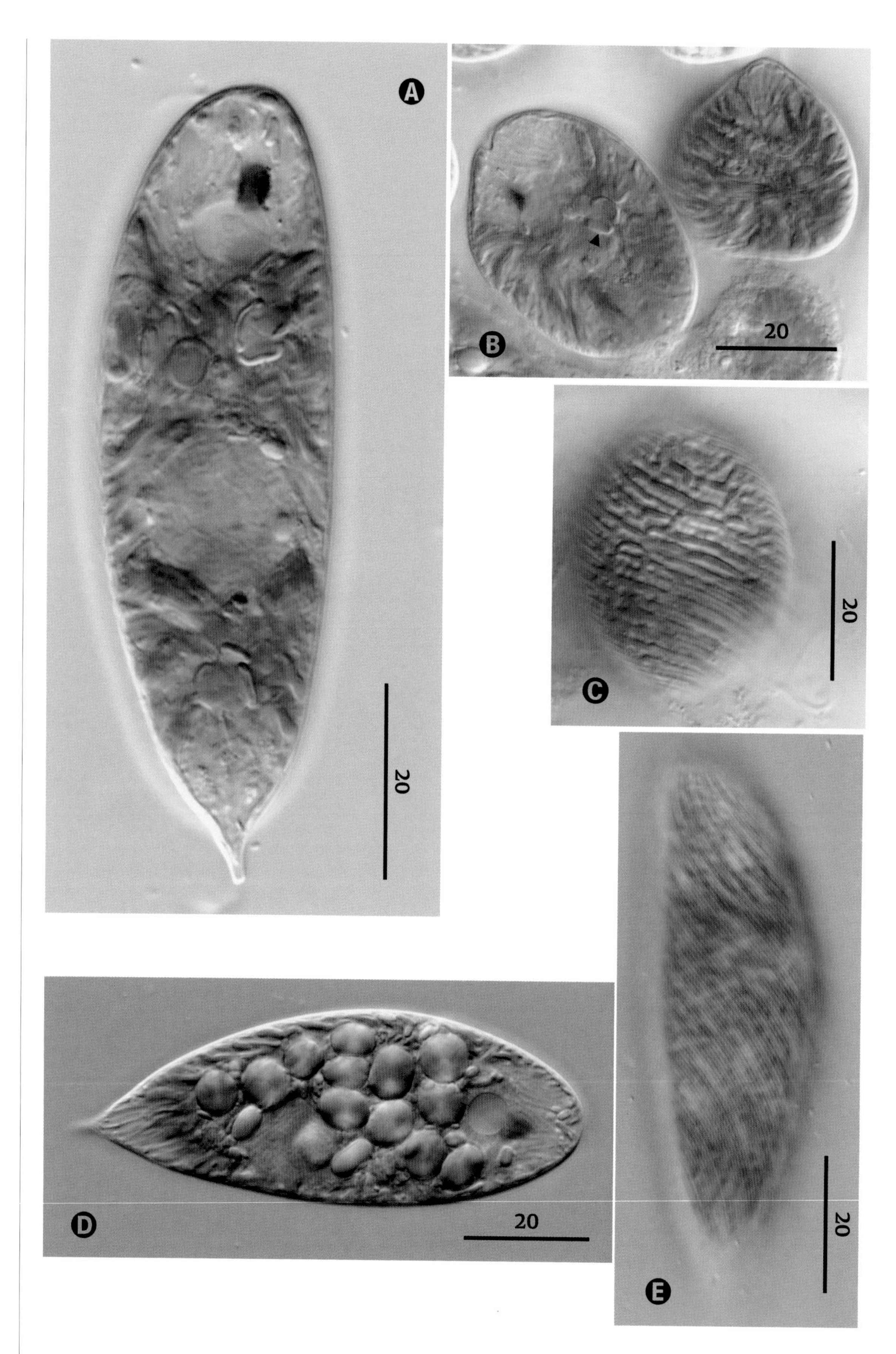

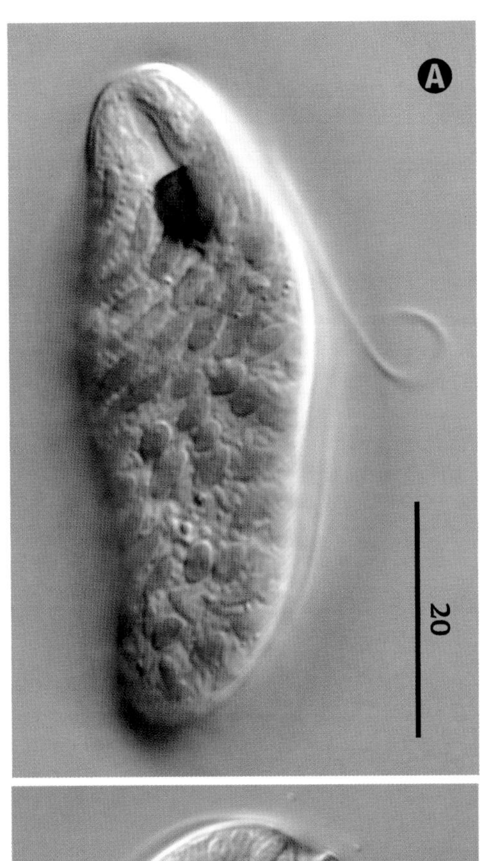

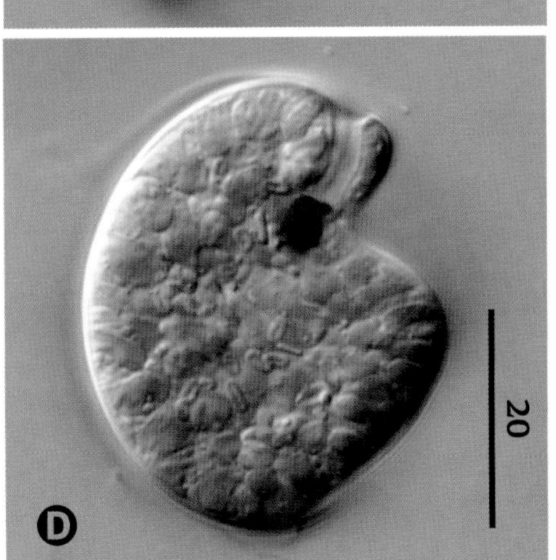

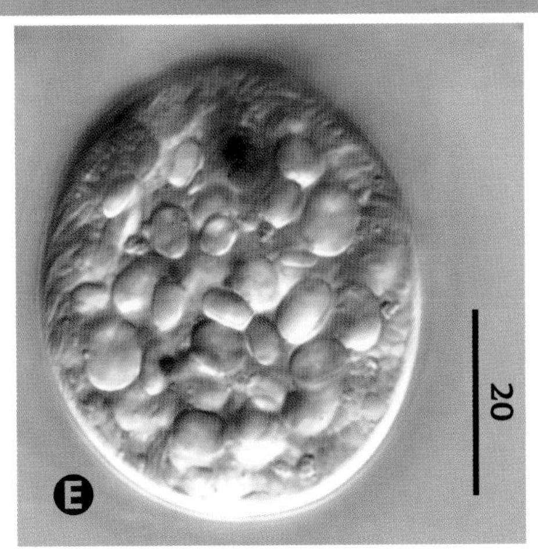

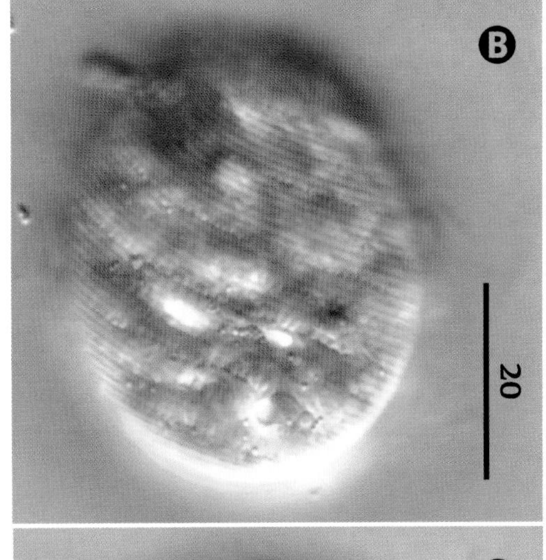

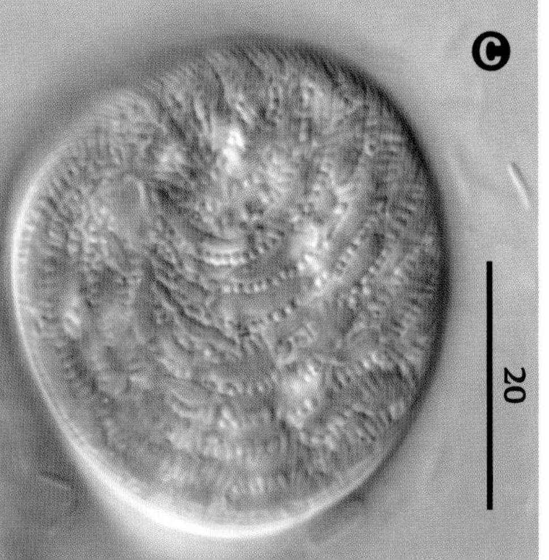

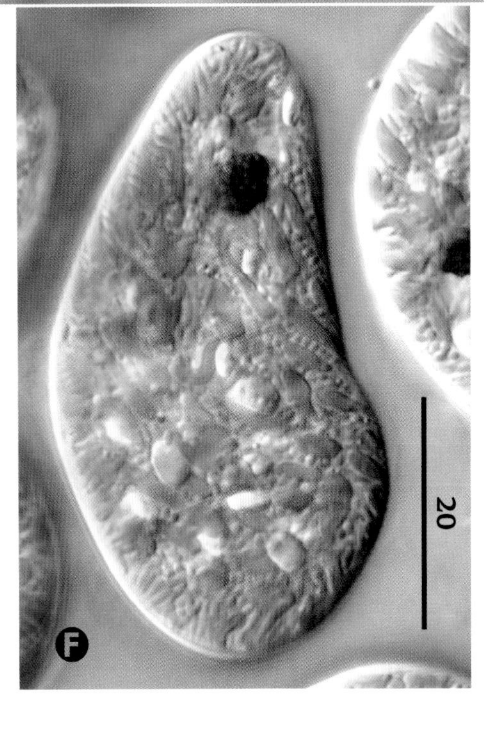

Euglena splendens

Dangeard 1901

SIZE: 60–120 µm long × 17–32 µm wide.

Broadly spindle-shaped cell with flagellum (body length), red eyespot, and disc-shaped, spirally arranged chloroplasts (**A**). Pellicle striations are fine and run spirally (**B**). Cell rounded through metaboly, with chloroplasts and spindle-shaped mucocysts spirally arranged; mucocysts appear spherical when viewed end on (**C**). Metabolic cell with basal part of flagellum visible in canal, above red eyespot (**D**). Metabolic cell with numerous ovoid paramylon bodies (**E**). Cell with chloroplasts and spindle-shaped mucocysts; several ovoid paramylon bodies are present (**F**).

Euglena stellata

Mainx 1926

SIZE: 25–55 µm long × 7–14 µm wide.

Cells showing flagellum (three-fourths to full body length), eyespot, ribbon-like extensions of the chloroplast, and small paramylon bodies (Ⓐ, Ⓑ, Ⓓ). Single, axial chloroplast with central pyrenoid (arrow); the deeply incised margin forms a stellate pattern (Ⓑ). Pellicular striations are fine (arrow); paramylon bodies and spindle-shaped mucocysts are present (Ⓒ).

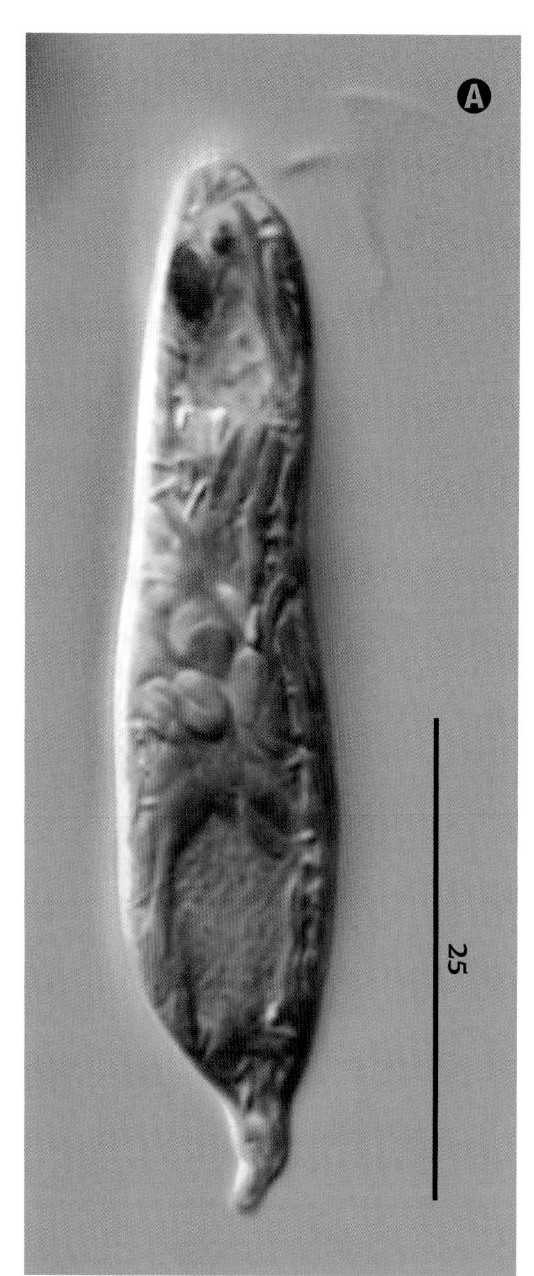

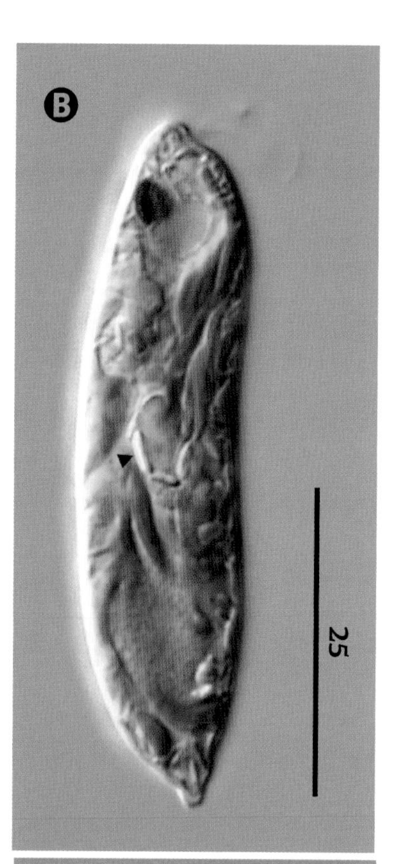

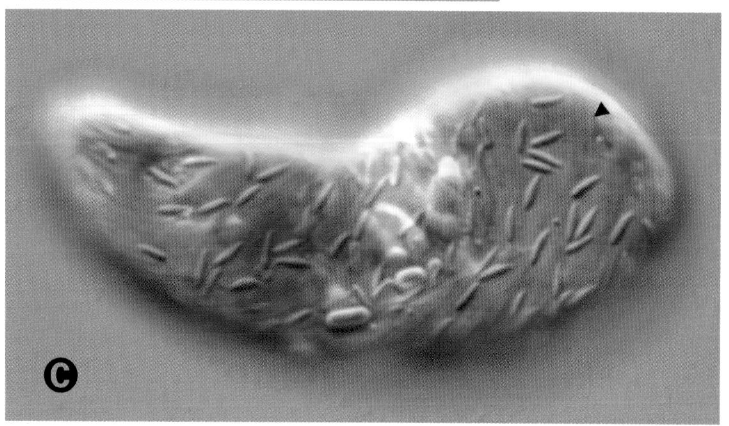

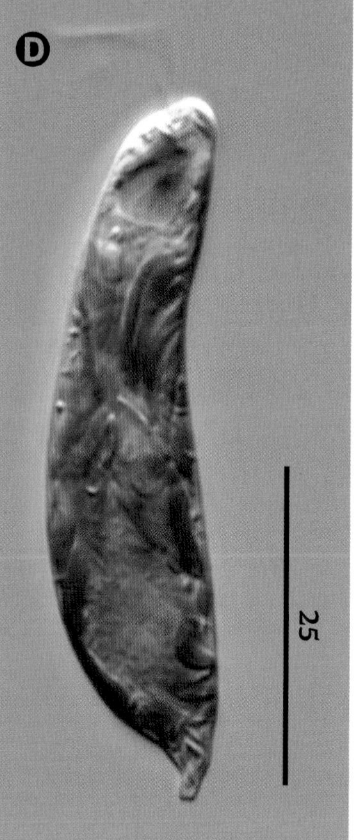

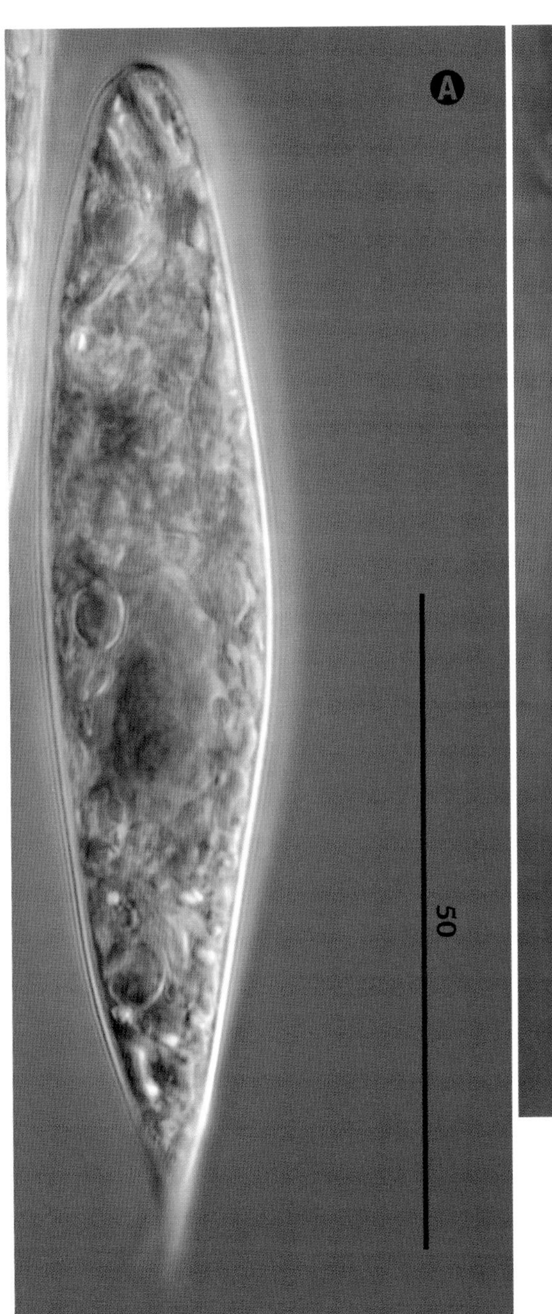

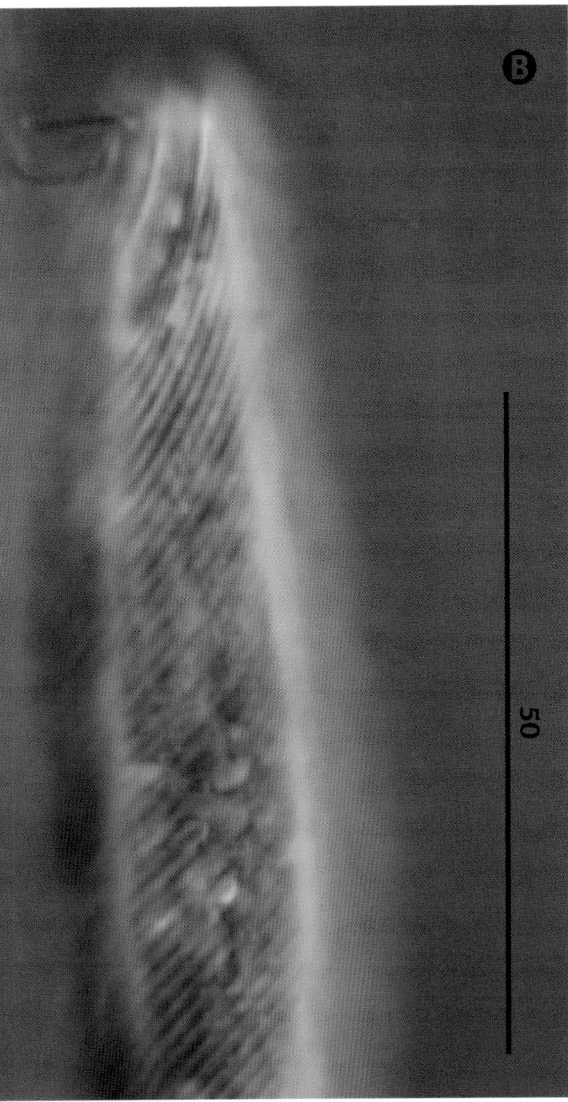

Euglena sulcifera

Chadefaud 1944

SIZE: 85–100 µm long × 25–30 µm wide.

Spindle-shaped cell showing eyespot, disc-shaped chloroplasts with diplopyrenoids, small paramylon grains, and groove at the anterior end (A). Detail of spiraling pellicular striations and anterior groove (B).

Euglena tristella

Chu 1946

SIZE: 53–73 µm long × 18–22 µm wide.

Spindle-shaped cell with flagellum (about one-and-one-fourth times body length), red eyespot, and three stellate chloroplasts with paramylon centers (arrows) (A). Cell with pyrenoid visible in one of the three paramylon centers (arrow); paramylon bodies and spindle-shaped mucocysts (B). Cells showing flagellum, eyespot, paramylon bodies, and nucleus (C, E). Cell with spindle-shaped mucocysts and spirally striated pellicle (arrow) (D).

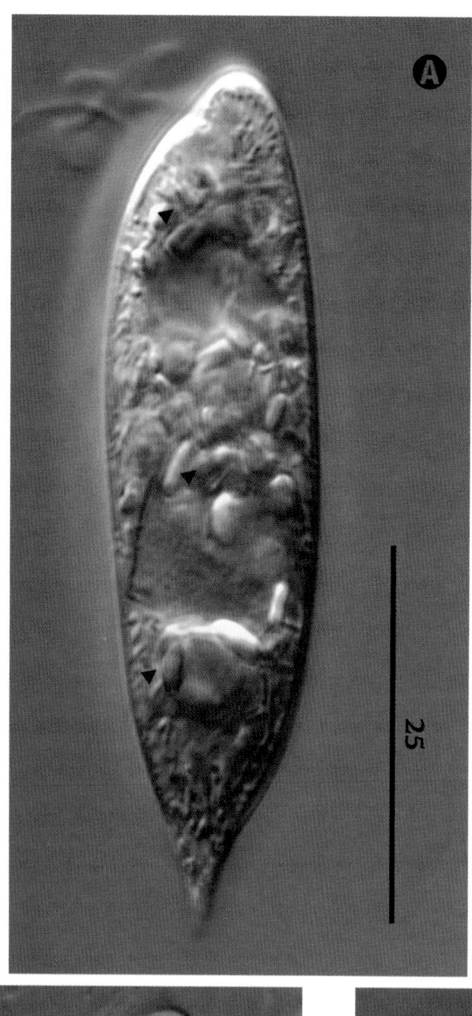

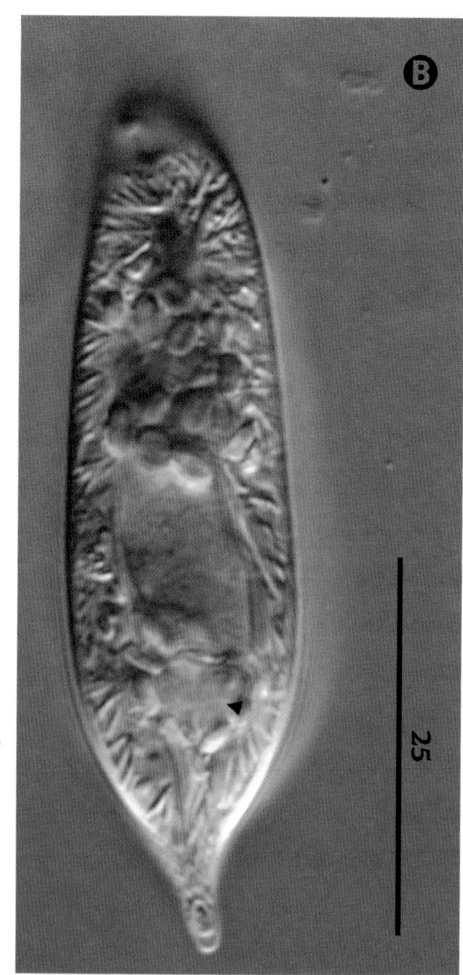

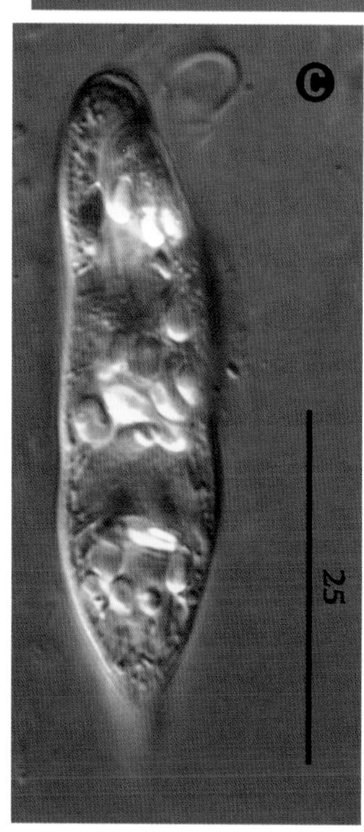

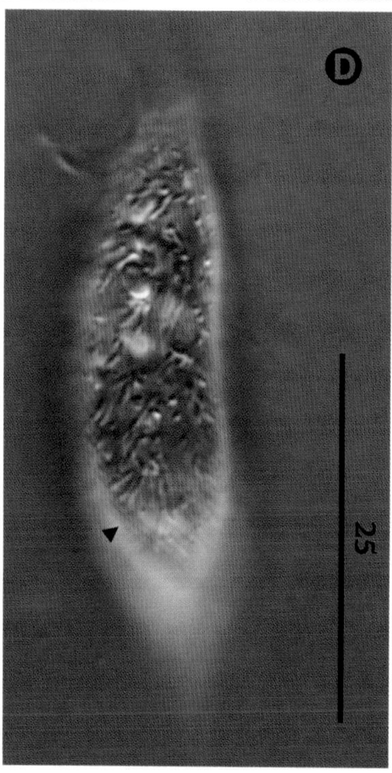

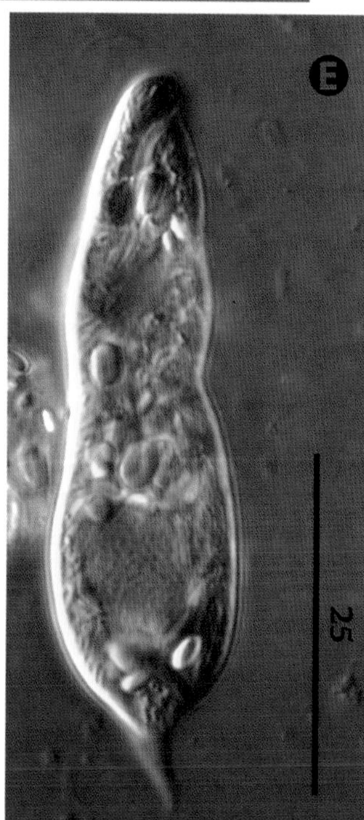

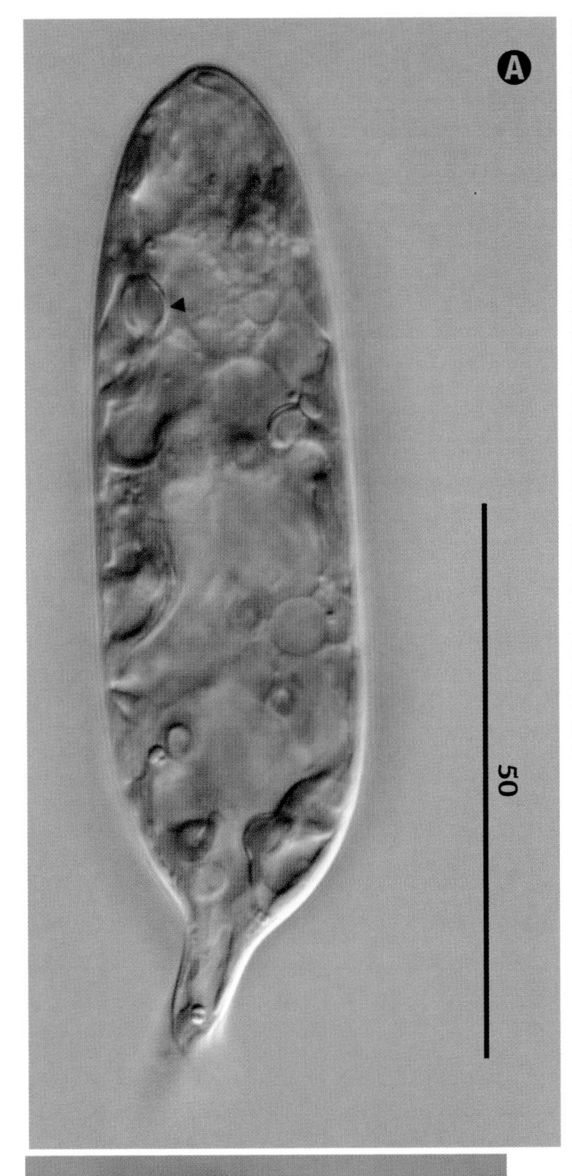

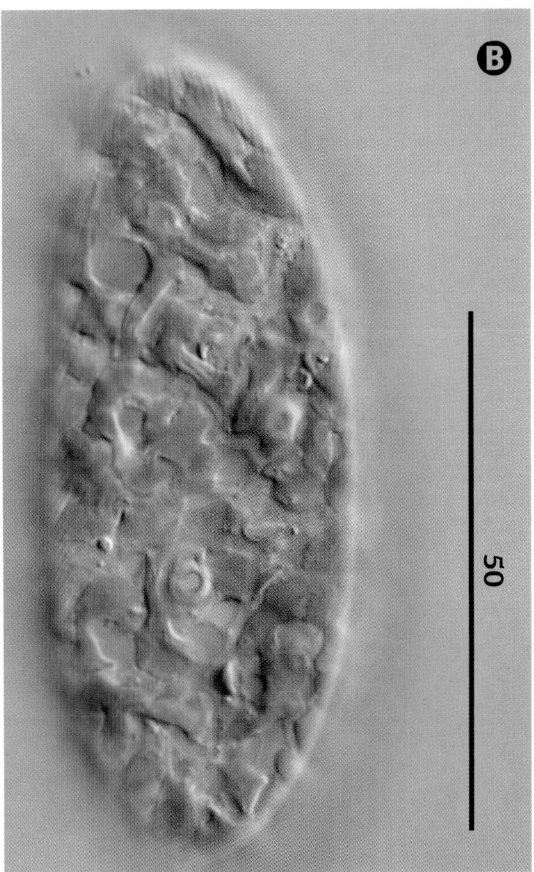

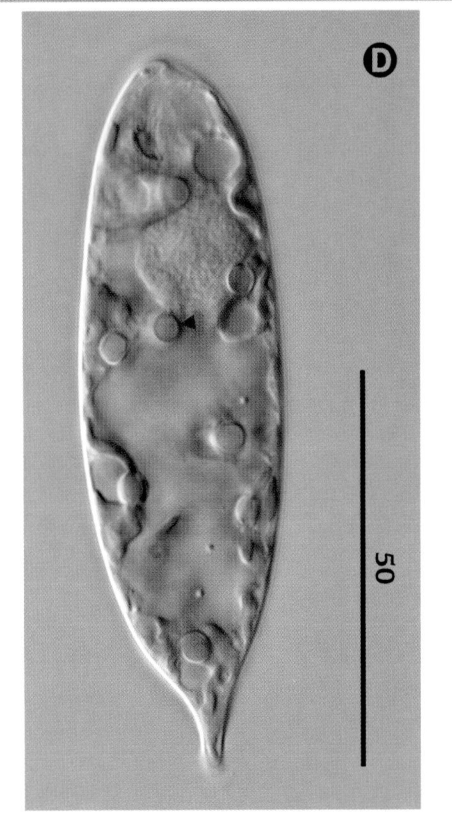

Euglena velata

Klebs 1883

SIZE: 94–124 μm long × 19–31 μm wide.

Spindle-shaped cells with eyespot and deeply incised chloroplasts with diplopyrenoids, in lateral view (arrow) (A) and face view (arrow) (D). Chloroplasts with deep and irregularly incised margins (B, C). Metabolic cell showing pellicle with fine, spiral striations (C). Flagellum is body length.

Euglena viridis

(O. F. Müller) Ehrenberg 1830

SIZE: 42–79 µm long × 7–20 µm wide.

Spindle-shaped cell with flagellum (one-third to full body length), red eyespot, contractile vacuole, small paramylon bodies, and stellate chloroplast with deeply lobed extensions. The pyrenoid (arrow) is surrounded by short paramylon rods (**A**). Cell showing spirally striated pellicle and chloroplast lobes (**B**).

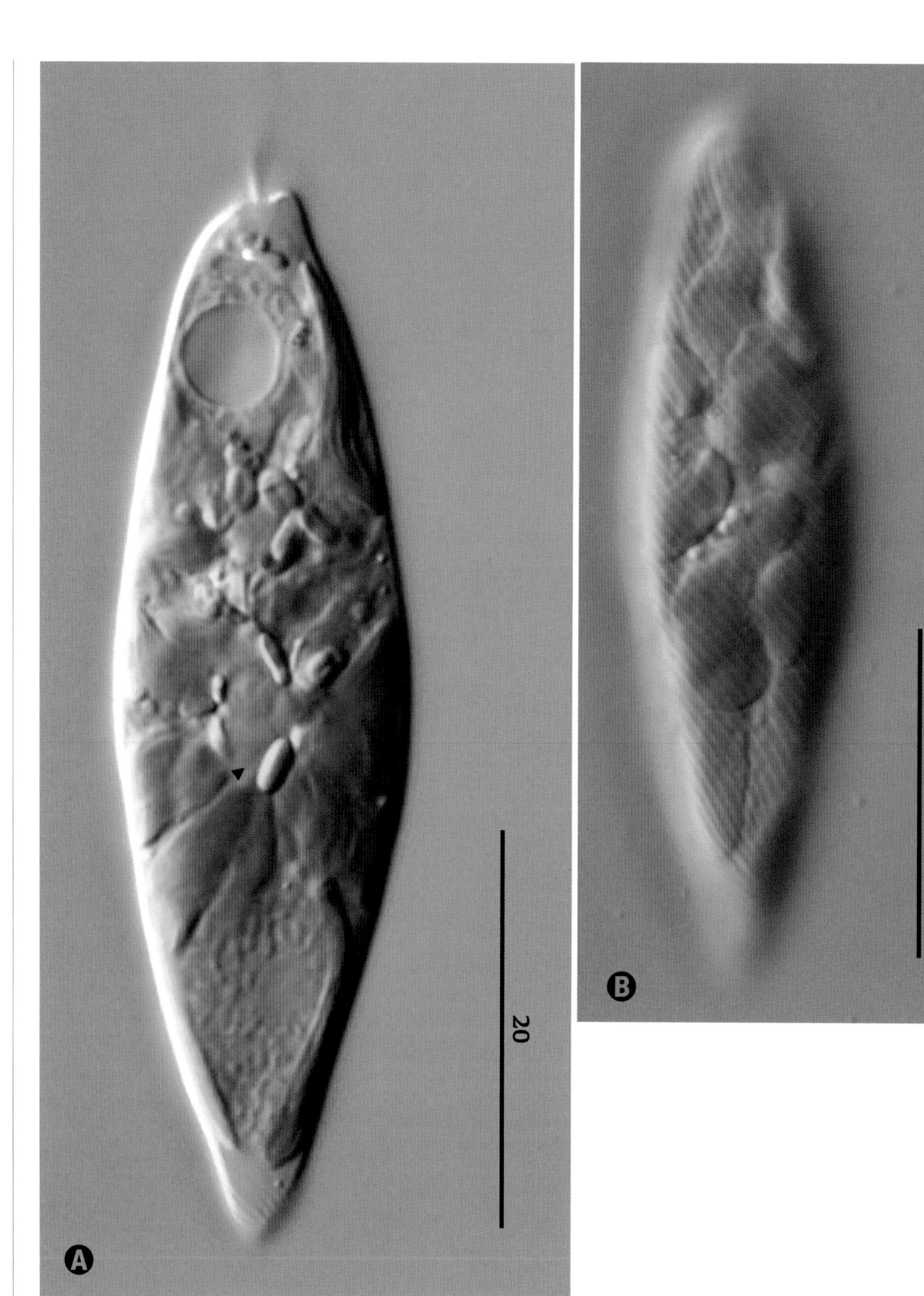

EUGLENARIA

KARNKOWSKA, LINTON ET KWIATOWSKI 2010

Cells free-living, solitary, with one emergent flagellum when swimming; club-shaped fusiform or cylindrically fusiform, narrowing to the posterior and tapering into a pointed tail; metabolic, display euglenoid movement; parietal, lobed chloroplasts, each with a single pyrenoid accompanied by bilateral paramylon caps located on both sides of the chloroplast (diplopyrenoid); mucocysts absent; freshwater.

Euglenaria anabaena var. *minima*

(Mainx) Karnkowska et Linton 2010

SIZE: 25–45 µm long × 8–11 µm wide.

Spindle-shaped cell, red eyespot, and plate-like chloroplasts (up to four) with diplopyrenoids (arrow) (**A**). Cell with flagellum (one-third to one-half body length), central nucleus, and small paramylon bodies (**B**). Detail of finely striated pellicle and plate-like chloroplasts (**C**, **D**).

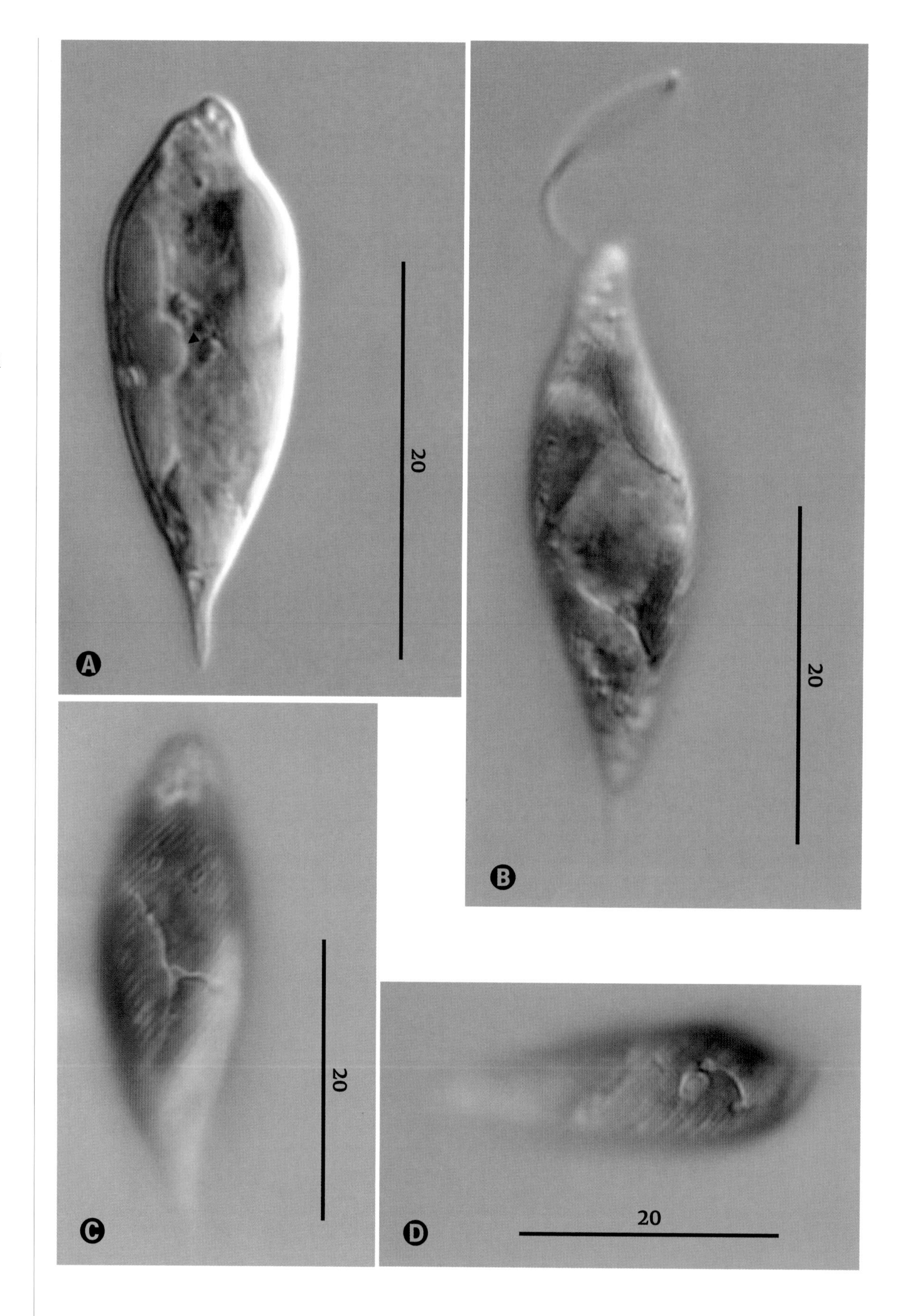

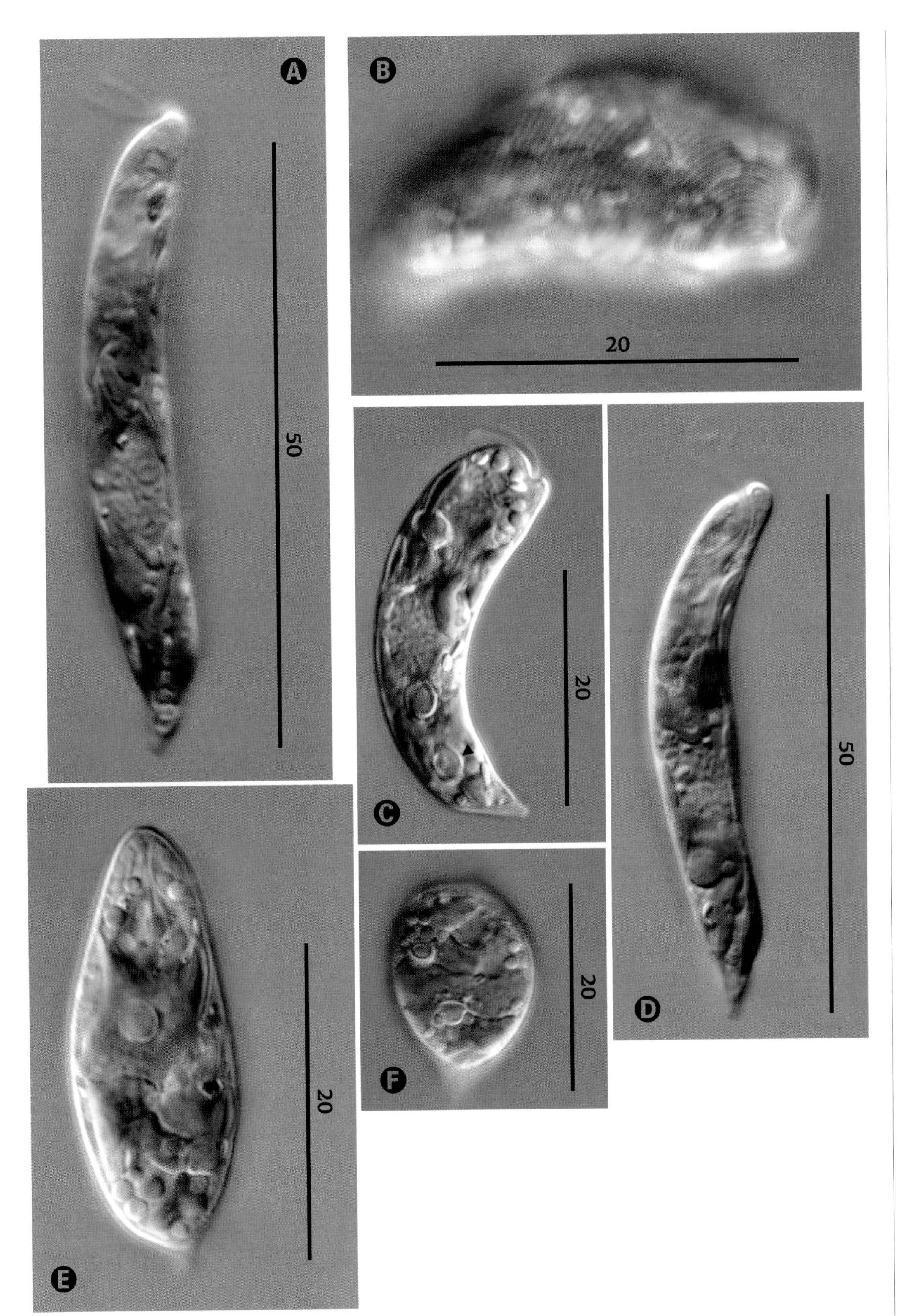

Euglenaria anabaena var. *minor*

(Mainx) Karnkowska et Linton 2010

SIZE: 44–59 µm long × 9–12 µm wide.

Cylindrical cell with a short caudus, flagellum, red eyespot, chloroplasts (up to eight), and small paramylon bodies (A, D). Pellicle is spirally striated (B). Cell with flagellum, paramylon grains, chloroplasts, and three diplopyrenoids (arrow) (C). Plate-like chloroplast with diplopyrenoid in face view (E). Metabolic cell showing large plate-like chloroplasts and two diplopyrenoids (F).

Euglenaria caudata

(Hübner) Karnkowska et Linton 2010

SIZE: 56–92 µm long × 11–20 µm wide.

Spindle-shaped cell with red eyespot, reservoir, disc-shaped chloroplasts with diplopyrenoids, and numerous small, ovoid paramylon bodies (A). Pellicle strips are fine and spirally arranged (B). Chloroplasts with diplopyrenoids seen in face view (arrow) (C). Metabolic cell with eyespot, chloroplasts, and diplopyrenoids (D). Flagellum is body length.

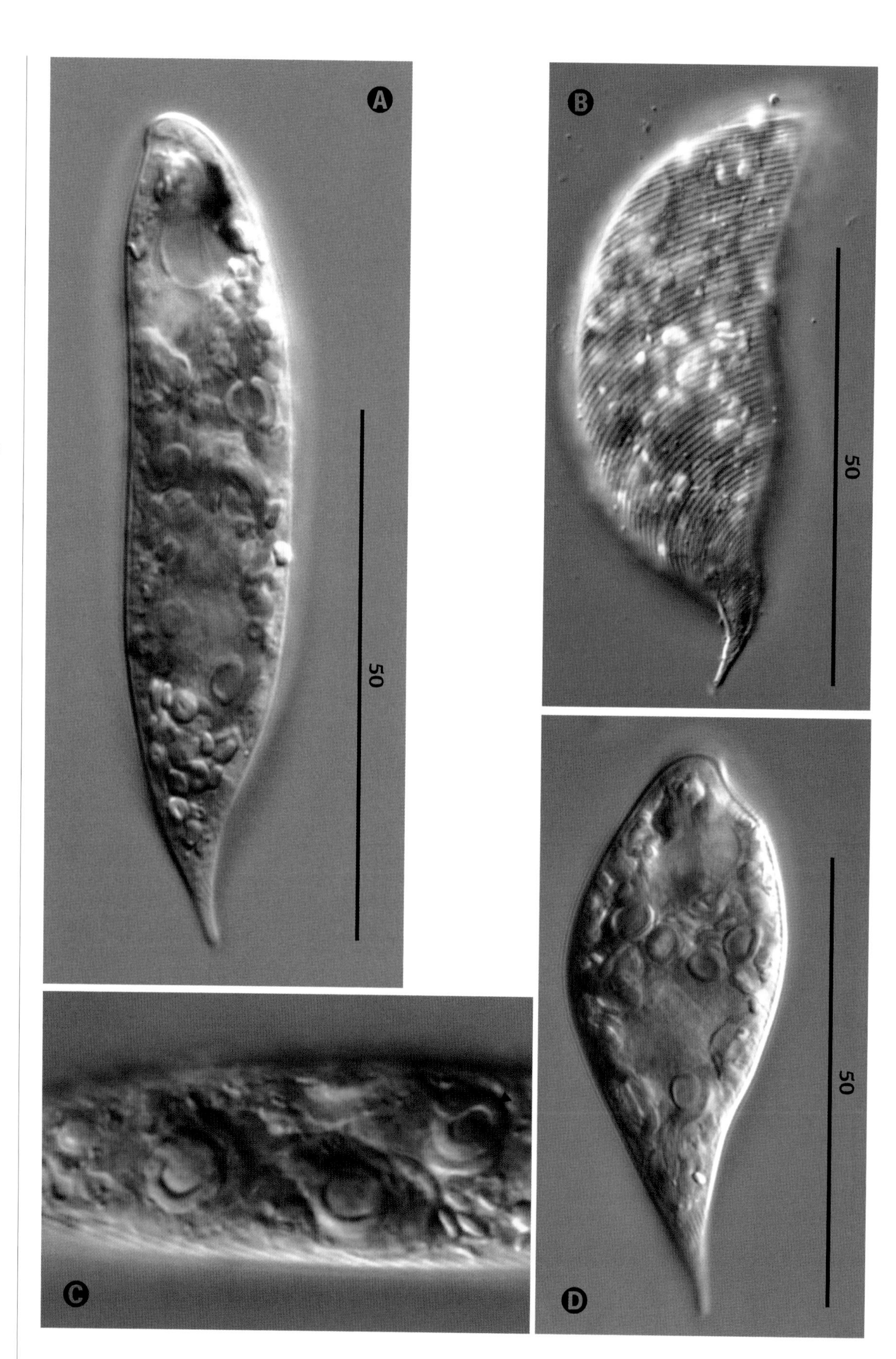

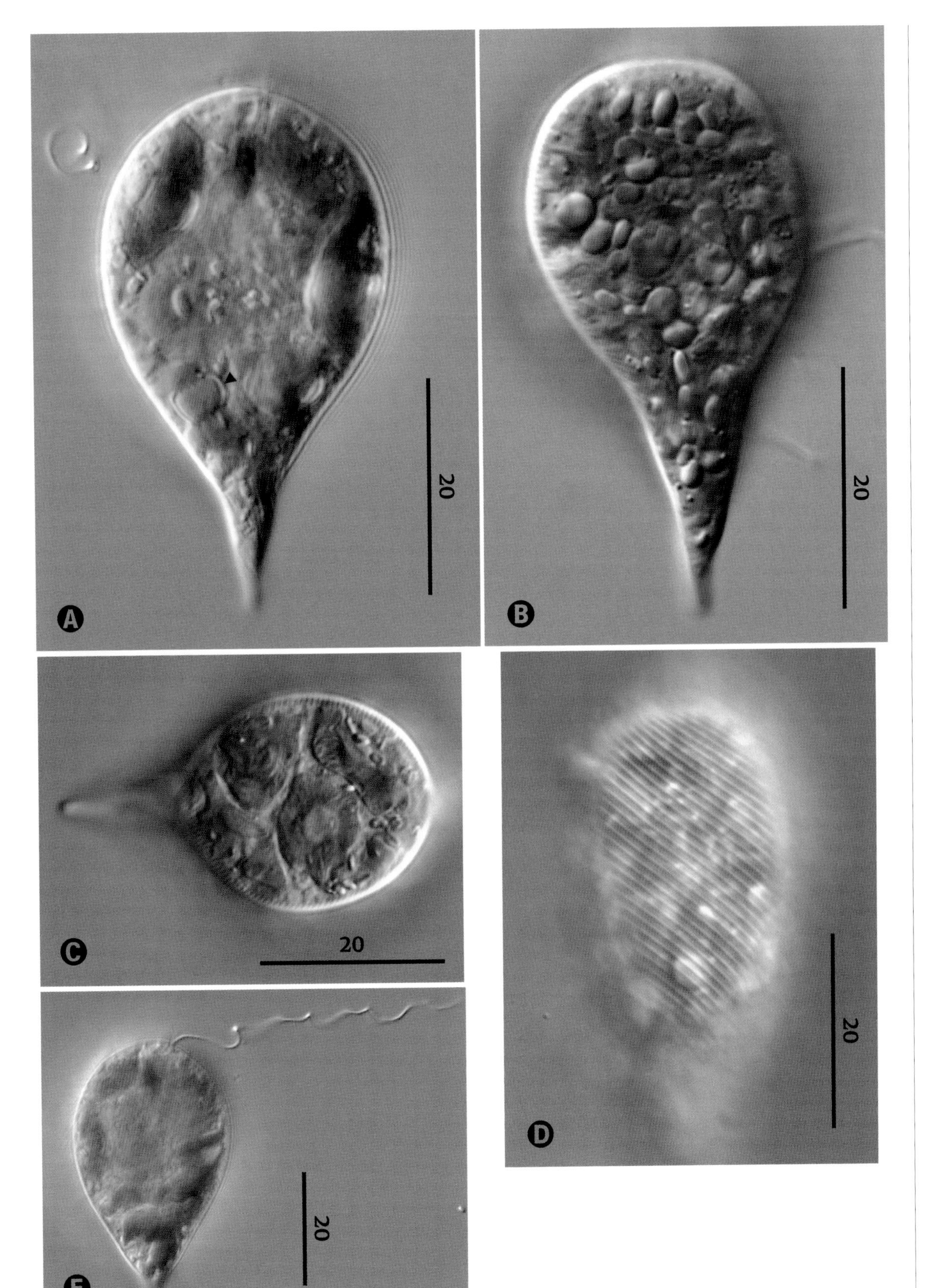

Euglenaria clavata

(Skuja) Karnkowska et Linton 2010

SIZE: 42–85 µm long × 20–35 µm wide.

Typical club-shaped cell showing eyespot, chloroplasts, and diplopyrenoids (arrow) (A). Osmium-fixed cell with numerous ovoid paramylon grains (B). Chloroplasts are disc-shaped and irregularly lobed (C). Pellicular striations are fine and spirally arranged (D). Flagellum up to two-and-a-half times body length (E).

EUTREPTIA

PERTY 1852

Cells ellipsoidal or spindle-shaped, solitary, free-swimming, with two visible, nearly equal flagella; pellicle finely striated; chloroplasts numerous, discoidal (in freshwater taxa, e.g., *Eutreptia viridis*), without pyrenoids or ribbon-like chloroplast with a central pyrenoid in marine species (e.g., *E. pertyi*); paramylon bodies cylindrical or round; eyespot present at the anterior end; strongly metabolic; cysts known; some freshwater but most species found in brackish and marine waters.

Eutreptia viridis

Perty 1852

SIZE: 49–66 µm long × 13 µm wide.

Broadly spindle-shaped cell showing two emergent flagella equal in size (about body length), red eyespot, paramylon center, and nucleus with nucleolus located at the posterior end (A). Two metabolic, rounded cells; paramylon center clearly visible (arrow) in the chloroplast; small paramylon bodies are present (B). Detail of pellicle striations (C). Surface view showing chloroplast and pellicle strips (D). Cell showing two emergent flagella, chloroplast, and nucleus (E). A dividing cell (F).

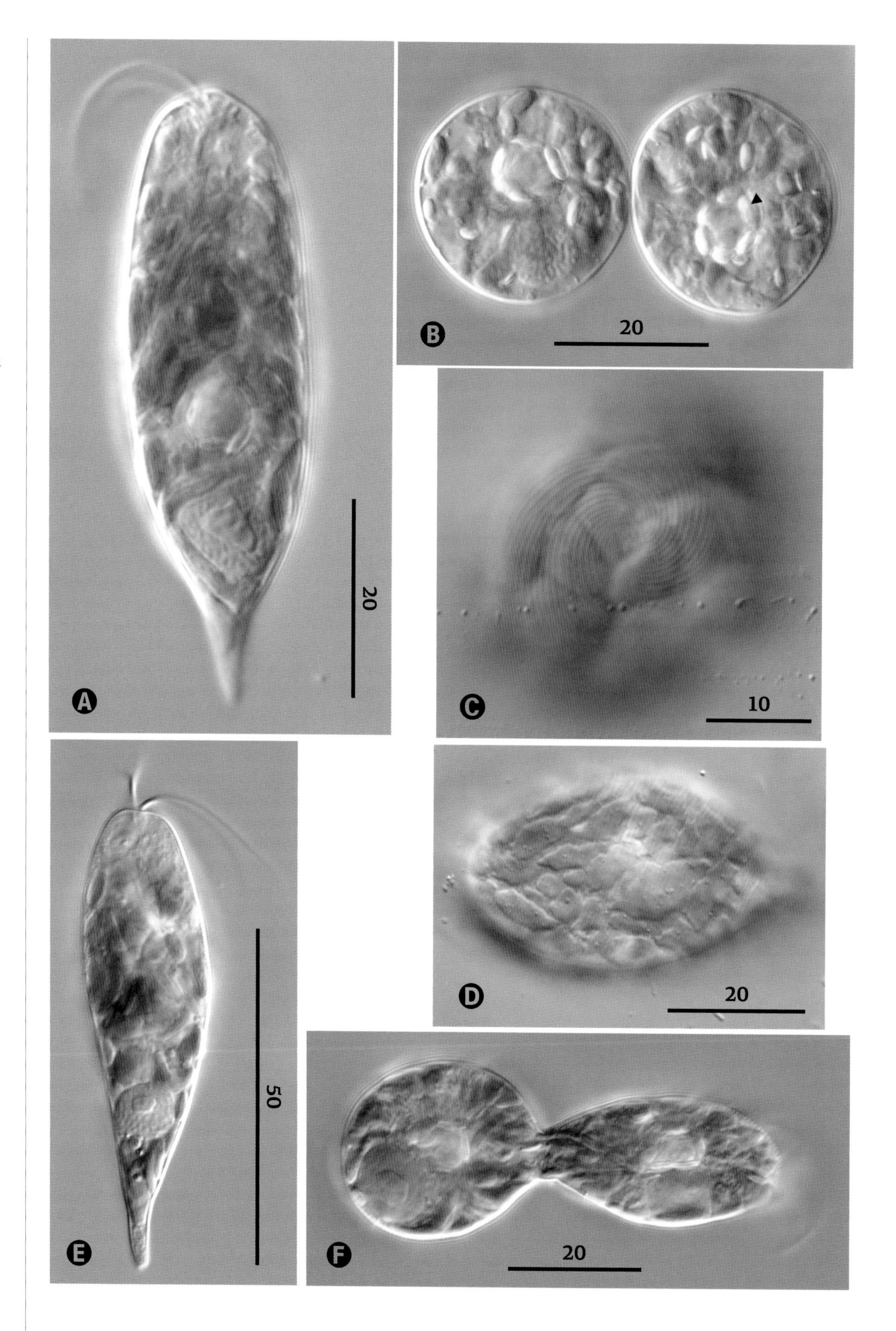

EUTREPTIELLA

DA CUNHA 1914

Cells often spindle-shaped, rounded at the anterior end and pointed at the posterior, free-swimming, with two unequal emergent flagella (rarely two pair of unequal flagella); pellicle finely striated; chloroplasts can be 6 to 10 leaf-like discs without pyrenoids (e.g., *Eutreptiella hirudoidea*), stellate clusters (two in *E. eupharyngea,* two or three in *E. braarudii*) with pyrenoids, or single, parietal, reticulate (e.g., *E. cornubiense*) with pyrenoids (e.g., *E. eupharyngea*); paramylon granules are small and elongate; eyespot present at the anterior end; strongly metabolic; cysts known; brackish and marine water.

Eutreptiella pomquetensis

(McLachlan, Seguel et Fritz) Marin et Melkonian 2003

SIZE: 70–100 µm long × 8–13 µm wide.

Spindle-shaped cell with four unequal flagella and central nucleus (A). Same cell showing red eyespot and ribbon-like chloroplasts arranged in two stellate clusters with pyrenoids located at the center (arrow) (B). Detail of spirally arranged pellicle strips (C).

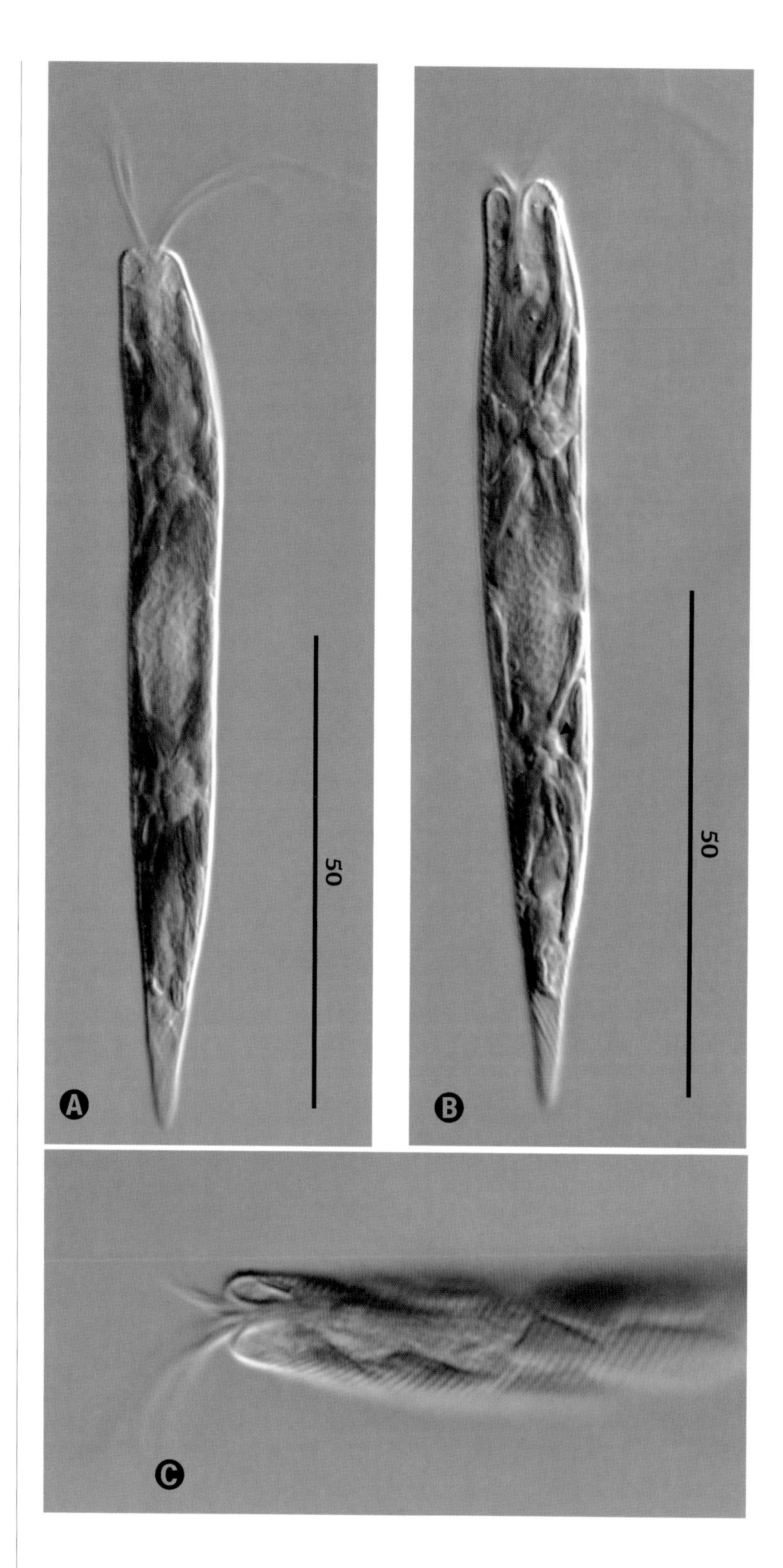

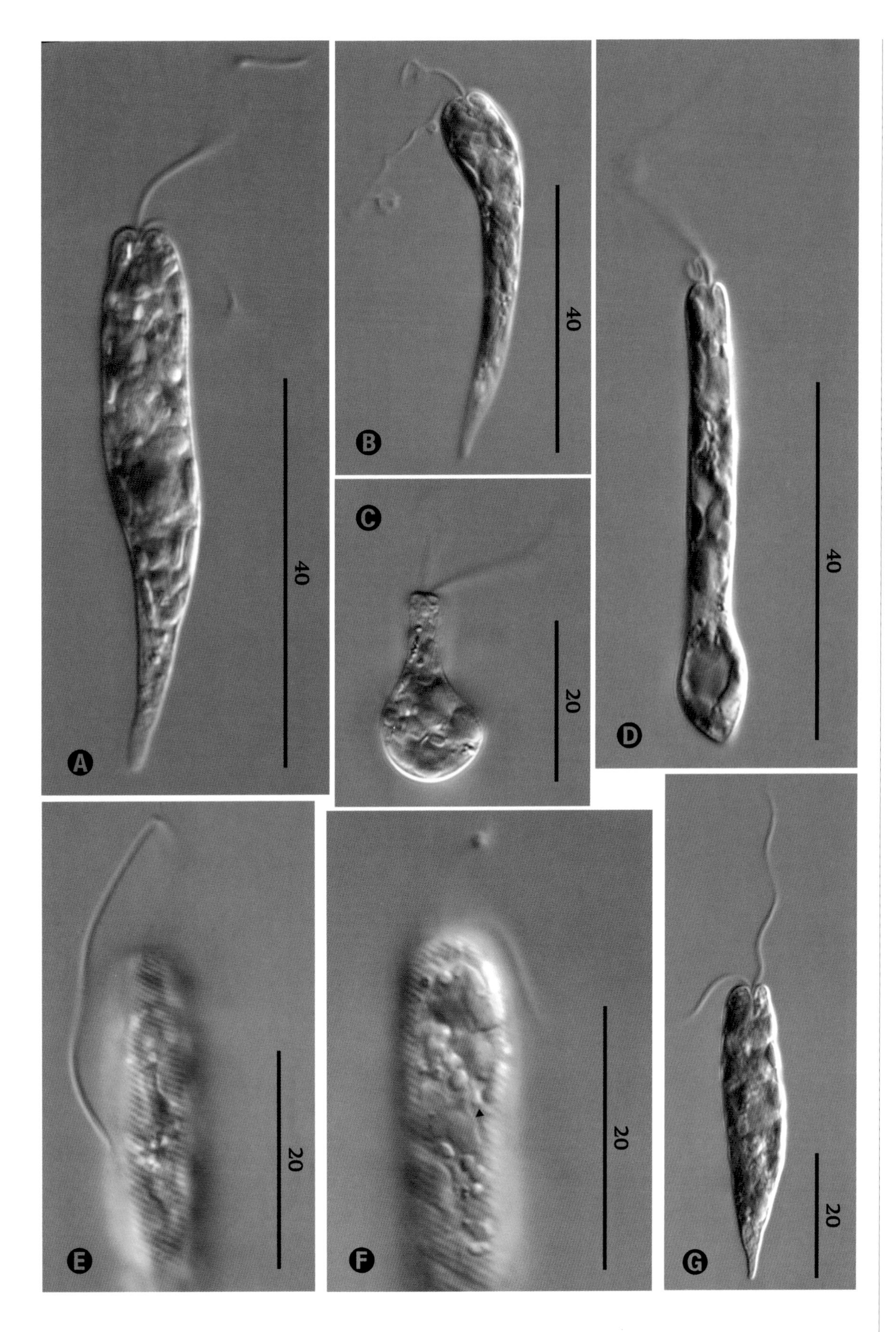

Eutreptiella sp.

SIZE: 46–58 µm long × 9–10 µm wide.

Spindle-shaped cells with two unequal flagella, red eyespot, chloroplasts, and small paramylon bodies (A, B, G). Metabolic cells (C, D). Pellicle striations are spirally arranged (E). Detail of chloroplast with irregular margins (arrow) (F).

LEPOCINCLIS

PERTY 1852

Cells ovoid or cylindrical, straight or twisted; a single emergent flagellum; pellicle rigid, some pellicular striations very pronounced, running almost longitudinally; chloroplasts small, discoid, with no pyrenoids; paramylon grains large and rod-shaped (mainly in elongated cells), disc or links; frequently one to two; some forms colorless (e.g., *Lepocinclis acus* var. *hyalina*); eyespot present at the anterior end; no euglenoid movement, although bending may occur; palmelloid stages and cysts unknown; typically freshwater.

Lepocinclis acus

(O. F. Müller) Marin et Melkonian 2003

SIZE: 67–230 µm long × 7–14 µm wide.

Spindle-shaped cells, narrowed at the anterior end showing red eyespot, flagellum (one-third body length), rod-like paramylon bodies, and numerous disc-shaped chloroplasts lacking pyrenoids (Ⓐ, Ⓒ). Pellicular striations are fine and follow the longitudinal axis of the body (Ⓑ).

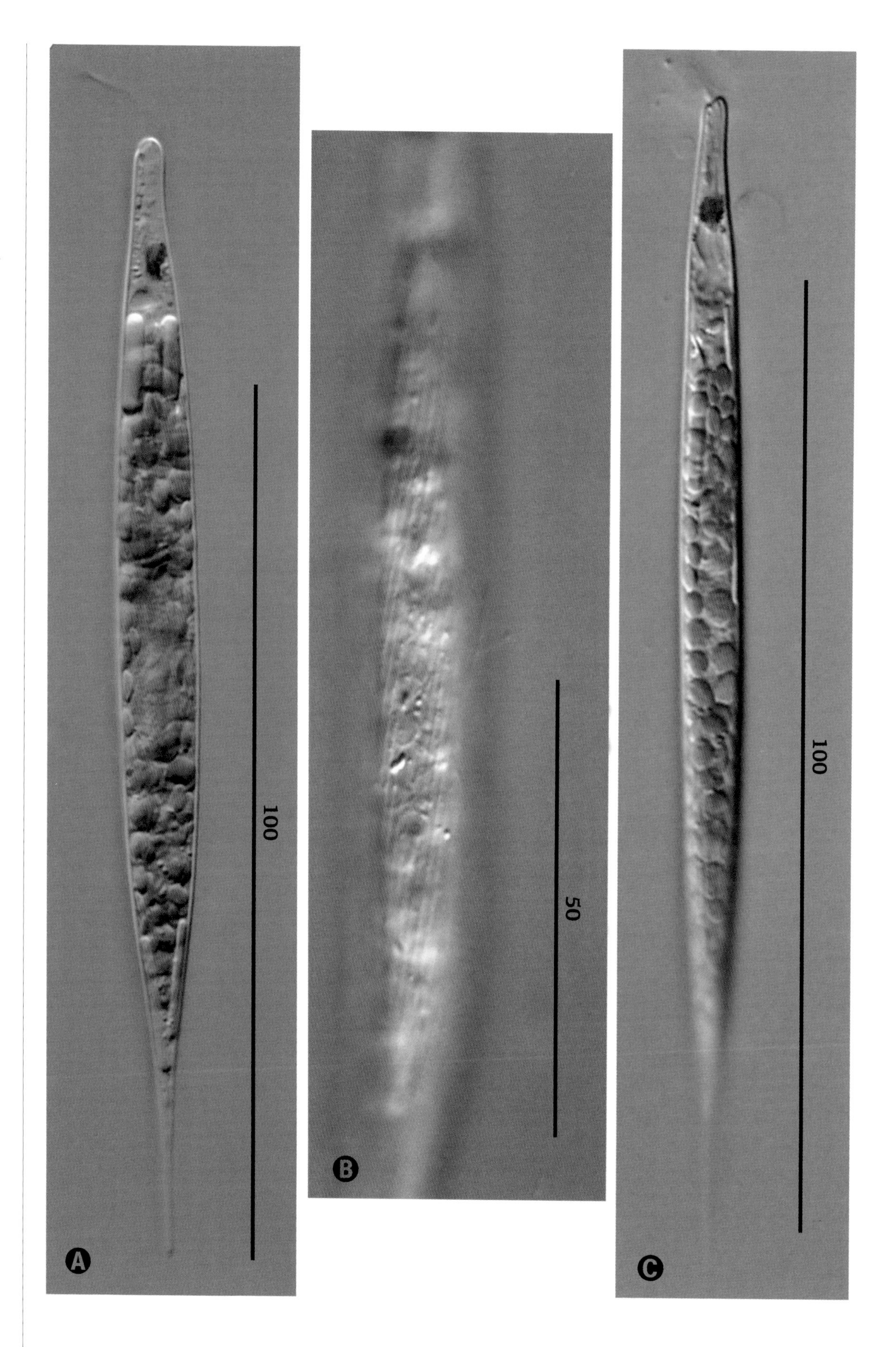

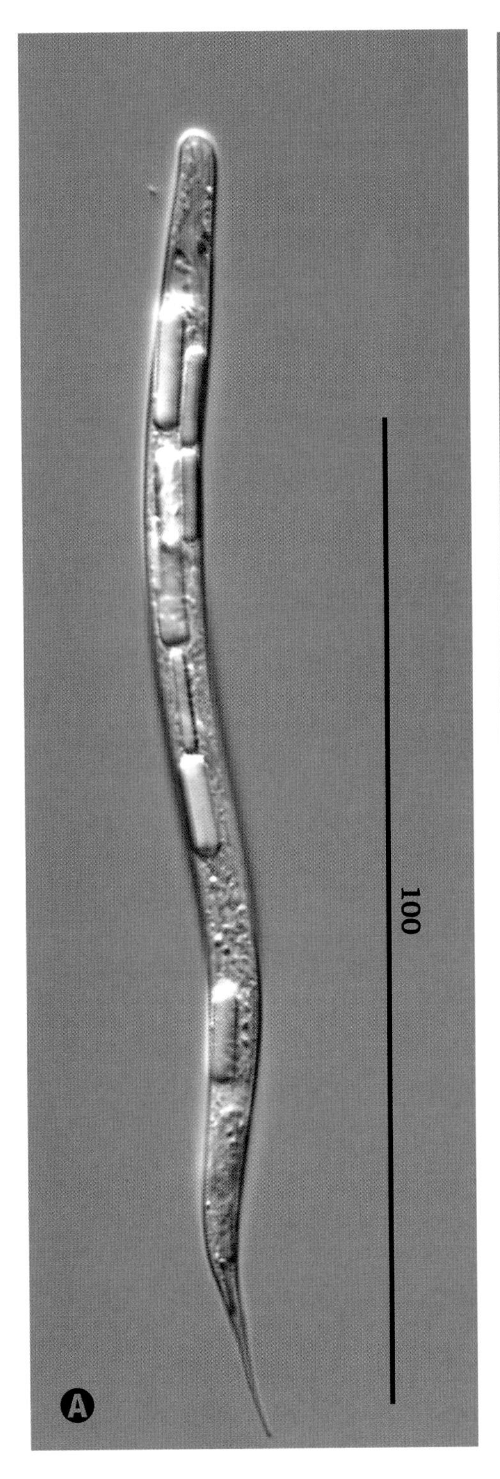

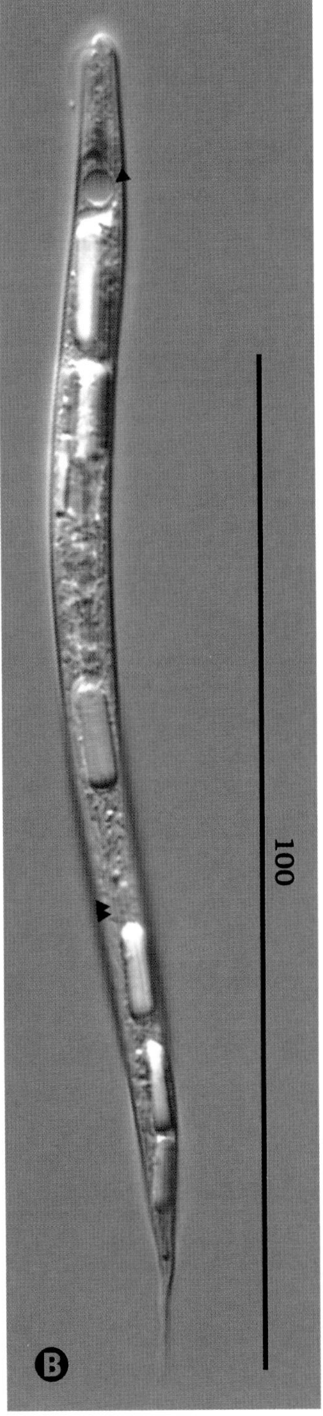

Lepocinclis acus var. *hyalina*

(= ***Euglena acus*** var. ***hyalina*** Klebs 1883)

SIZE: 97–190 µm long × 6–11 µm wide.

Colorless, cylindrical cells showing the eyespot and rod-like paramylon bodies (**A**, **B**). The contractile vacuole is visible at the anterior of the cell (arrow), and the pellicle strips are very fine (double arrow) (**B**).

Lepocinclis antefossa

(= ***Euglena antefossa*** Johnson 1944)

SIZE: 295–336 µm long × 22–28 µm wide.

Cells rigid, elongate, with a short groove at the anterior end and tapering to form a prominent, colorless tail (Ⓐ, Ⓒ, Ⓓ). The numerous disc-shaped chloroplasts lack pyrenoids (Ⓑ). Large rod-like paramylon bodies positioned anterior and posterior to the nucleus (Ⓐ, Ⓒ). Pellicle striations parallel to the longitudinal axis (Ⓑ).

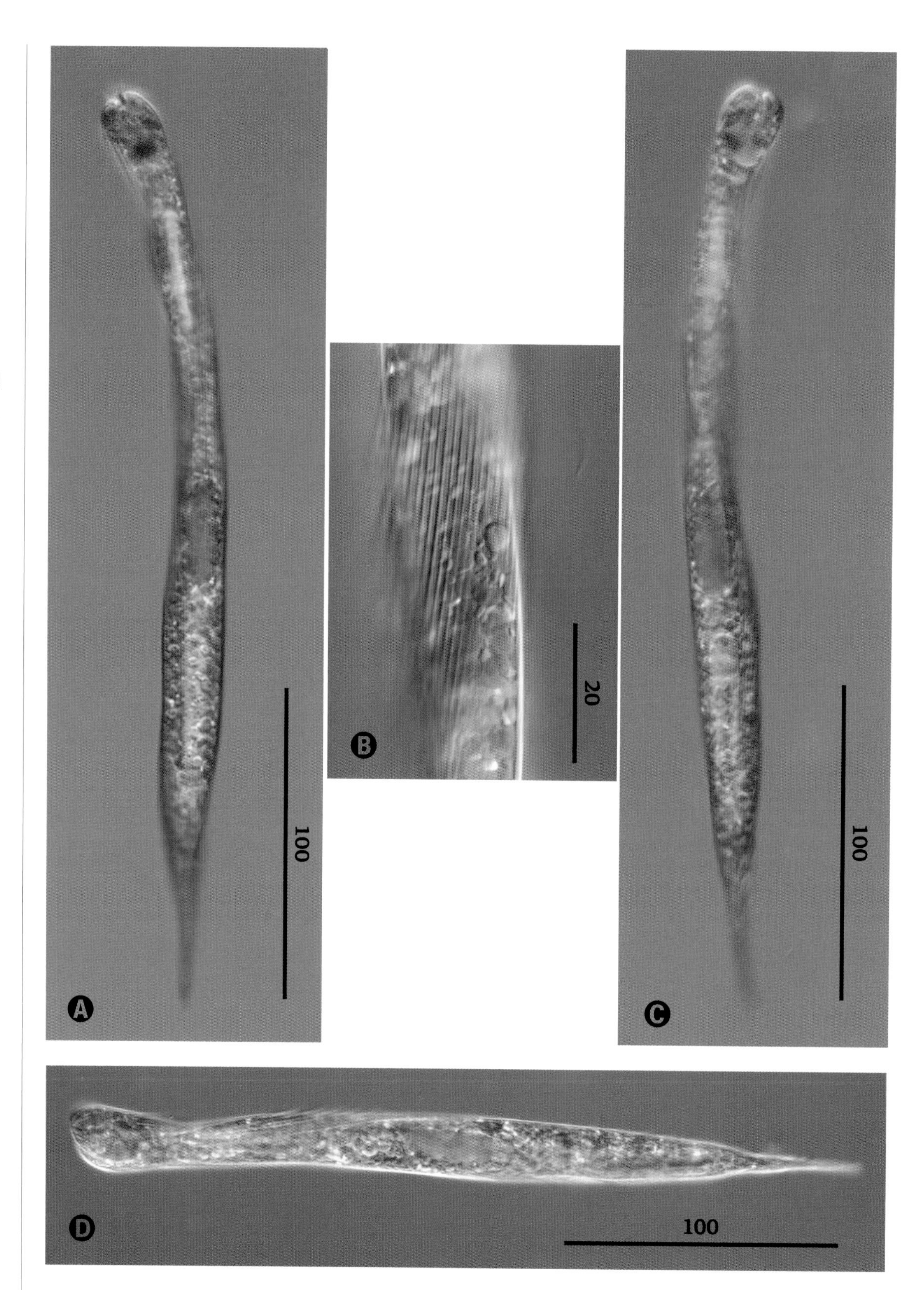

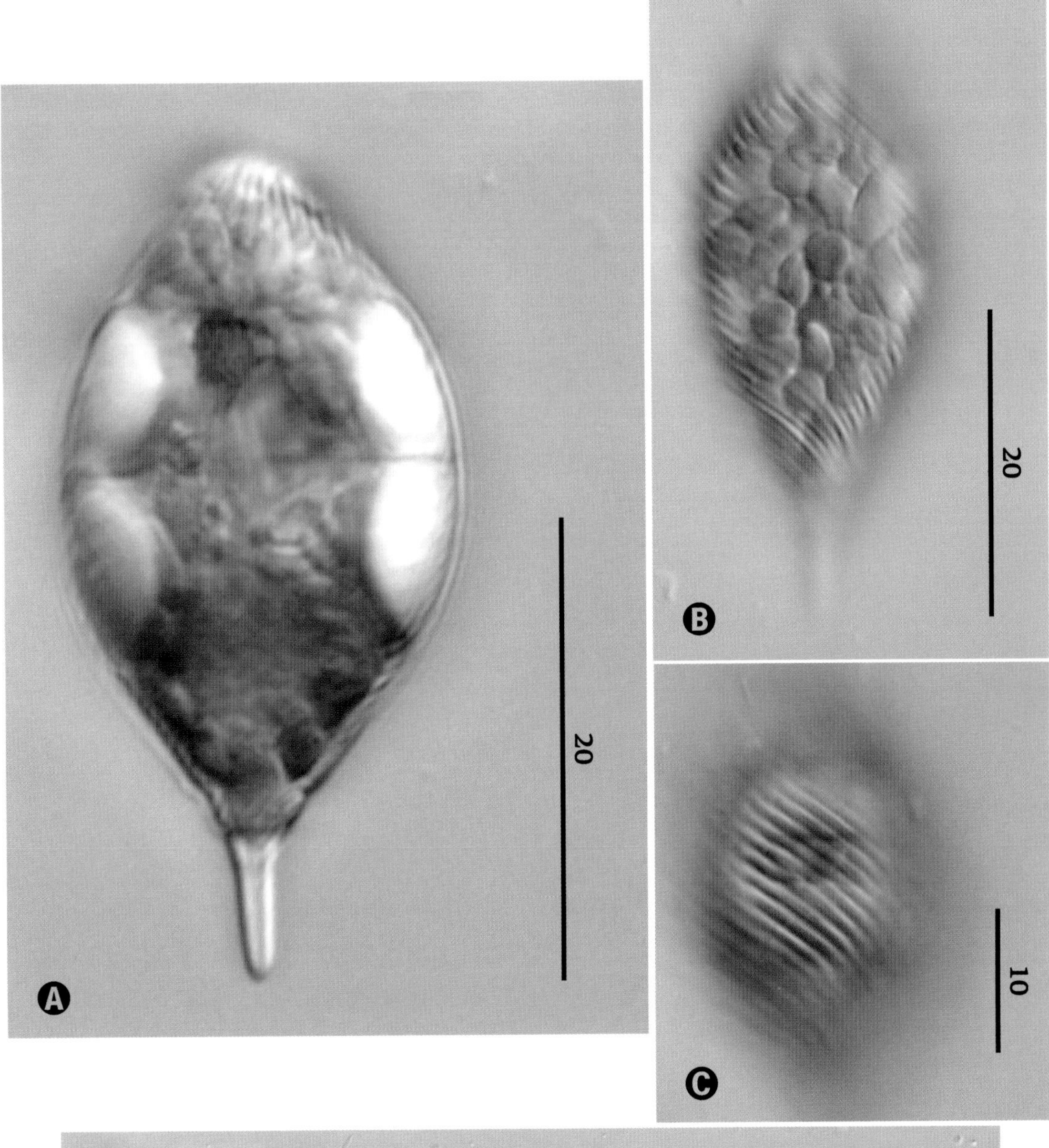

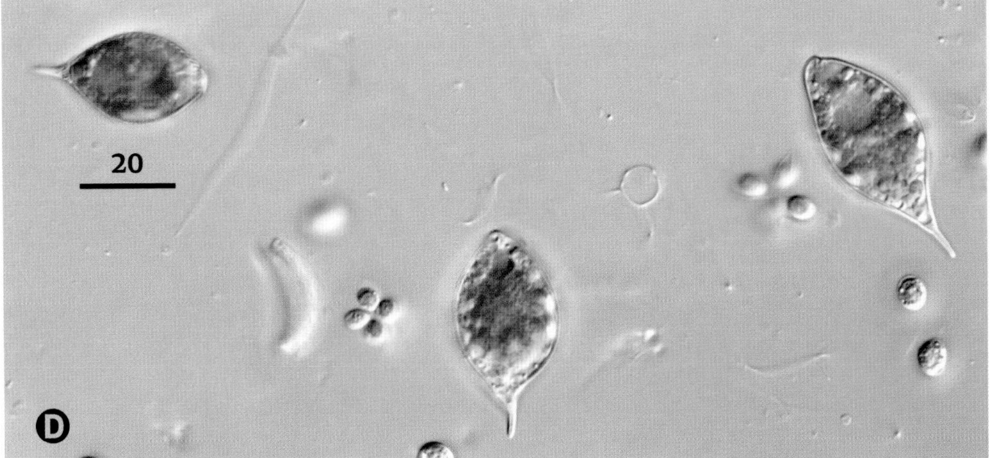

Lepocinclis bütschlii

Lemmermann 1901

SIZE: 30–42 µm long × 17–24 µm wide.

Ovoid cell with hyaline tail; red eyespot located anterior; ends of the large paramylon link are visible on sides (**A**). Cells contain numerous disc-shaped chloroplasts (**B**). Detail of yellow-brownish pellicle striations with left-handed orientation and portion of the large paramylon link in face view (**C**). Comparison showing one cell of *L. bütschlii* (*upper left corner*) and two cells of *L. playfairiana;* notice the differences in size and color (**D**).

Lepocinclis capito

Wehrle 1939

SIZE: 32–38 µm long × 11–24 µm wide.

Broadly pyriform cell showing flagellum, eyespot, and a portion of the lateral paramylon link (arrow) (A). Surface view of cell showing thick left-handed pellicular striations and numerous small disc-shaped chloroplasts (B).

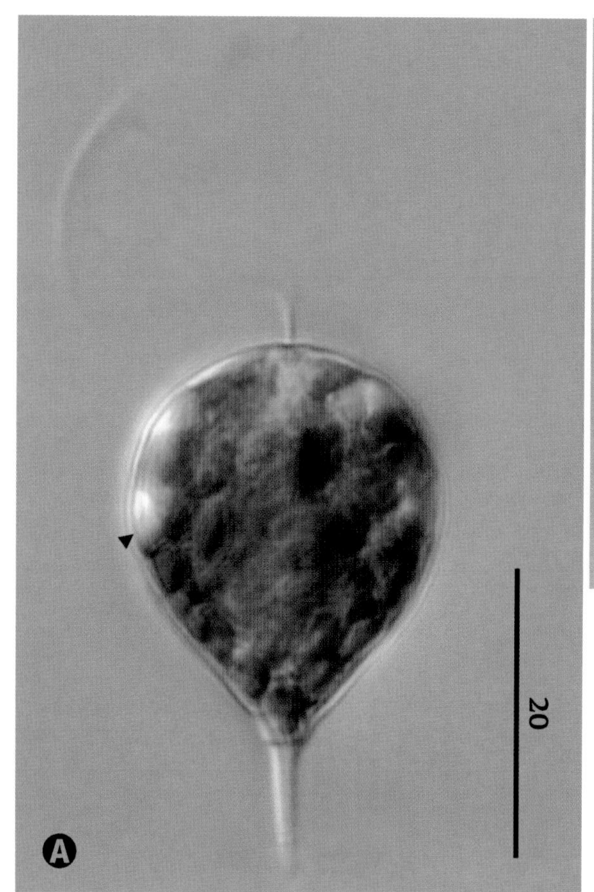

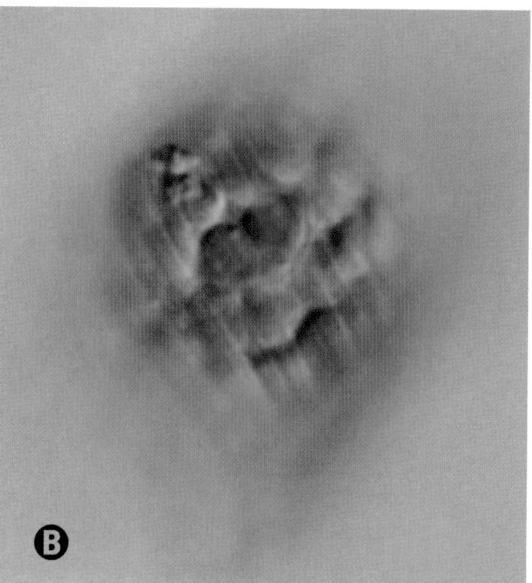

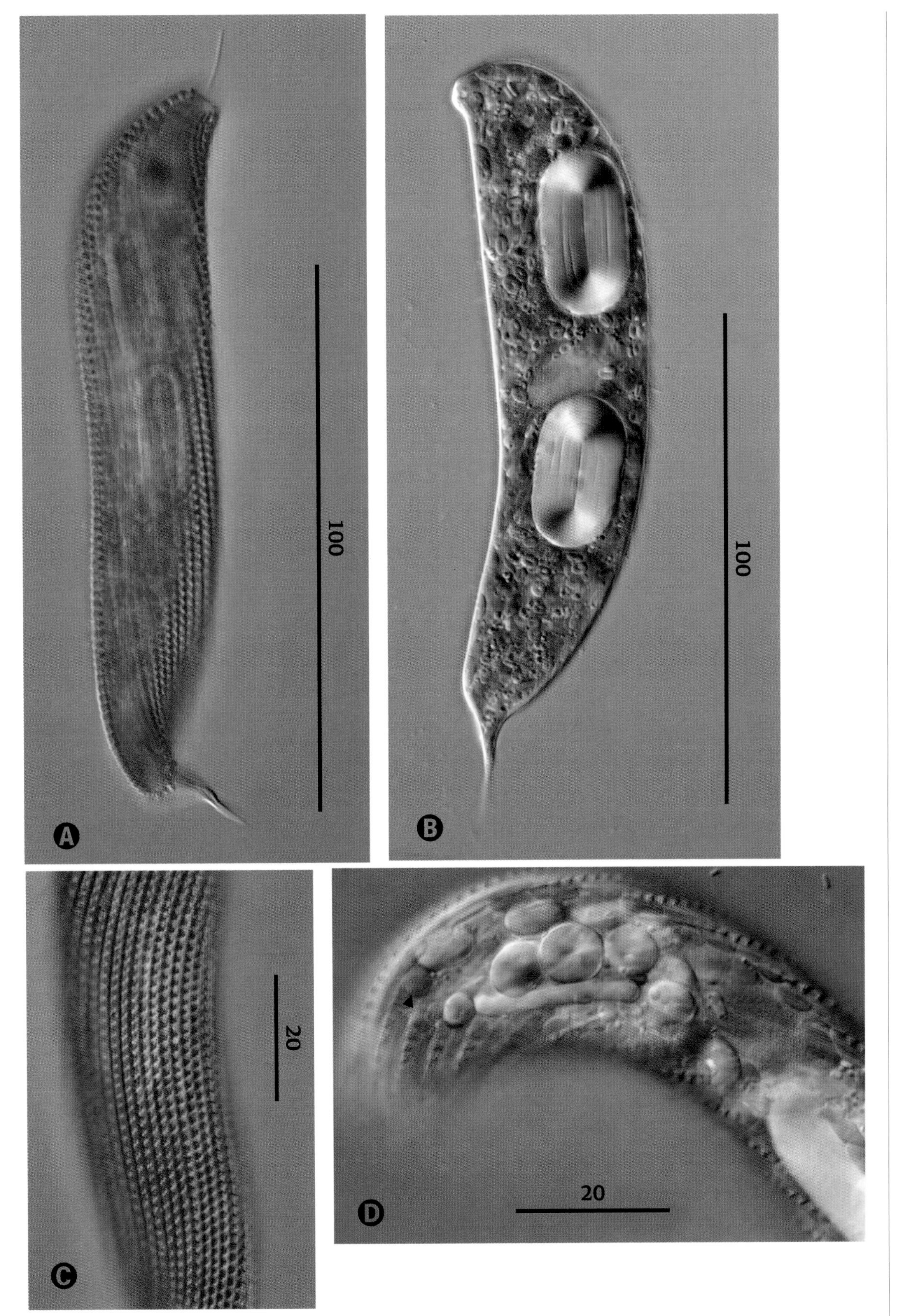

Lepocinclis fusca

(Klebs) Kosmala et Zakryś 2005

SIZE: 140–241 µm long × 11–24 µm wide.

Cylindrical, brownish cell with flagellum (Ⓐ). Eyespot, two large paramylon links (Ⓑ), and numerous smaller paramylon grains (Ⓒ, Ⓓ). Pellicular striations exhibit pyramid-like ornamentations (Ⓒ). Chloroplasts are small and disc-shaped (arrow) (Ⓓ).

Lepocinclis fusiformis

(Carter) Lemmermann 1901

SIZE: 28–40 µm long × 17–24 µm wide.

Citriform cells with flagellum (one to one-and-a-half times body length) (A), eyespot, and nucleus with nucleolus (A, B). Pellicle striations oriented in a left-handed spiral and numerous small disc-shaped chloroplasts (C). Paramylon is in the form of a large curved link, seen in face view (D).

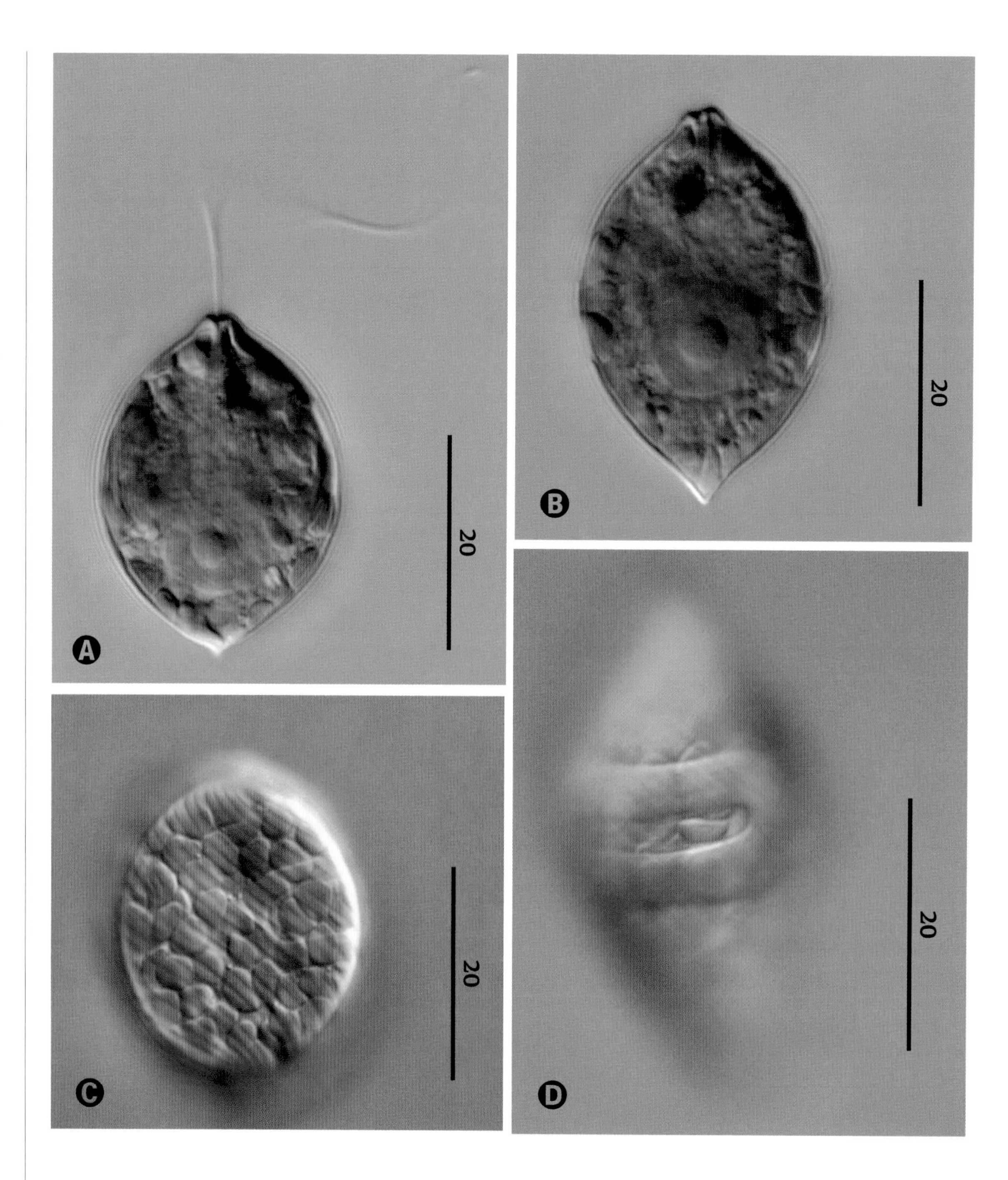

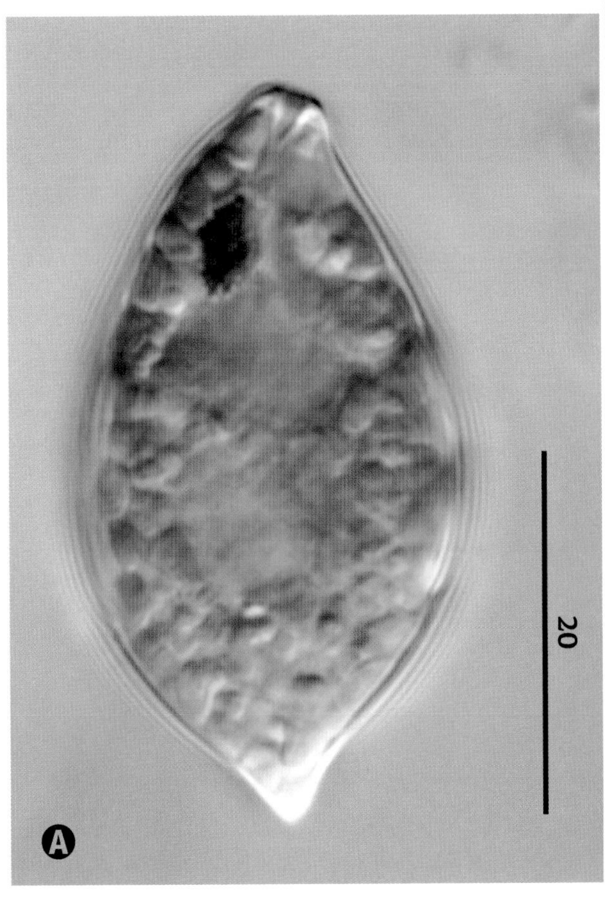

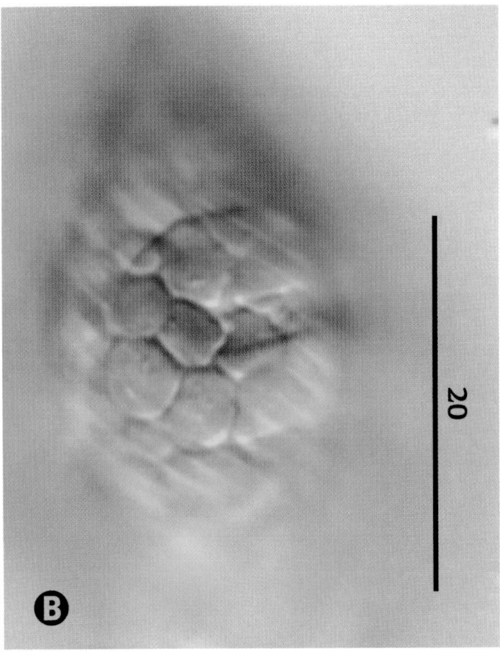

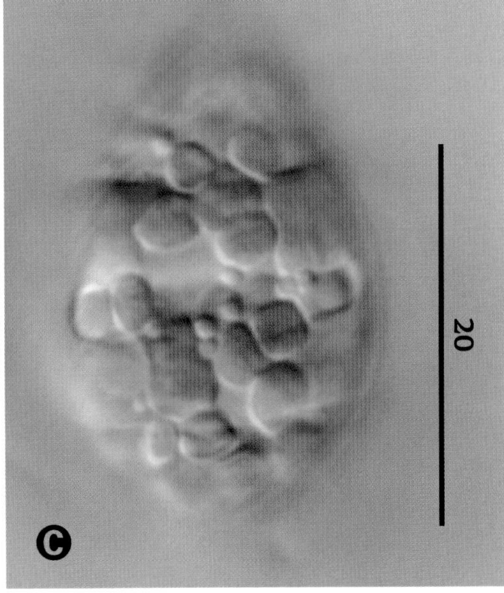

Lepocinclis fusiformis var. *amphirhynchus*

Nygaard 1949

SIZE: 30–42 µm long × 23–26 µm wide.

Cell showing short caudus, red eyespot, chloroplasts, and ends of large paramylon link on lateral sides (A). Paramylon link in face view; disc-shaped chloroplasts and pellicle strips arranged in a left-handed spiral (B). Large paramylon link seen in face view and several small disc-shaped chloroplasts (C).

Lepocinclis helicoideus

(= ***Euglena helicoideus*** (Bernard) Lemmermann 1910)

SIZE: 103–485 µm long × 18–47 µm wide.

Large cylindrical cell with red eyespot and numerous rod-like paramylon bodies (**A**). Pellicular striations run longitudinally; numerous disc-shaped chloroplasts present (**B**). Large curved cell surrounding a smaller *E. rubra* (**C**). Twisted cell body seen during swimming (**D**).

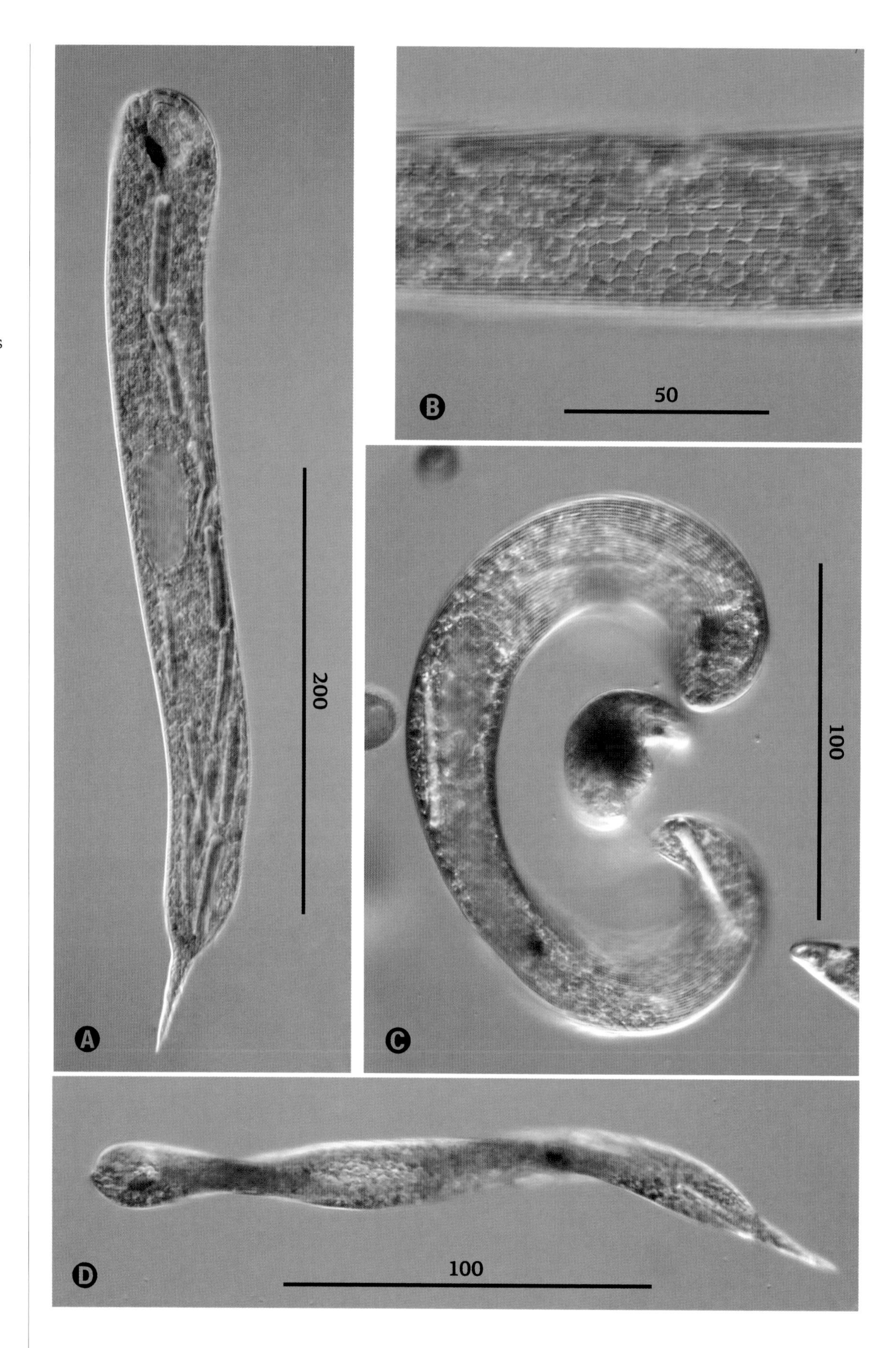

Lepocinclis lefèvrei

Conrad 1934

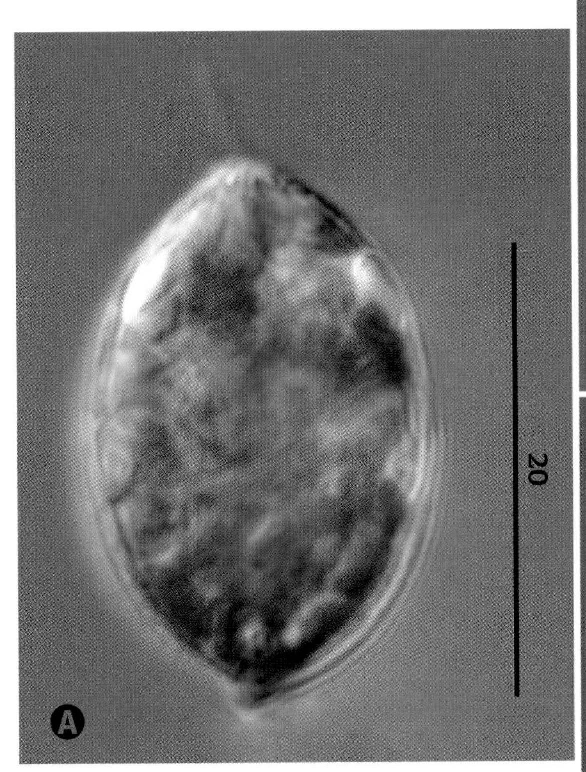

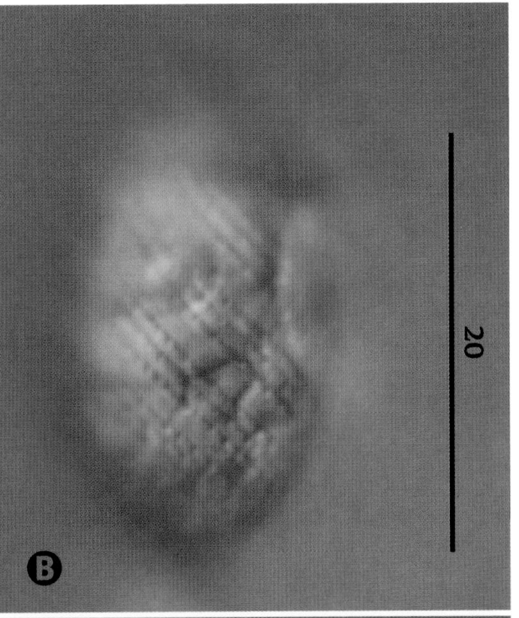

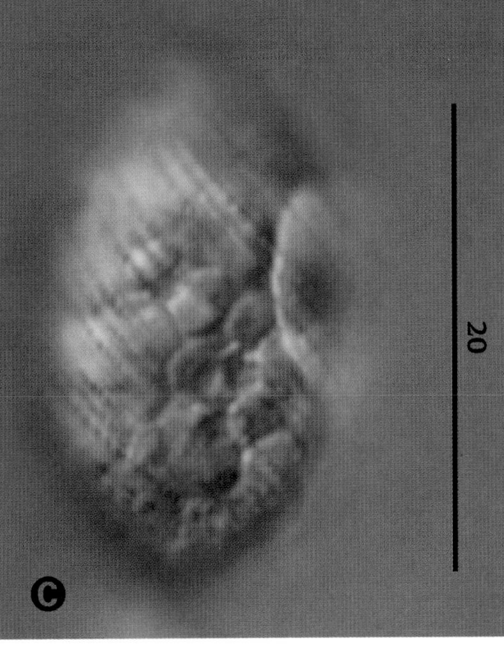

SIZE: 21–30 μm long × 12–17 μm wide.

Obovoid cell showing flagellum, eyespot, lateral link-like paramylon body, and short blunt caudus (A). The large, curved paramylon body is parietally positioned; pellicle strips are left-handed and finely granulated (B, C). Chloroplasts are numerous small discs (C).

Lepocinclis ovum

(Ehrenberg) Lemmermann 1910

SIZE: 20–38 µm long × 13–24 µm wide.

Broadly ovoid cells showing flagellum (about body length) (A), eyespot, disc-shaped chloroplasts, nucleus, and short caudus (A, B). Pellicular striations are left-handed (C). One large link-like paramylon body is seen in face view (C) and lateral view (B—arrows, D).

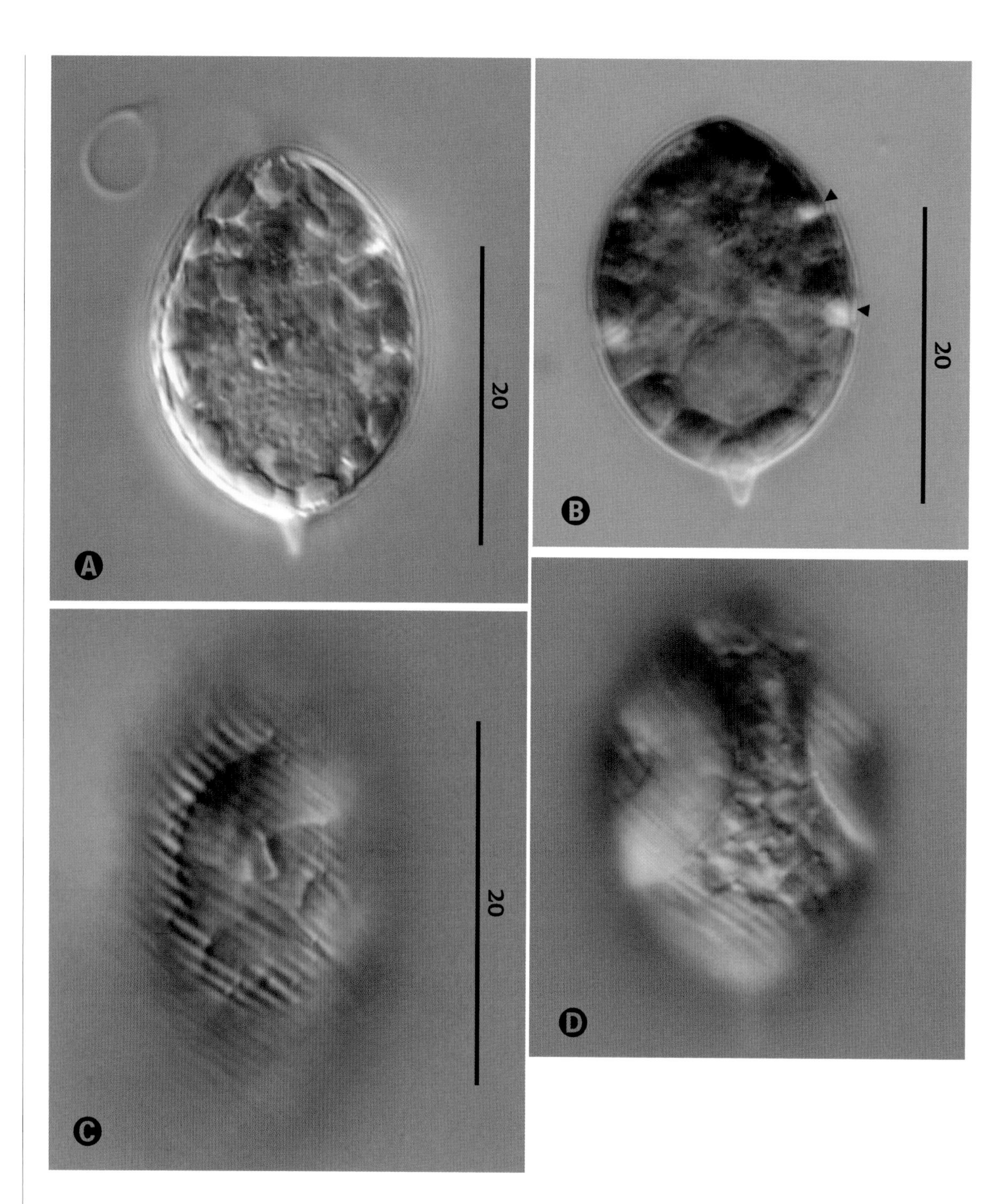

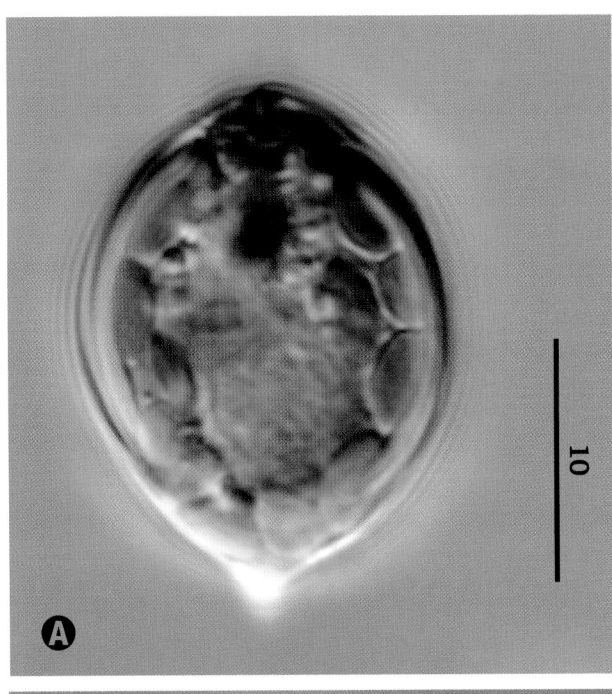

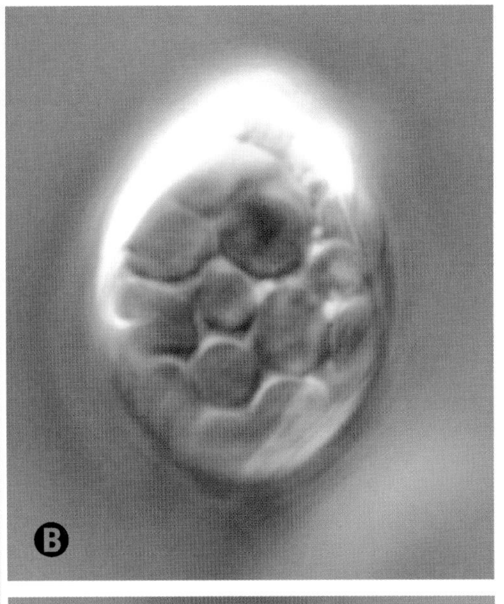

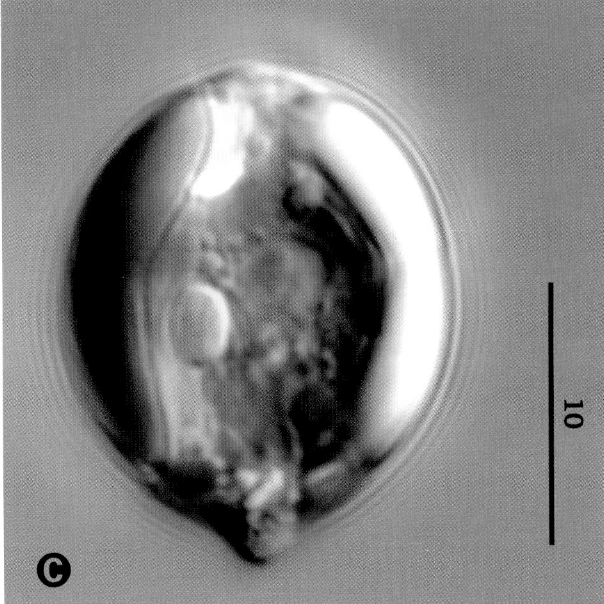

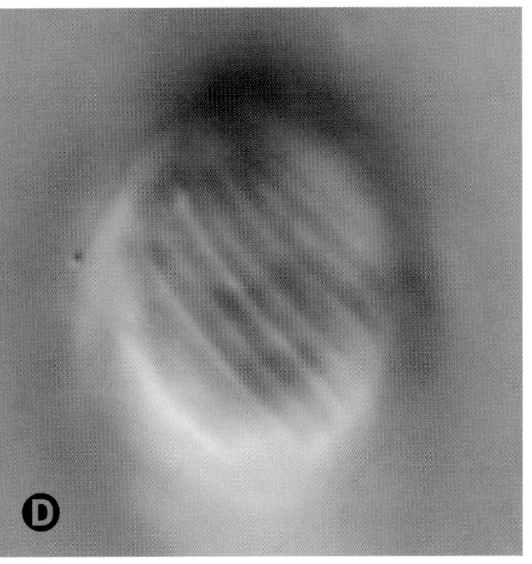

Lepocinclis ovum var. *globula*

(Perty) Lemmermann 1913

SIZE: 13–27 µm long × 10–21 µm wide.

Ovoid cell with short caudus, red eyespot, disc-shaped chloroplasts; the ends of the large link-like paramylon body can be seen on lateral sides (**A**). Disc-shaped chloroplasts lacking pyrenoids (**B**). Ends of paramylon body and smaller ovoid paramylon grain (**C**). Pellicle striations are left-handed (**D**). Flagellum is three to four times body length.

Lepocinclis ovum var. *ovata*

Swirenko 1915

SIZE: 28–40 µm long × 23–39 µm wide.

Ovoid cell with short caudus showing eyespot, reservoir, and large lateral paramylon body (A). Numerous disc-shaped chloroplasts (B, C) and flagellum (one-half body length) (C). The link-like nature of the paramylon body is shown in A–C. Pellicular striations running left-handed (C, D).

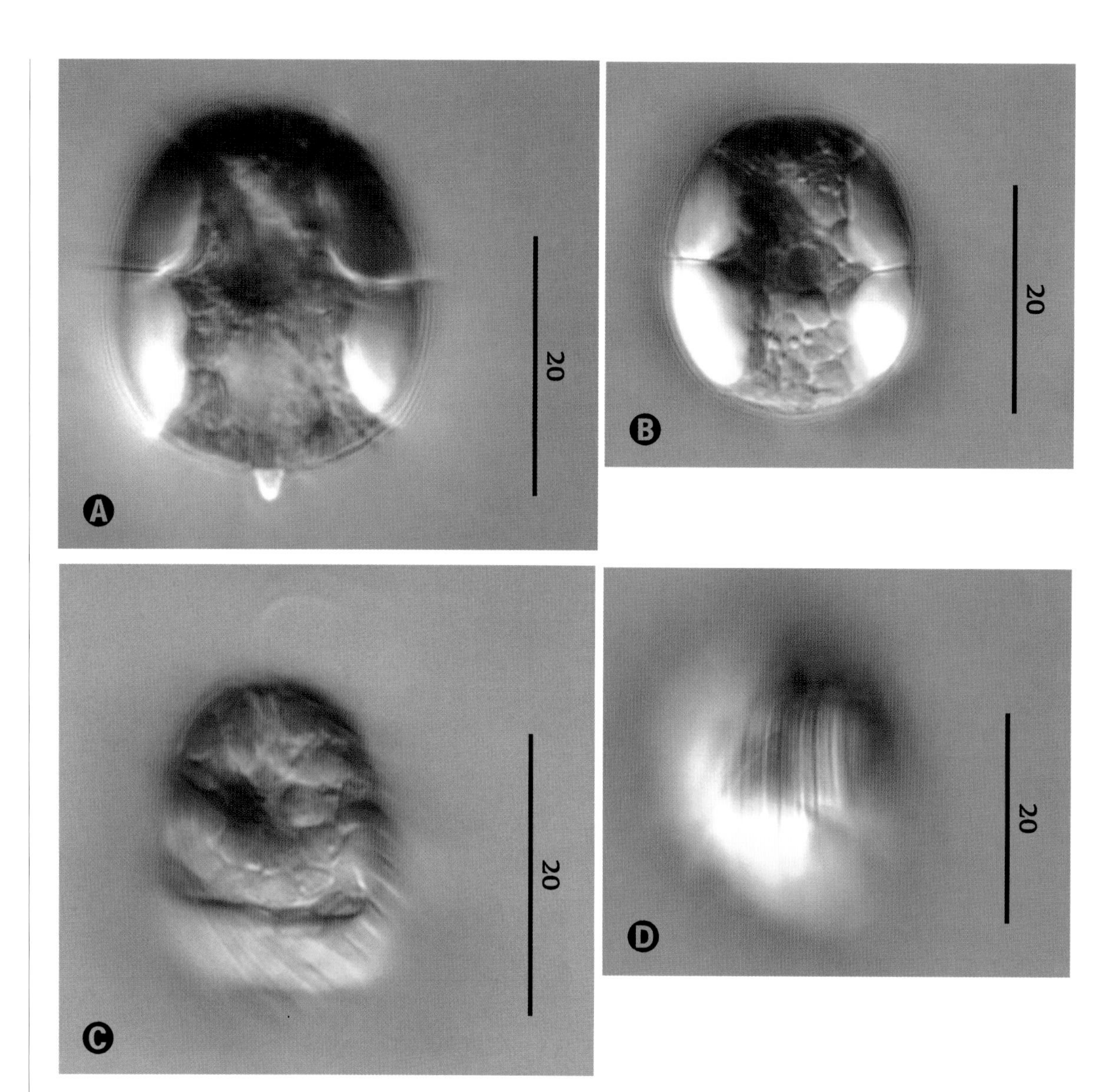

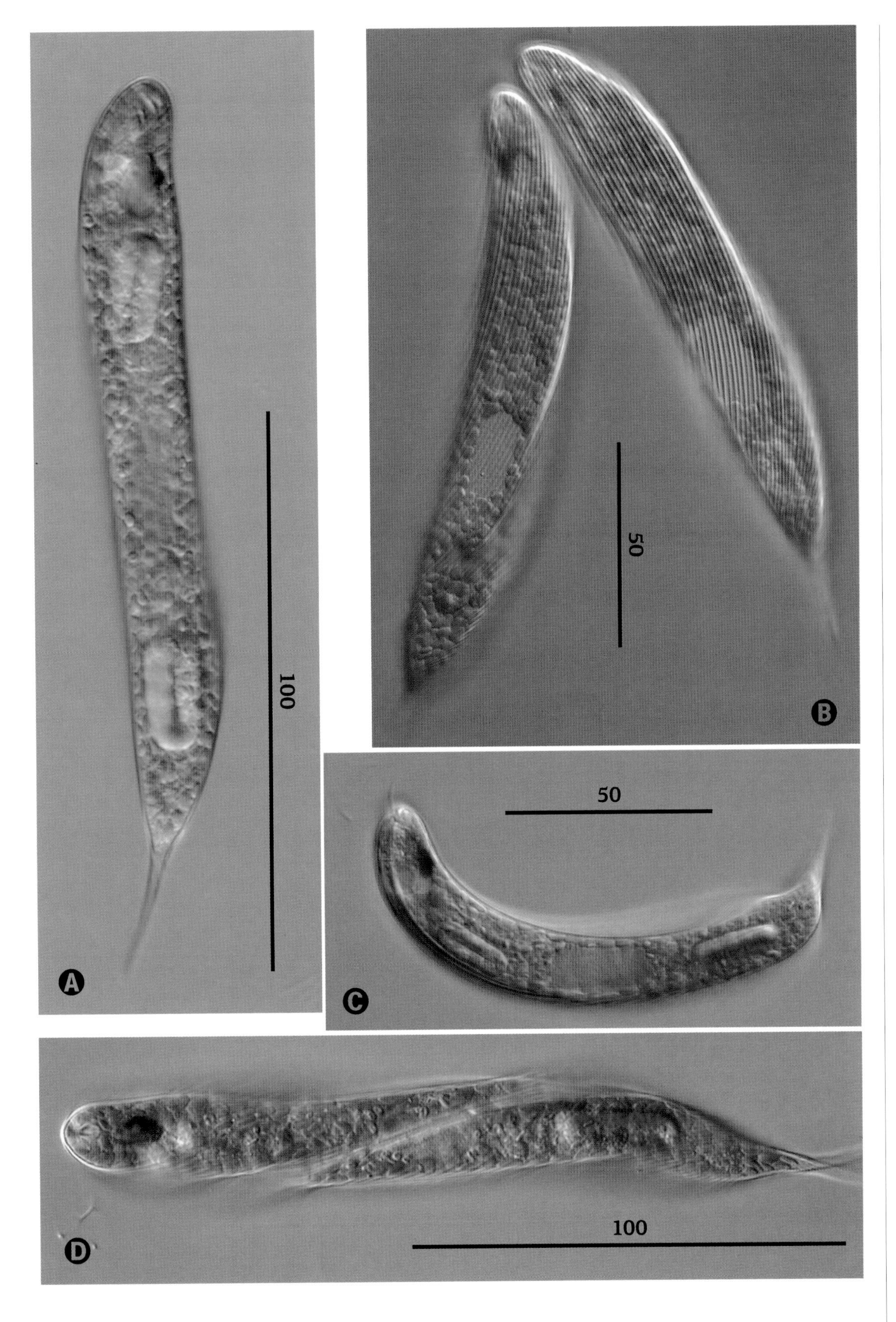

Lepocinclis oxyuris

(Schmarda) Marin et Melkonian 2003

SIZE: 152–225 µm long × 16–26 µm wide.

Cylindrical cell with hyaline caudus, red eyespot, two large paramylon links, and numerous small disc-shaped chloroplasts (**A**). Two cells with pellicular striations running longitudinally; large nucleus visible toward the posterior (**B**). Curved cell with flagellum (one-third to one-half body length) and two large link-like paramylon bodies placed anterior and posterior to the nucleus (**C**). Swimming cell showing the characteristic twisted body (**D**).

Lepocinclis playfairiana

Deflandre 1932

SIZE: 40–50 µm long × 17–30 µm wide.

Broadly ovoid cell with long caudus showing eyespot, reservoir, basal part of both flagella (arrow), disc-shaped chloroplasts, and nucleus with nucleolus (double arrow) (**A**). Cell showing large paramylon link (arrow) (**B**). Pellicular striations are fine and left-handed (**C**). Paramylon link seen in face view; small ovoid paramylon bodies also present (**D**). Flagellum about body length.

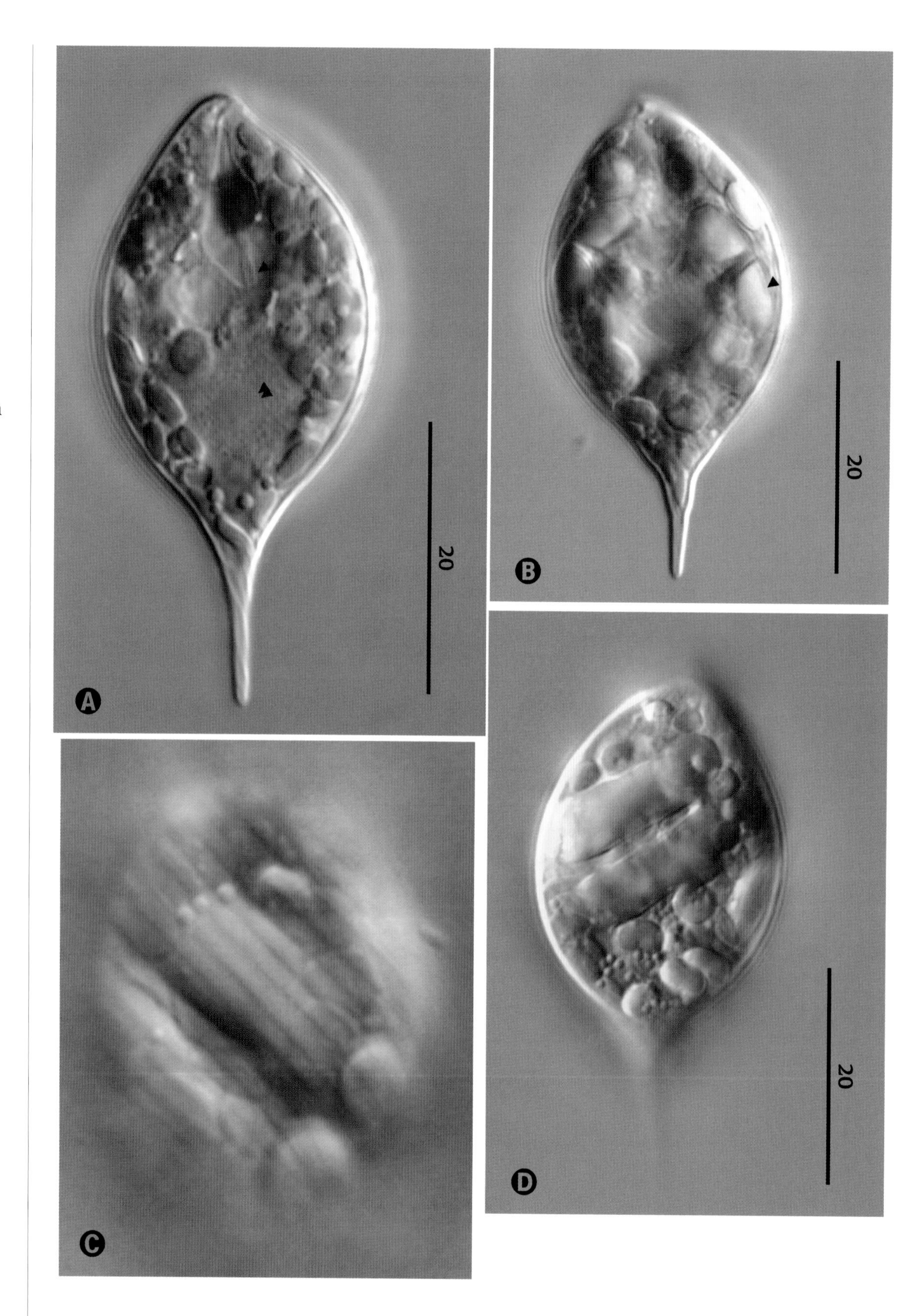

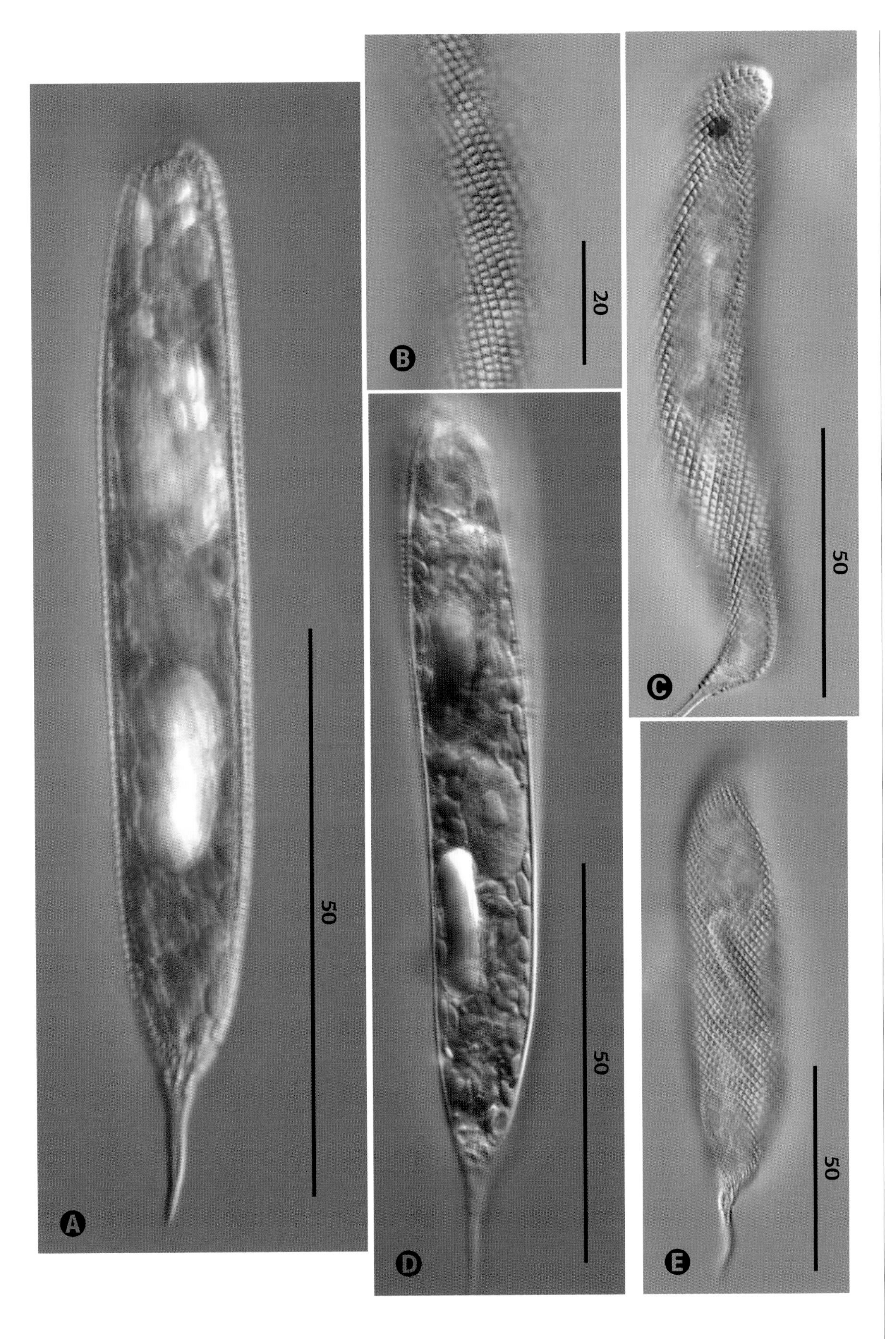

Lepocinclis spirogyroides

(Ehrenberg) Marin et Melkonian 2003

SIZE: 80–197 µm long × 6–30 µm wide.

Cylindrical cell with hyaline tail and two large paramylon bodies (A). Detail of pellicle strips with brick-like ornamentations (B). Slightly twisted cell showing pellicle ornamentation and red eyespot (C). Cell with central nucleus with nucleolus, two large paramylon bodies, and numerous small disc-shaped chloroplasts (D). Surface view of yellow-brown pellicle with brick-like ornamentations (E).

Lepocinclis steinii

Lemmermann 1901

SIZE: 22–30 μm long × 7–17 μm wide.

Fusiform cells with short conical caudus showing flagellum (twice body length), large paramylon link on the lateral sides (arrows) (**B**), numerous disc-shaped chloroplasts (**A**, **B**), and eyespot (**A**). Details of longitudinally arranged pellicular striations and paramylon link in face view (**C**, **D**).

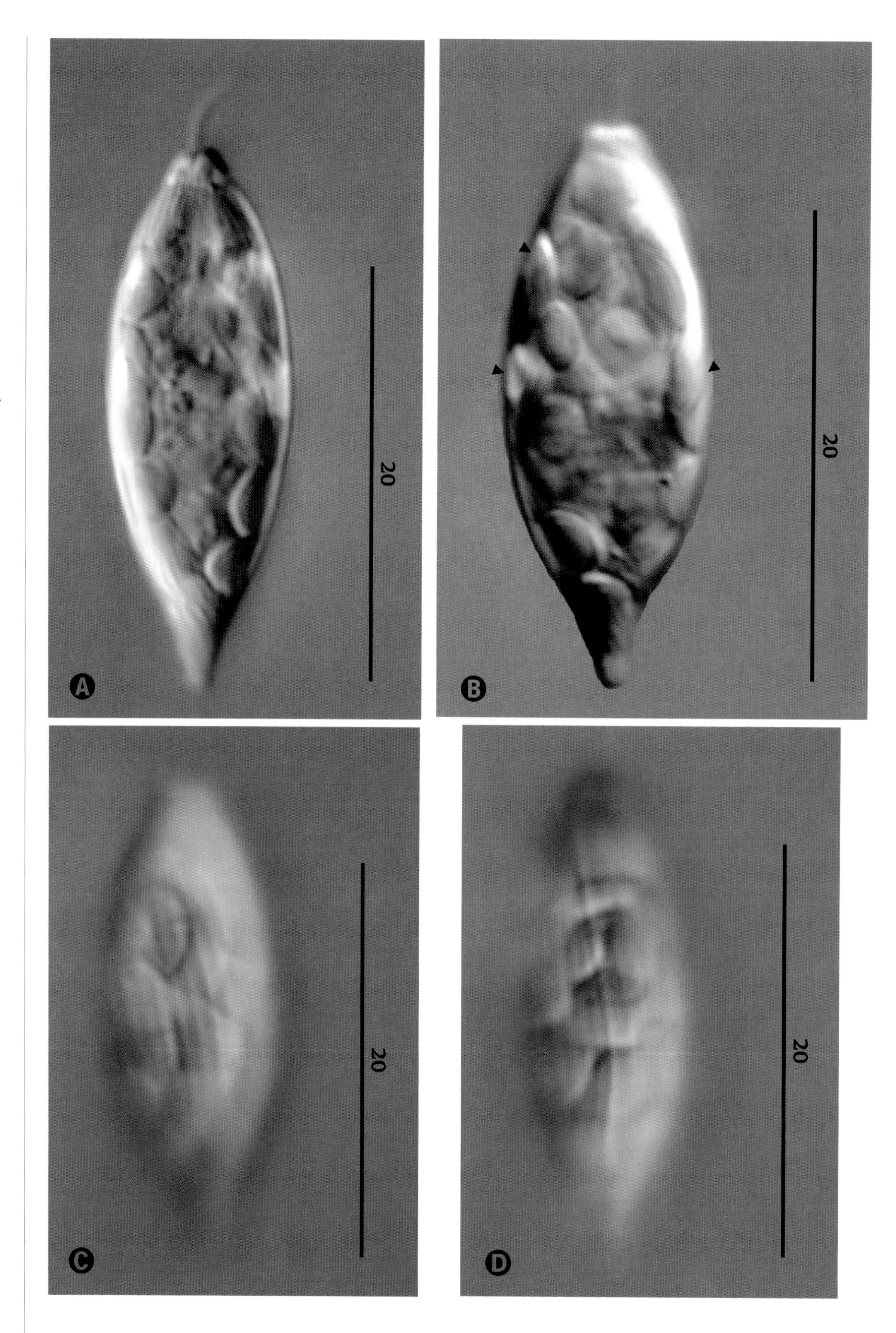

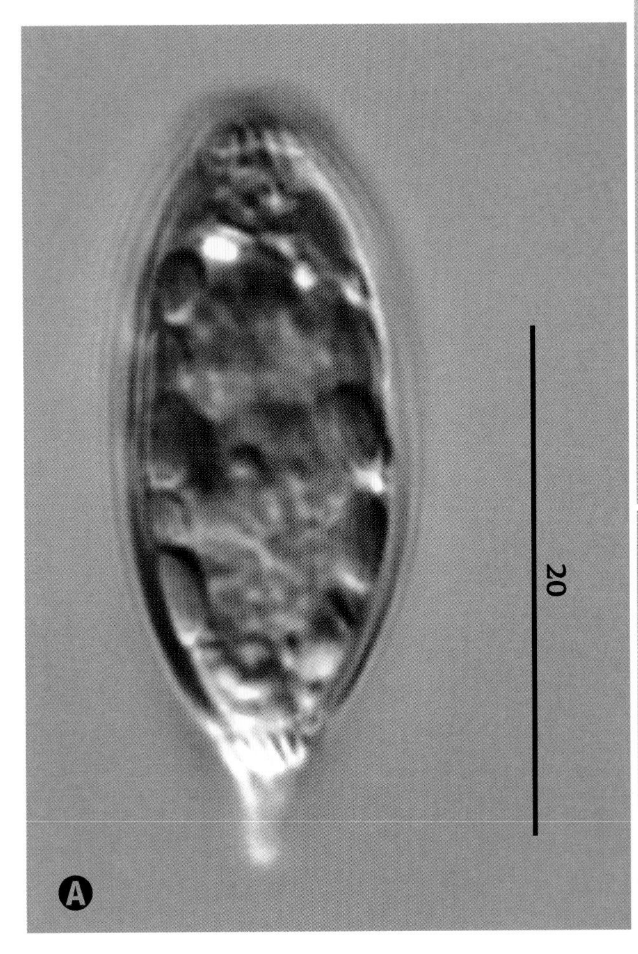

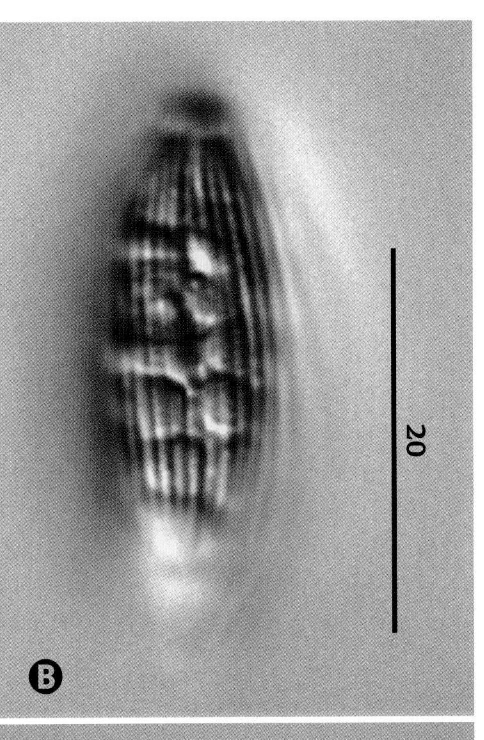

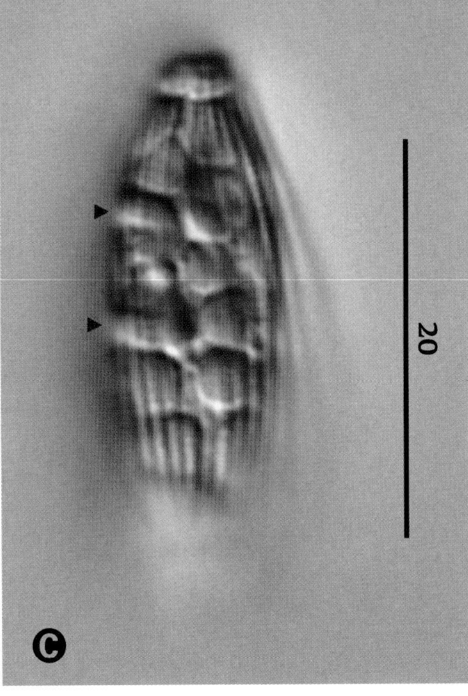

Lepocinclis steinii var. *suecica*

Lemmermann 1901

SIZE: 20–30 µm long × 10–15 µm wide.

Fusiform cell with caudus expanded at connection with body cell; eyespot, disc-shaped chloroplasts, and small paramylon grains are visible (**A**). Cell with longitudinally arranged pellicle strips, chloroplasts, and large link-like paramylon body (**B**, **C**—arrows).

Lepocinclis tripteris

(Dujardin) Marin et Melkonian 2003

SIZE: 60–135 µm long × 12–18 µm wide.

Spindle-shaped cells showing two large paramylon rods (Ⓐ), red eyespot, and numerous small disc-shaped chloroplasts (Ⓐ–Ⓒ). Cells are spirally twisted when swimming (Ⓑ, Ⓒ) and may bend (Ⓑ). Flagellum is one-half body length.

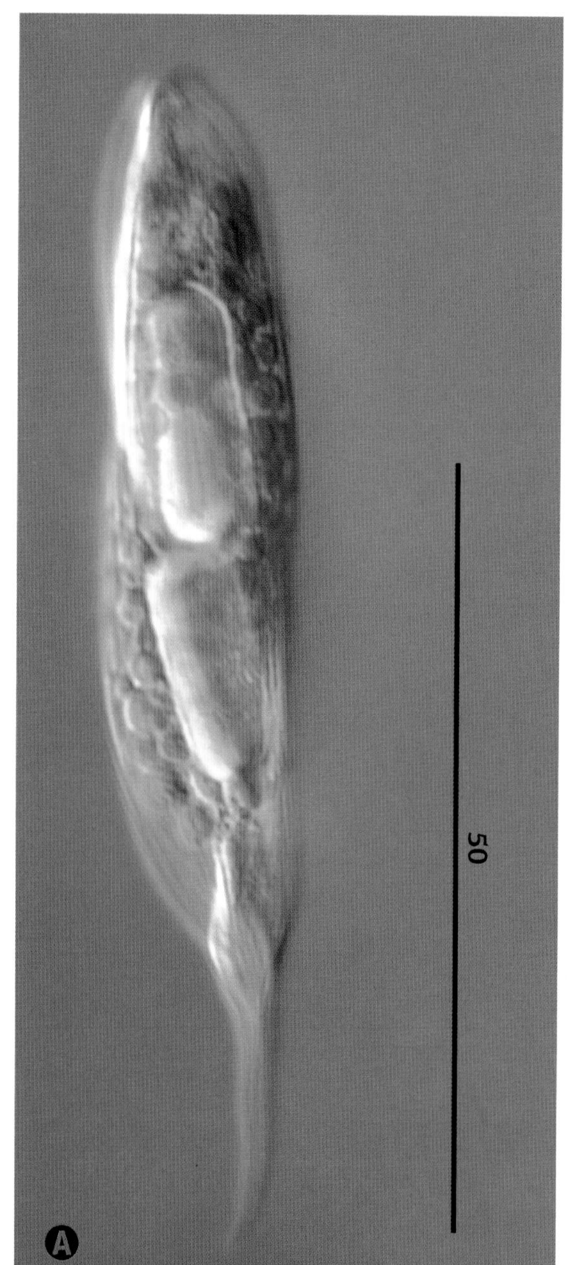

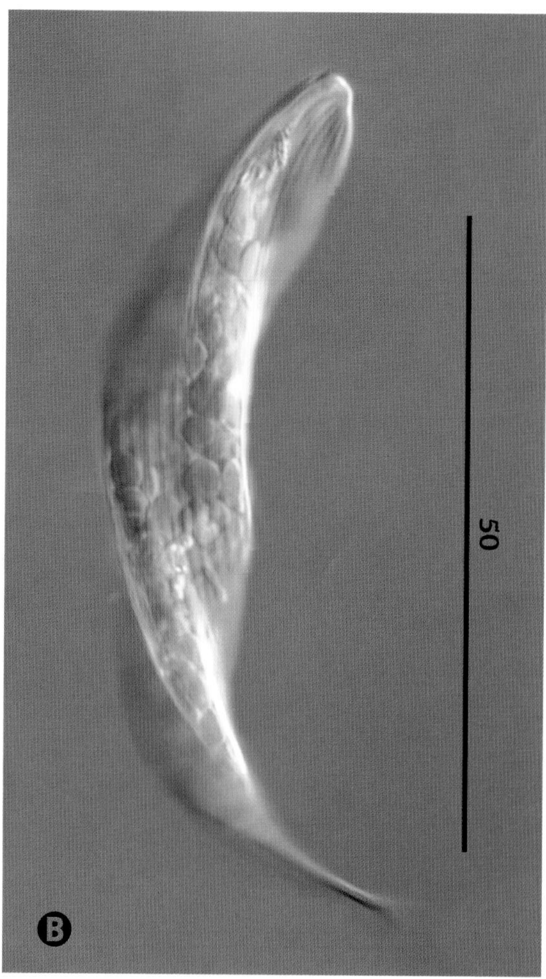

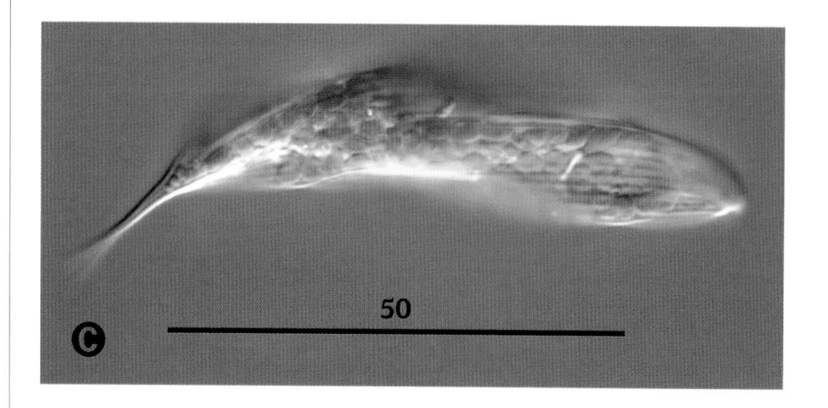

MONOMORPHINA

MERESCHKOWSKY EMEND. KOSMALA AND ZAKRYŚ 2007

Rigid euglenoids, pyriform in lateral view, with a hyaline spine at the posterior end; a single emergent flagellum; pellicle with clearly visible, helically arranged ribs; one large parietal chloroplast, typically without pyrenoids (however, haplopyrenoids have been reported in *M. aenigmatica*); usually two (rarely three) lateral plate-like paramylon bodies; eyespot at the anterior end; freshwater.

Monomorphina aenigmatica

(Dreżepolski) Nudelman et Triemer emend. Kosmala et Zakryś 2007

SIZE: 14–40 µm long × 5–15 µm wide.

Cell with flagellum (body length), large orbicular chloroplast, one haplopyrenoid (arrow), and large plate-like lateral paramylon bodies (double arrows) (A). The pellicle is markedly striated with few spirally oriented pellicle strips (B). Cells showing haplopyrenoid (arrow) (C) and curved plate-like paramylon bodies (C, D—arrow).

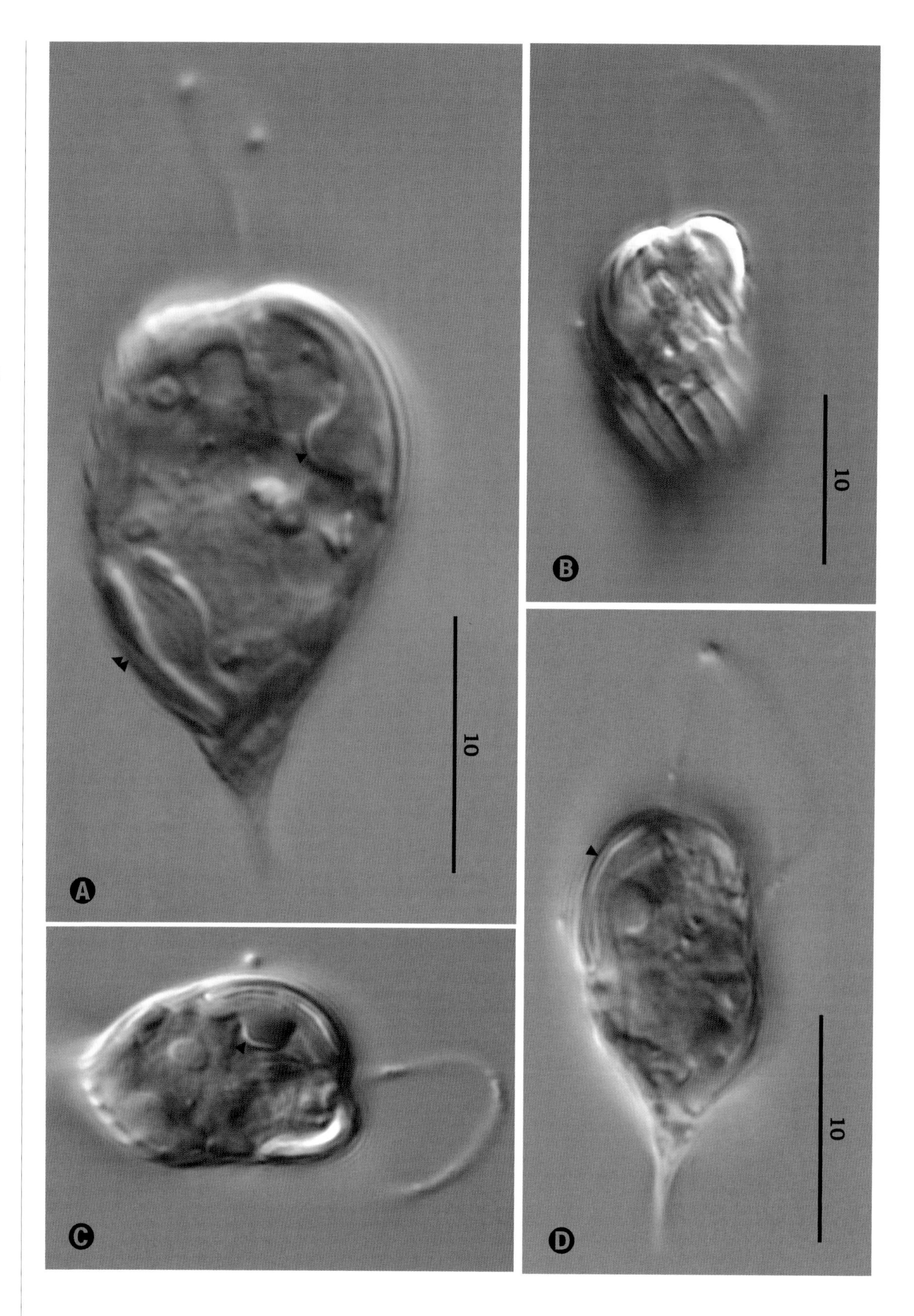

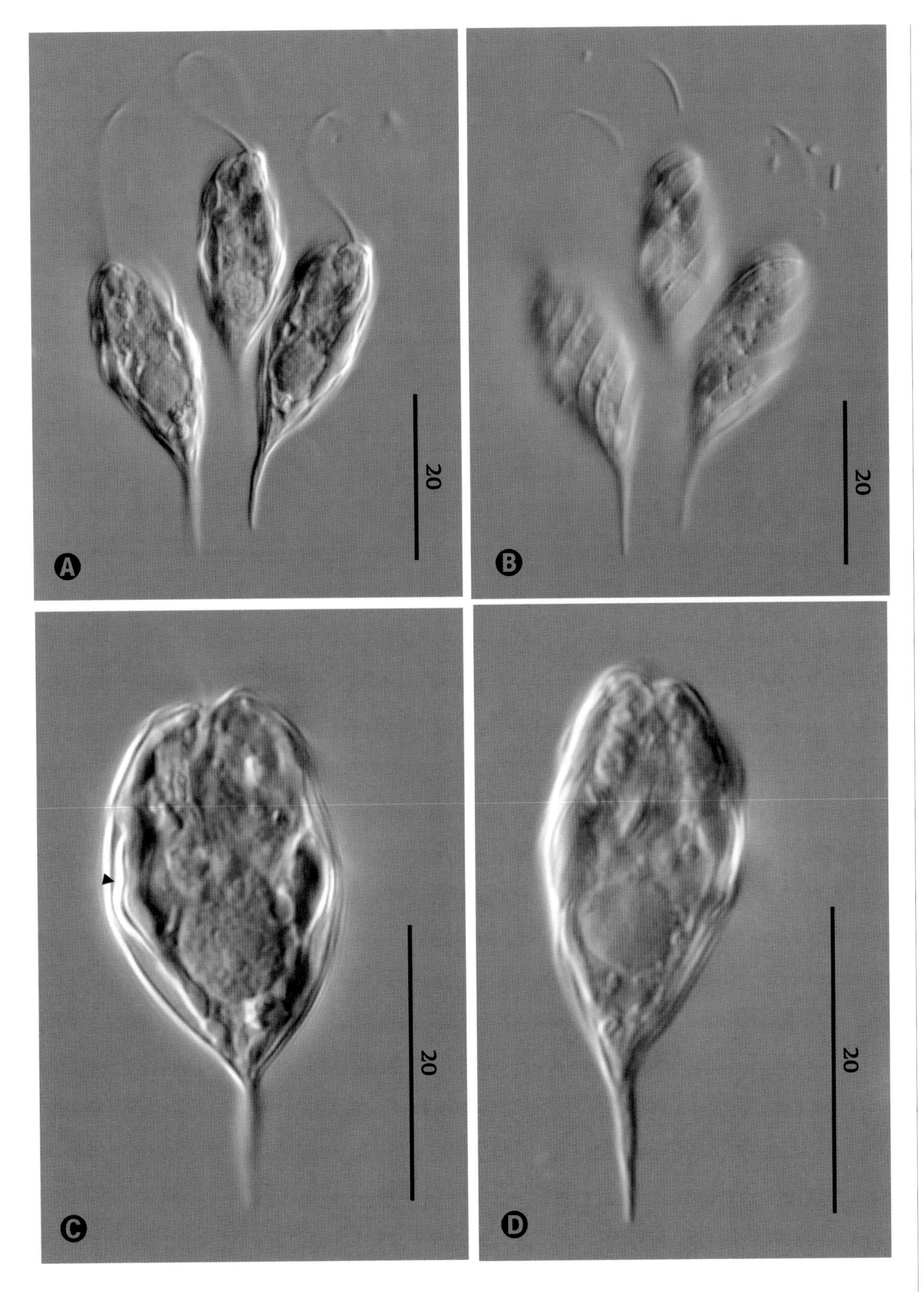

Monomorphina pseudopyrum

Kosmala, Milanowski, Brzóska, Pękala, Kwiatowski et Zakryś 2007

SIZE: 26–37 µm long × 10–16 µm wide.

Three spindle-shaped cells with flagellum, red eyespot, one orbicular perforated chloroplast, lateral paramylon plates, nucleus (A), and hyaline pellicle keels (B). Pear-shaped cells with eyespot, lateral paramylon plate (arrow) (C), and nucleus (C, D).

Monomorphina pyrum

(Ehrenberg) Mereschkowsky emend. Kosmala et Zakryś 2007

SIZE: 25–46 μm long × 9–19 μm wide.

Ovoid cell showing red eyespot and one orbicular perforated chloroplast and paramylon (arrow) (**A**). Cell in lateral view with eyespot, chloroplast, lateral paramylon plates (arrows), and nucleus (**B**). Cell with flagellum and pellicle (**C**). Two cells differing in shape (**D**). Detail of pellicle keels (ribs) (**E**).

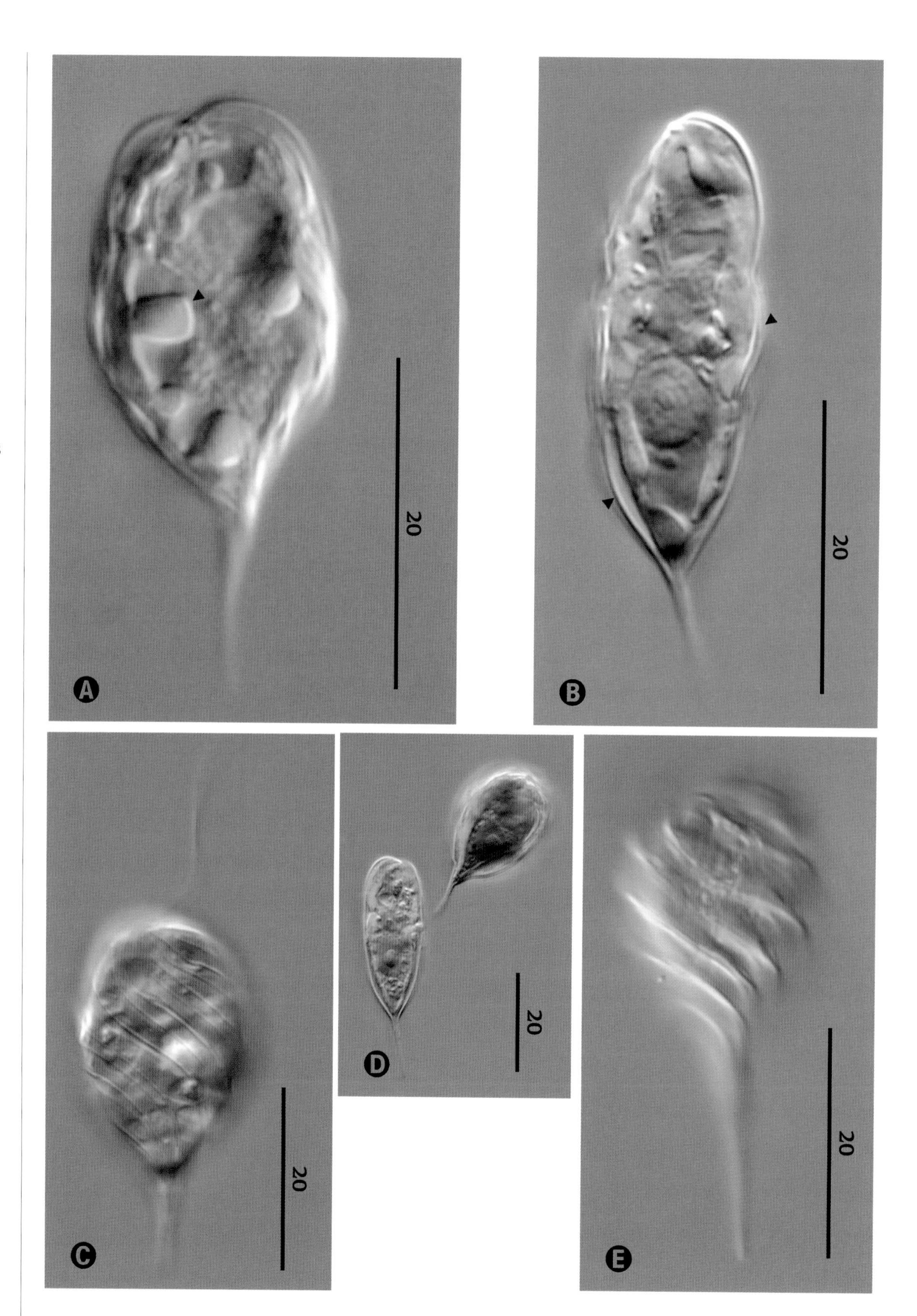

Phacus acuminatus

Stokes 1885

SIZE: 25–30 µm long × 18–27 µm wide.

Broadly ovoid cell with flagellum (body length or longer), eyespot, numerous disc-shaped chloroplasts lacking pyrenoids, and a small ring-like paramylon grain (A). Cells possess two paramylon bodies of different size (B, D). Cell exhibiting a median groove following the longitudinal axis (C). Pellicular striations are longitudinally disposed (D).

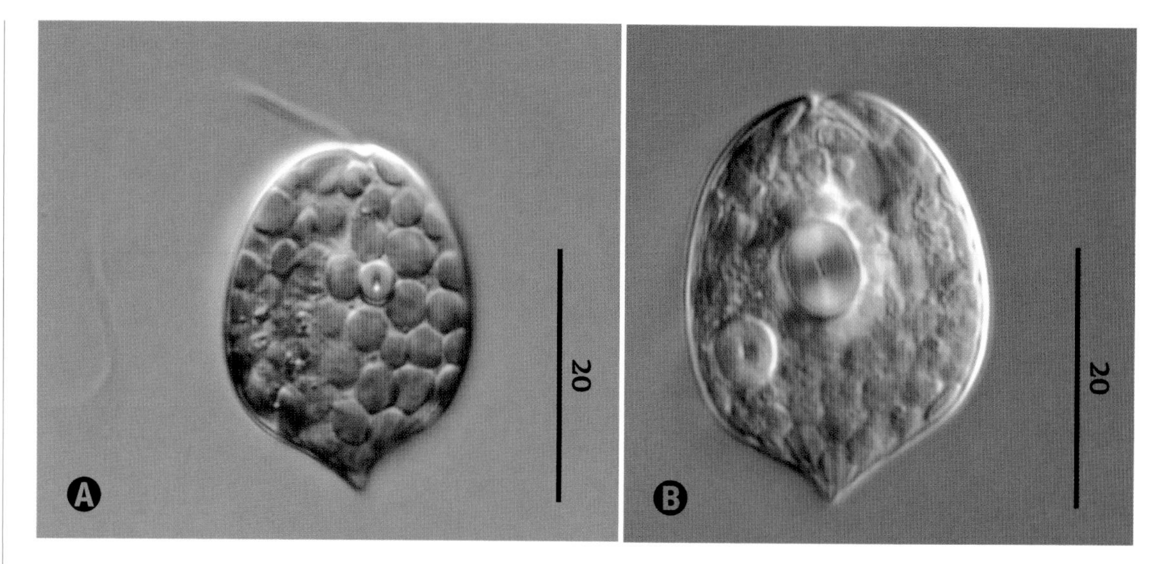

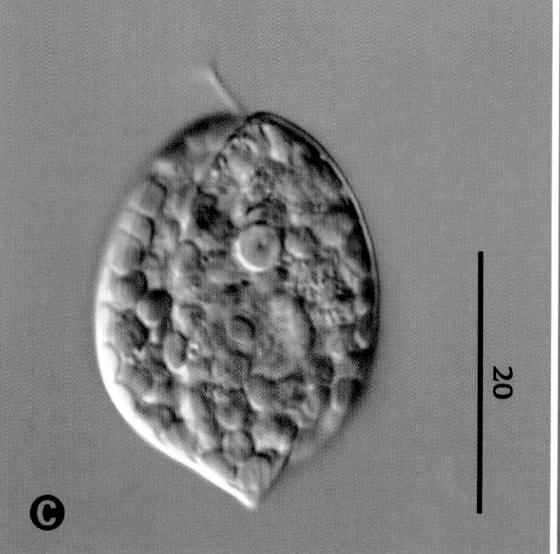

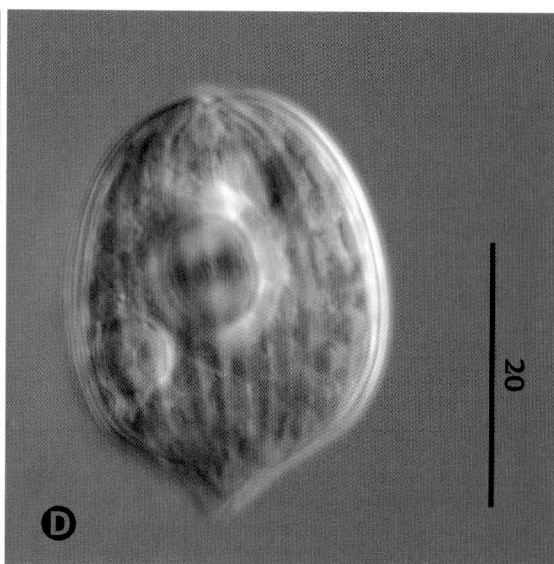

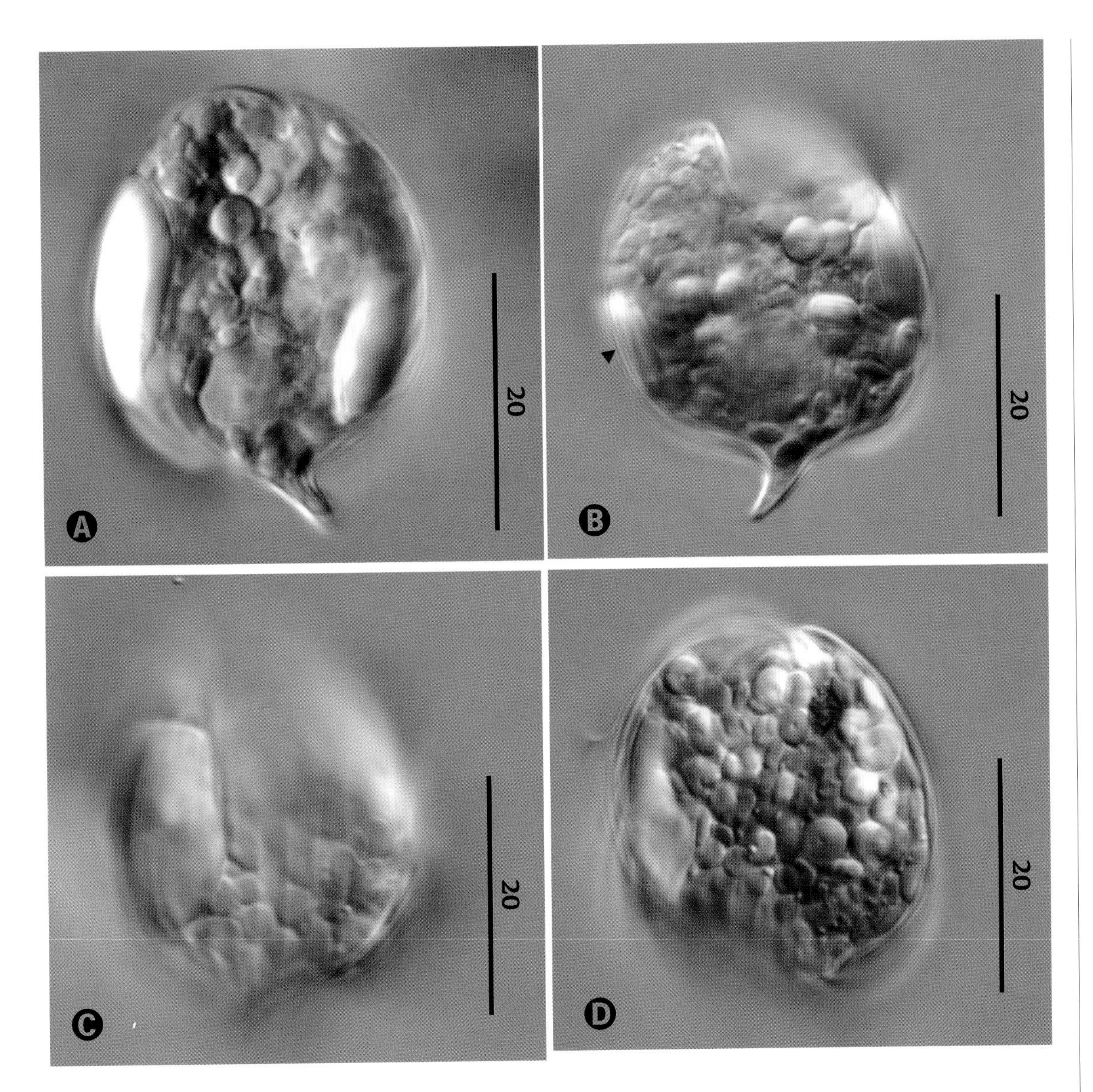

Phacus alatus

Klebs 1883

SIZE: 19–36 µm long × 16–22 µm wide.

Ovoid cell with short bent tail, red eyespot, disc-shaped chloroplasts, small paramylon bodies, and two lateral large paramylon discs (A). Detail of median indentation, nucleus posteriorly located, numerous chloroplasts, two lateral large paramylon discs (arrow), and several small paramylon bodies (B). Pellicle strips longitudinally arranged (C). Cell with red eyespot, chloroplasts, and paramylon bodies (D). Flagellum is body length.

Phacus applanatus

Pochmann 1942

SIZE: 36 µm long × 18 µm wide.

Ovoid cell with eyespot and one large paramylon body (A). Longitudinal pellicular striations (B). Group of cells (C). Lateral view showing eyespot, nucleus, numerous disc-shaped chloroplasts lacking pyrenoids, and paramylon body (D). Flagellum is body length.

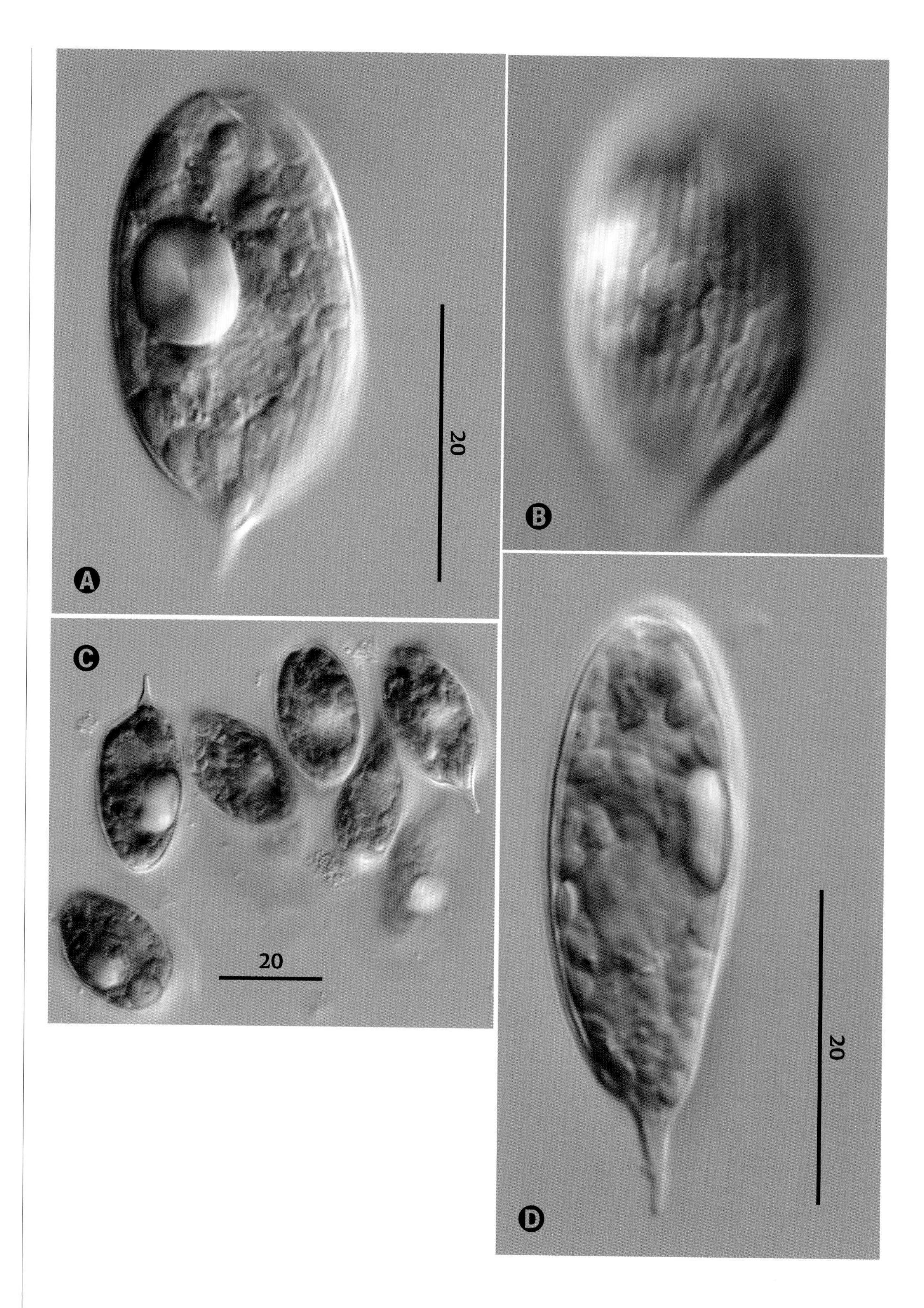

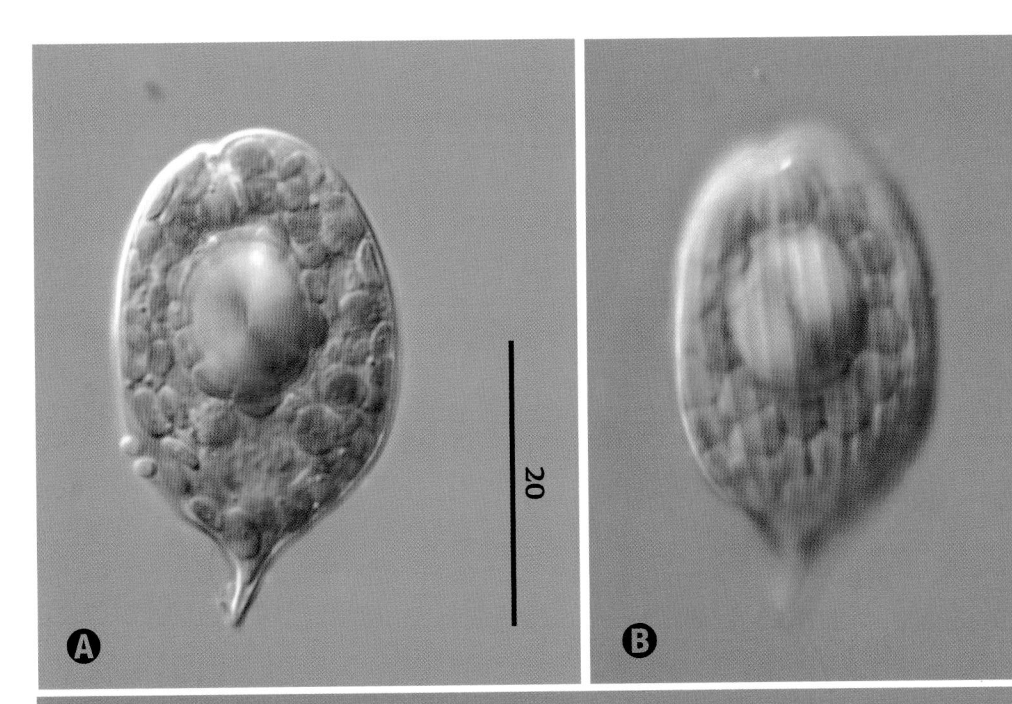

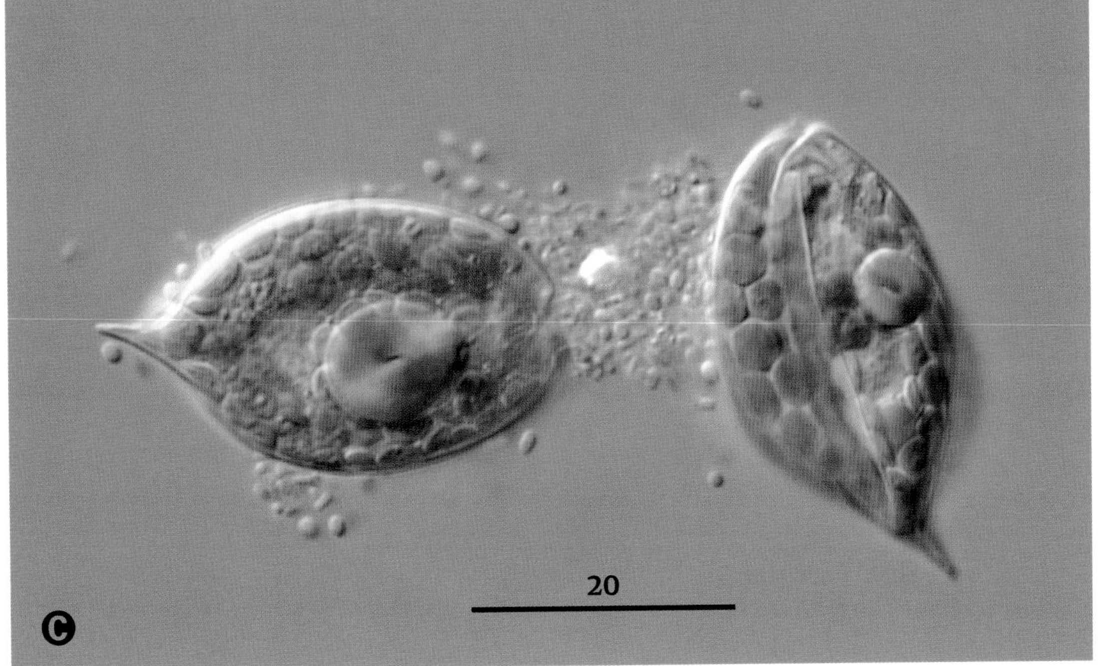

Phacus caudatus

Hübner 1886

SIZE: 31–50 µm long × 15–27 µm wide.

Ovoid cell with eyespot, paramylon disc, and numerous disc-shaped chloroplasts (A). Pellicle striations run longitudinally (B). Two cells, one exhibiting a ventral groove (C, *right*). Flagellum about body length.

Phacus curvicauda

Swirenko 1915

SIZE: 20–35 μm long × 18–25 μm wide.

Broadly ovoid cells with longitudinal groove visible at the apical end, red eyespot, disc-shaped chloroplasts, two large paramylon discs, smaller paramylon bodies, and short bent caudus (Ⓐ, Ⓒ). Detail of pellicle with longitudinal strips (Ⓑ). Cell showing longitudinal groove, large paramylon discs, and small chloroplasts (Ⓓ).

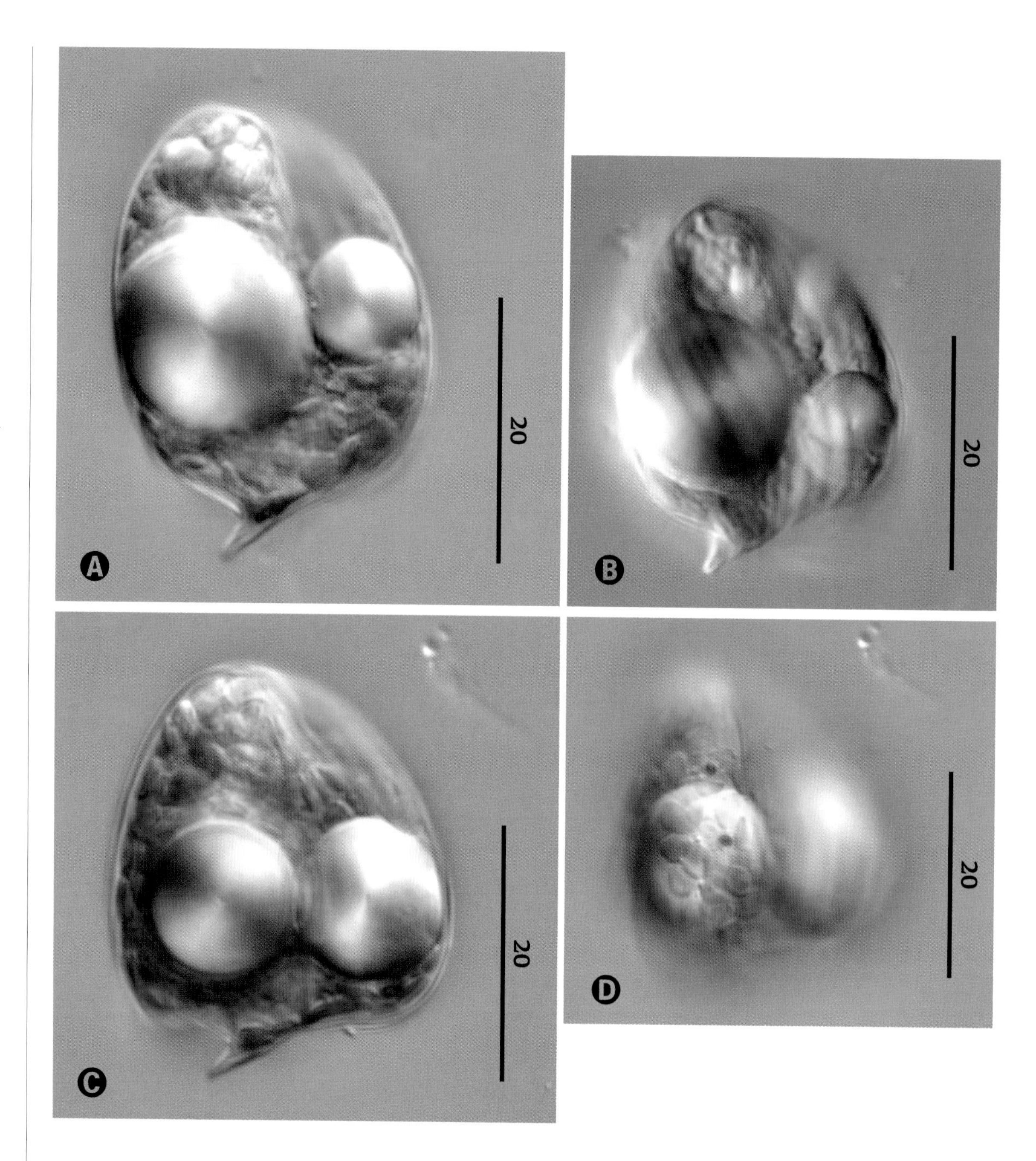

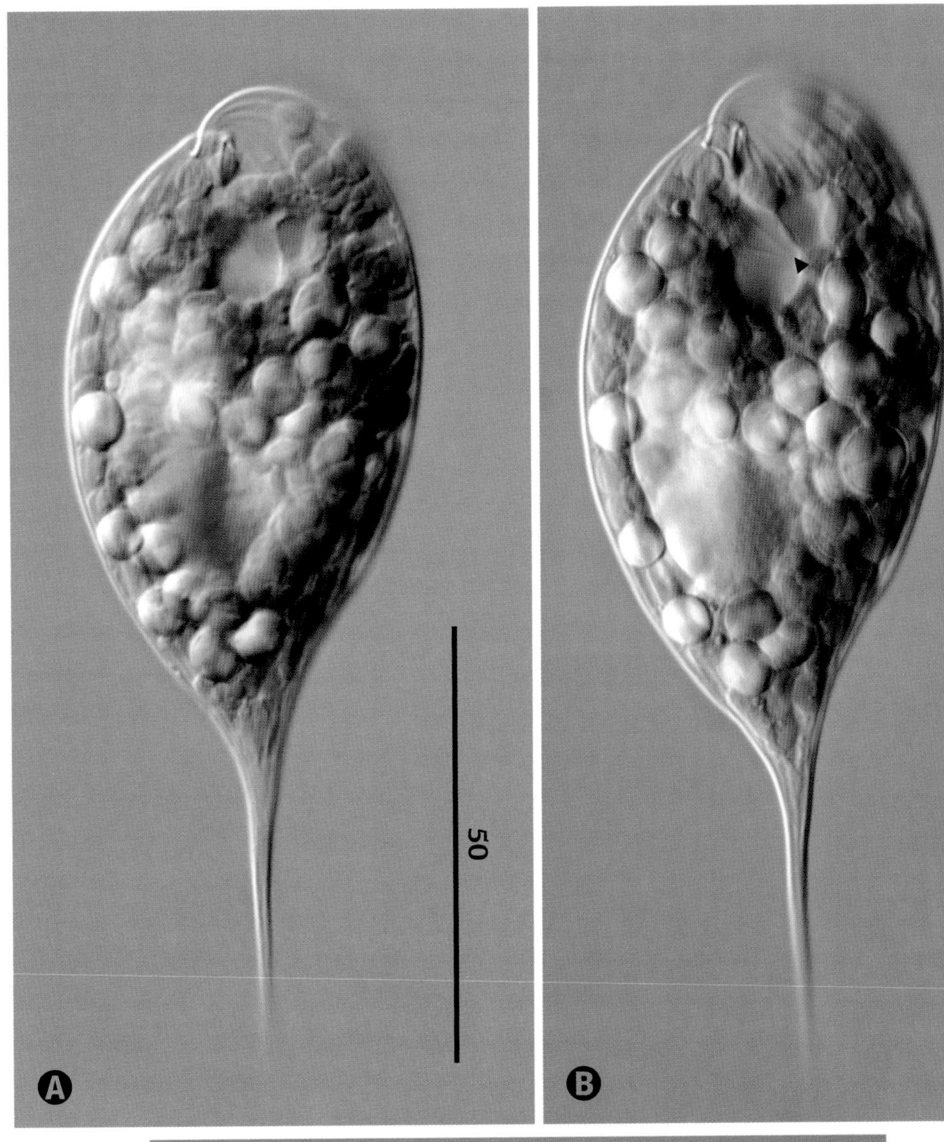

Phacus elegans

Pochmann 1942

SIZE: 106–147 µm long × 33–40 µm wide.

Elongated ovoid cell showing eyespot, numerous disc-shaped choloroplasts, several ovoid paramylon bodies (Ⓐ, Ⓑ), and the two flagella present in the reservoir adjacent to the eyespot (arrow) (Ⓑ). Pellicular striations run longitudinally (Ⓒ). Flagellum is one-third to one-half body length.

Phacus gigas

Da Cunha 1913

SIZE: 100–115 µm long × 70–80 µm wide.

Broadly ovoid cell with bent caudus showing eyespot, numerous small disc-shaped chloroplasts, and one large and numerous small paramylon discs (**A**). Longitudinally disposed pellicular striations (**B**). Detail showing red eyespot, reservoir and basal part of one of the two flagella (arrow), chloroplasts, central nucleus, and one large paramylon body (**C**). Flagellum is body length.

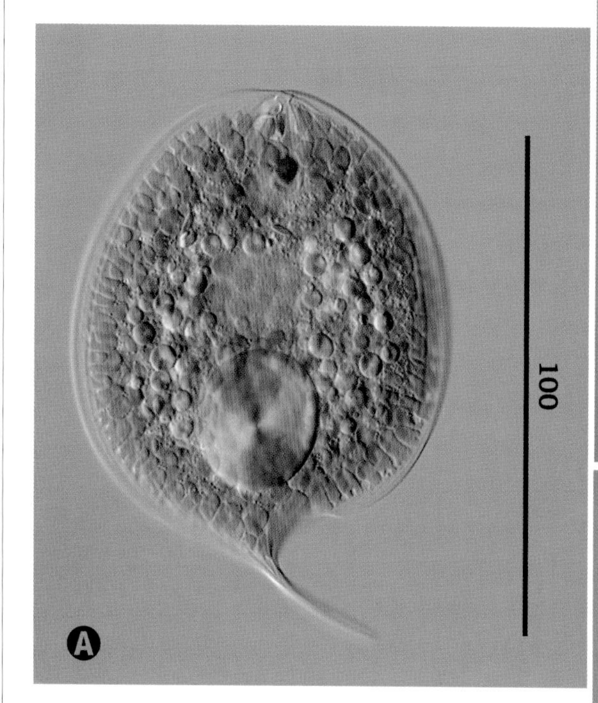

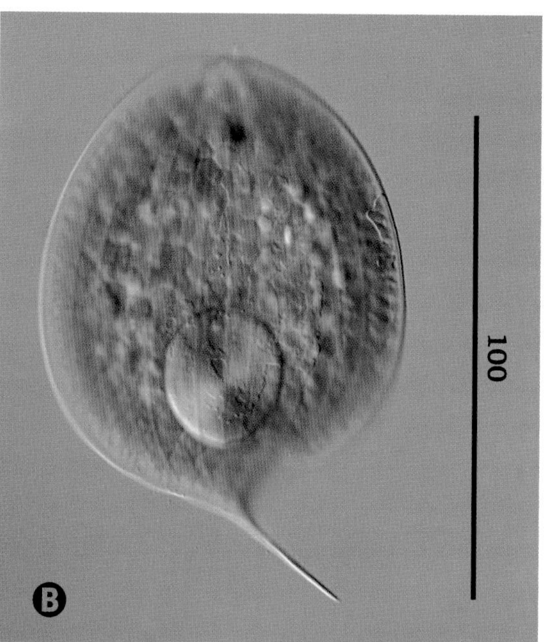

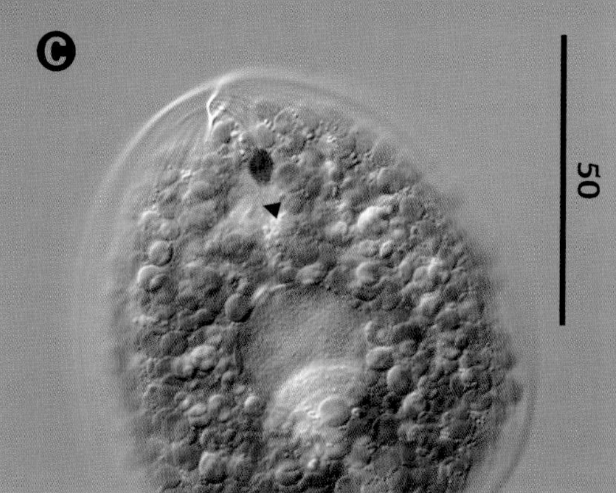

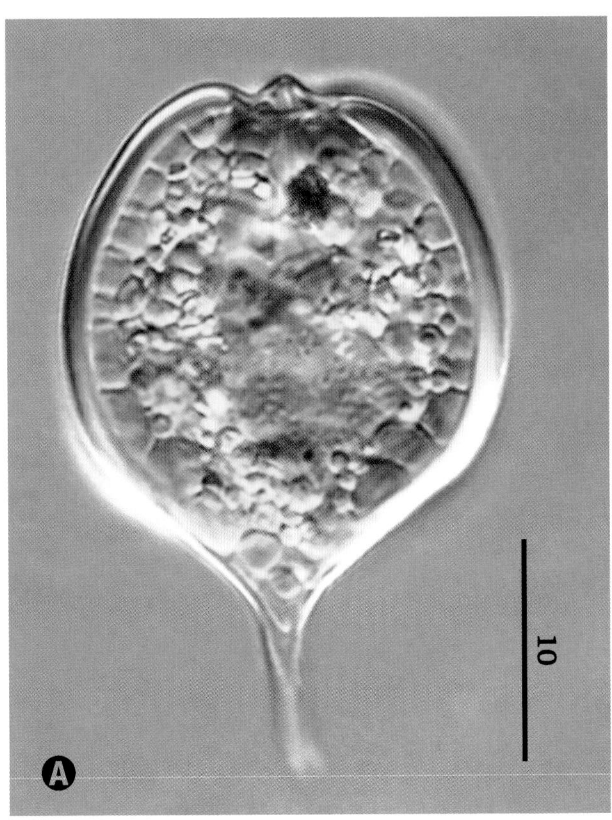

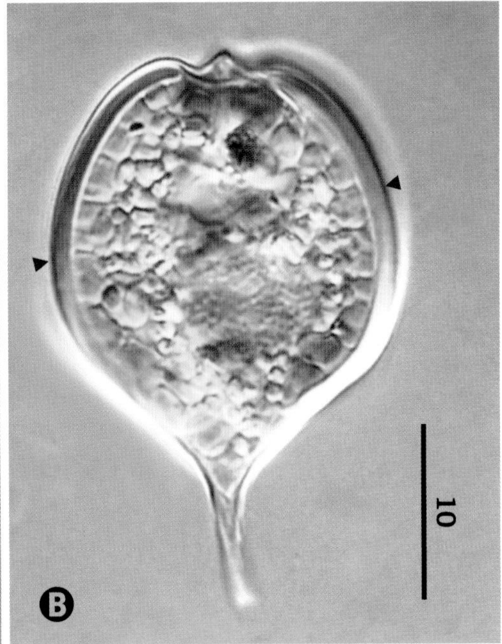

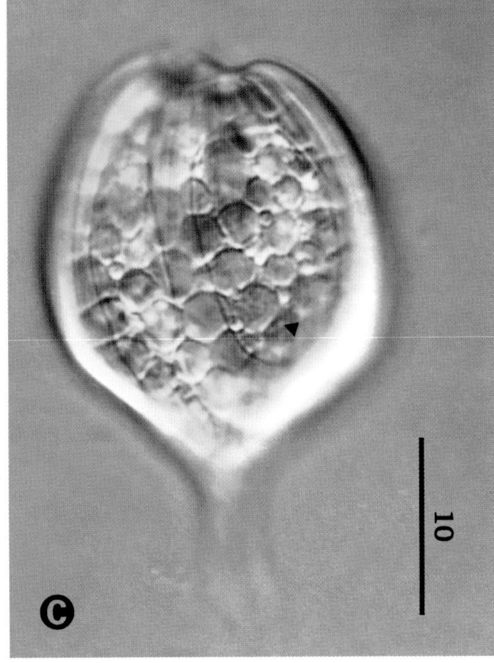

Phacus glaber

(Deflandre) Pochmann 1942

SIZE: 25–34 µm long × 15–22 µm wide.

Two broadly ovoid cells showing median papilla at the anterior end, eyespot, and cross sectional view of the large paramylon links (Ⓐ, Ⓑ—arrows). Surface view of paramylon link (Ⓒ). Numerous small disc-shaped chloroplasts are present (Ⓐ–Ⓒ), and pellicular striations are fine and longitudinally disposed (arrow) (Ⓒ).

Phacus hamatus

Pochmann 1942

SIZE: 38–55 μm long × 25–35 μm wide.

Broadly ovoid cell with bent caudus showing eyespot, numerous chloroplasts, and one large paramylon disc (A). Cell with flagellum and small disc-shaped chloroplasts (B). Paramylon body, nucleus, and nucleolus (C). Detail of longitudinal pellicle strips, chloroplasts, and paramylon (D).

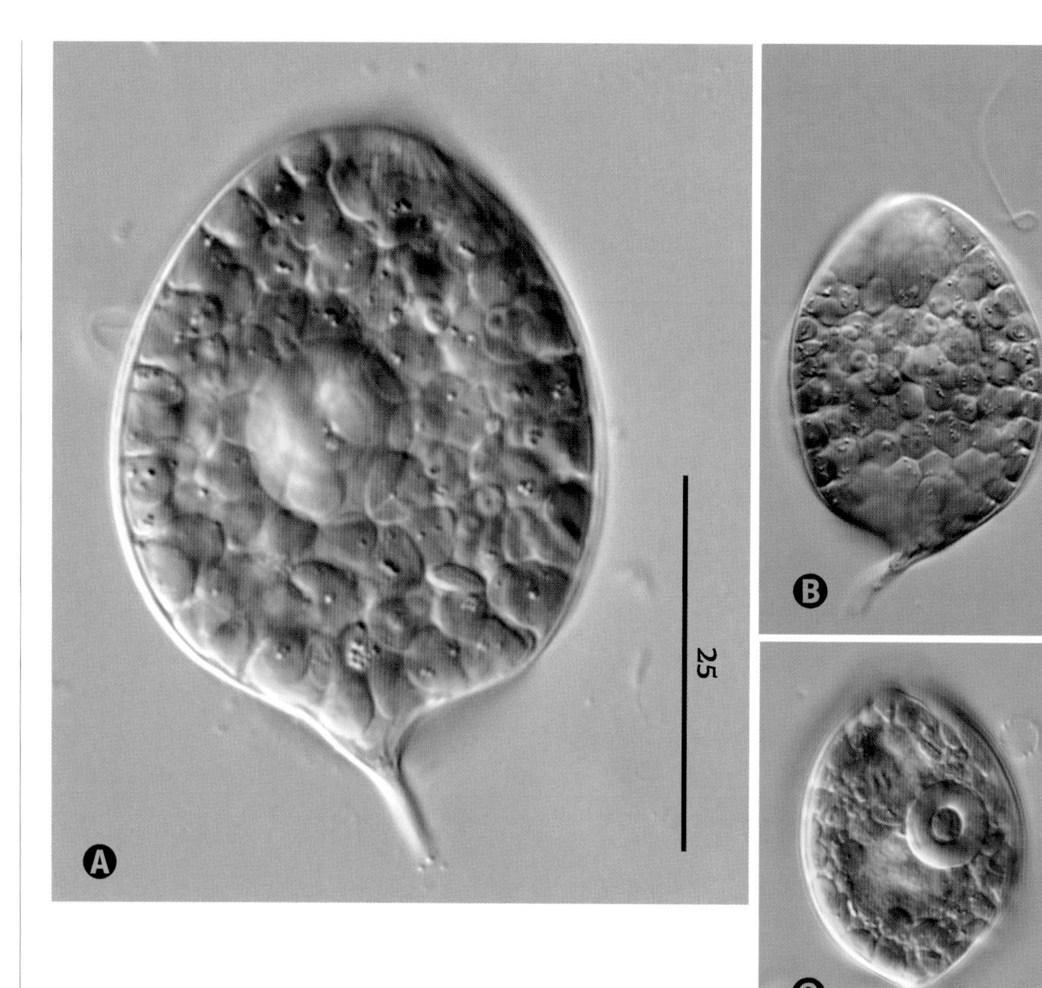

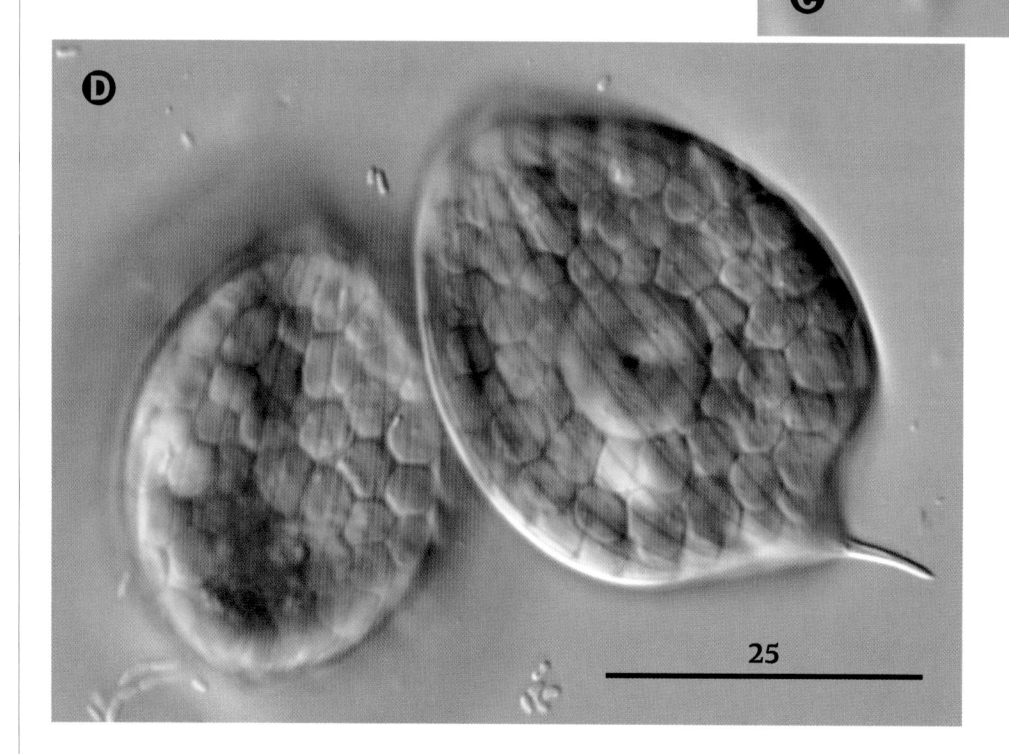

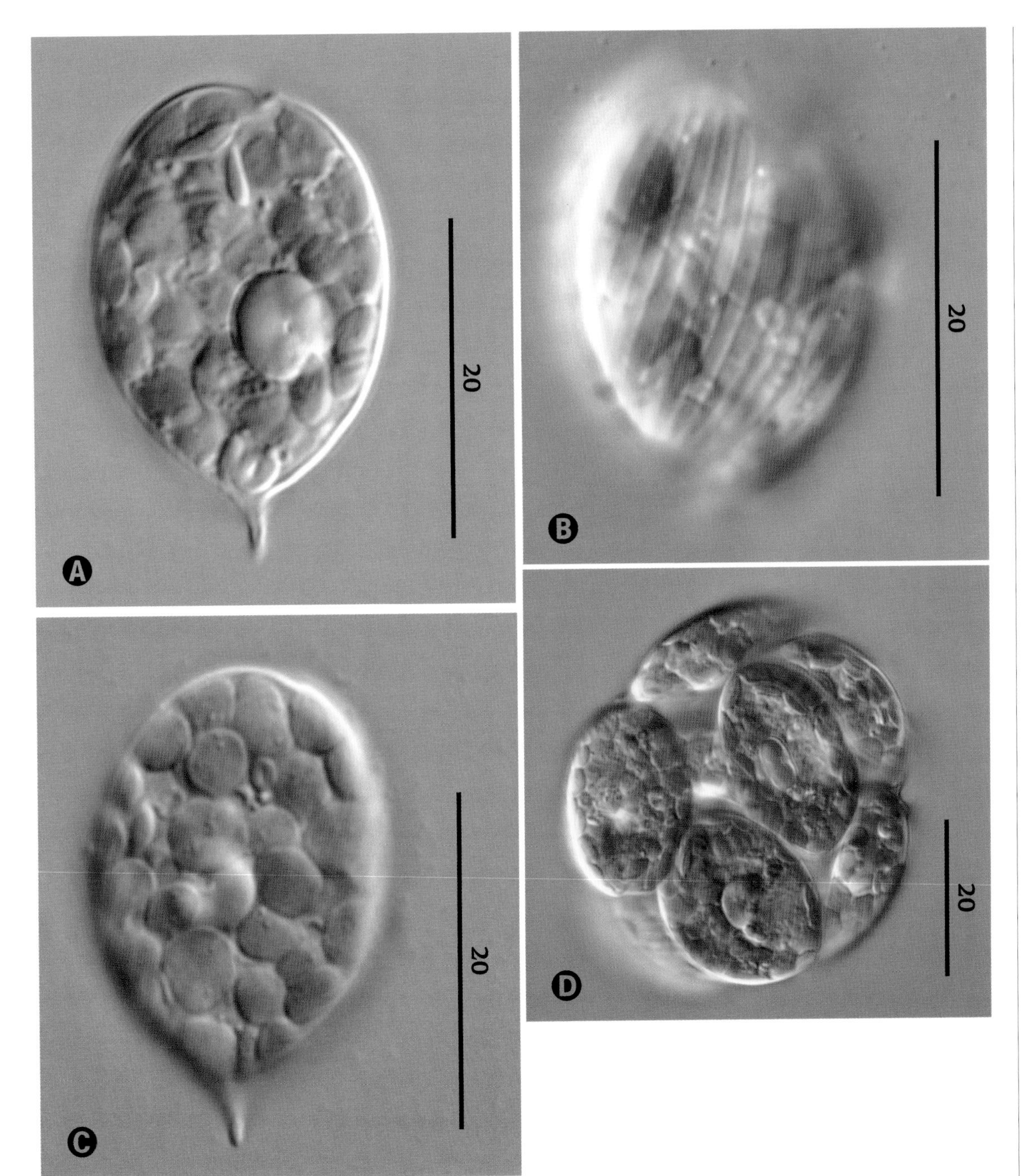

Phacus hamelii

Allorge et Lefèvre emend. Zakryś et Kosmala 2007

SIZE: 24–37 µm long × 12–21 µm wide.

Ovoid cells with short straight caudus showing numerous disc-shaped chloroplasts and one large and several smaller paramylon bodies (A, C). Pellicular striations form a right-handed spiral (B). Group of cells showing eyespot and chloroplasts (D).

Phacus helikoides

Pochmann 1942

SIZE: 70–120 µm long × 30–54 µm wide.

Triple-twisted cell with long straight caudus showing red eyespot and large paramylon disc (A). Longitudinally disposed pellicle strips with cross-striations (arrow) and large paramylon disc (B). Triple-twisted cells (C, D) with numerous small disc-shaped chloroplasts, large paramylon disc (C), and smaller paramylon grains (D).

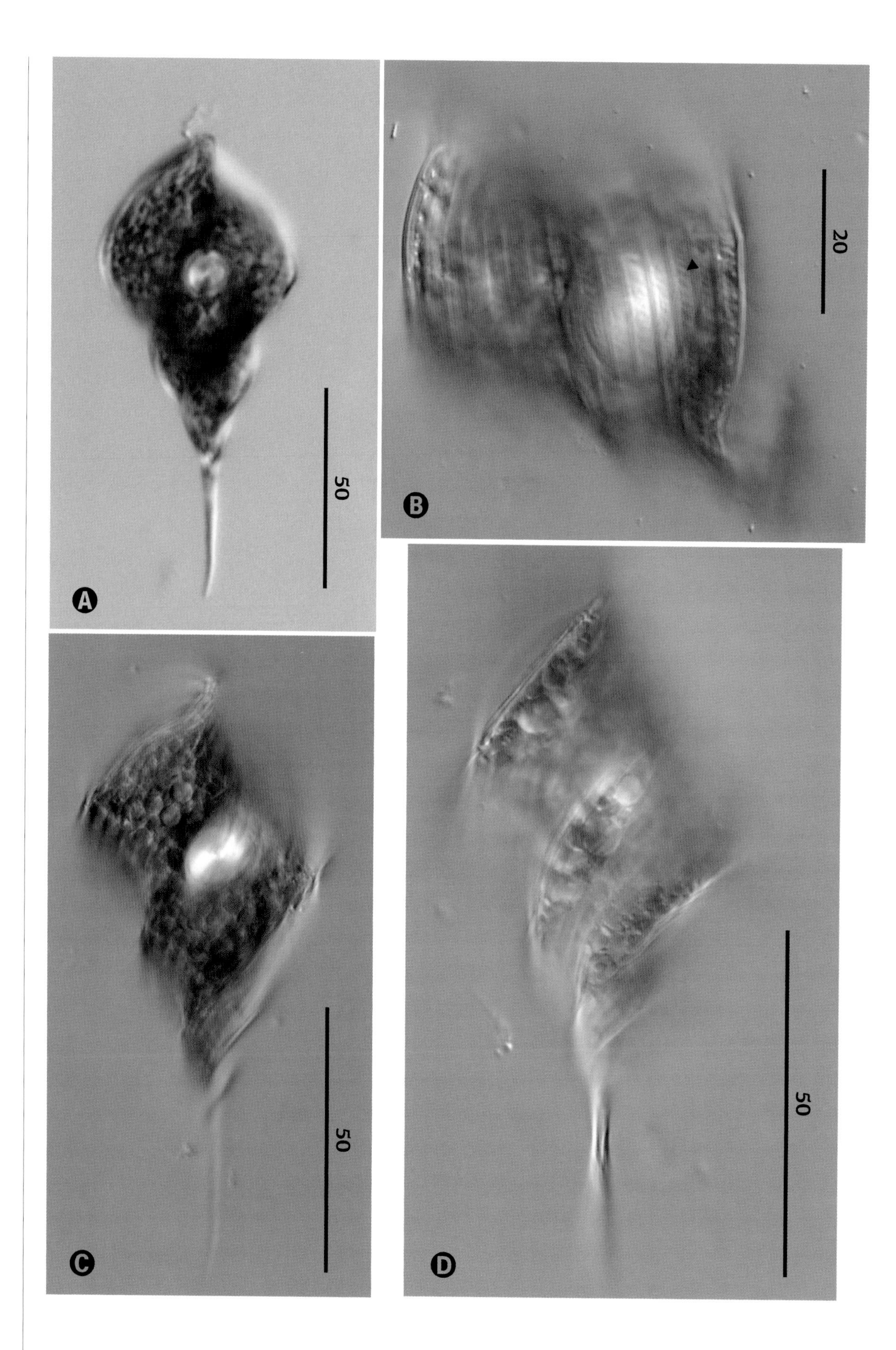

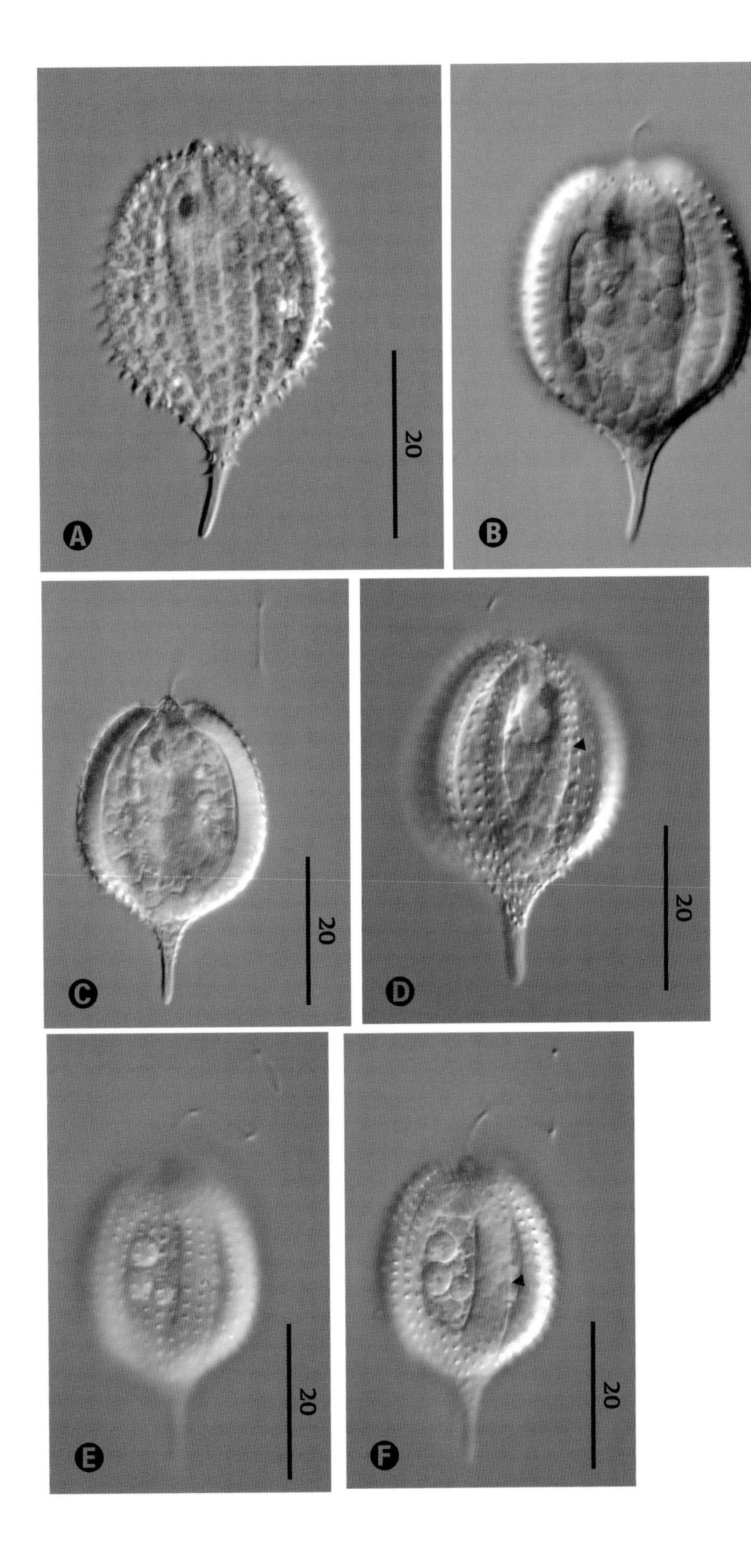

Phacus horridus

Pochmann 1942

SIZE: 30–57 µm long × 22–36 µm wide.

Ovoid cell with a median protuberance at the anterior end showing eyespot, small disc-shaped chloroplasts, and pellicle with rows of small spines oriented posteriorly (A). Cell with flagellum (one to one-and-a-half times body length), eyespot, and chloroplasts (B). Two large lateral paramylon bodies (C). Cells showing flagellum, eyespot, large link-like paramylon bodies (arrows) (D, F), and small disc-like paramylon bodies (E, F).

Phacus inflexus

(Kisselew) Pochmann emend. Zakryś et Karnkowska 2010

SIZE: 18–31 μm long × 7–11 μm wide.

Flattened cell with eyespot, small disc-shaped chloroplasts, and nucleus (A). Pellicular striations run longitudinally (B). Twisted cell with flagellum, eyespot, and chloroplasts (C). Cell showing typical body shape and possessing two dissimilarly sized paramylon bodies (D).

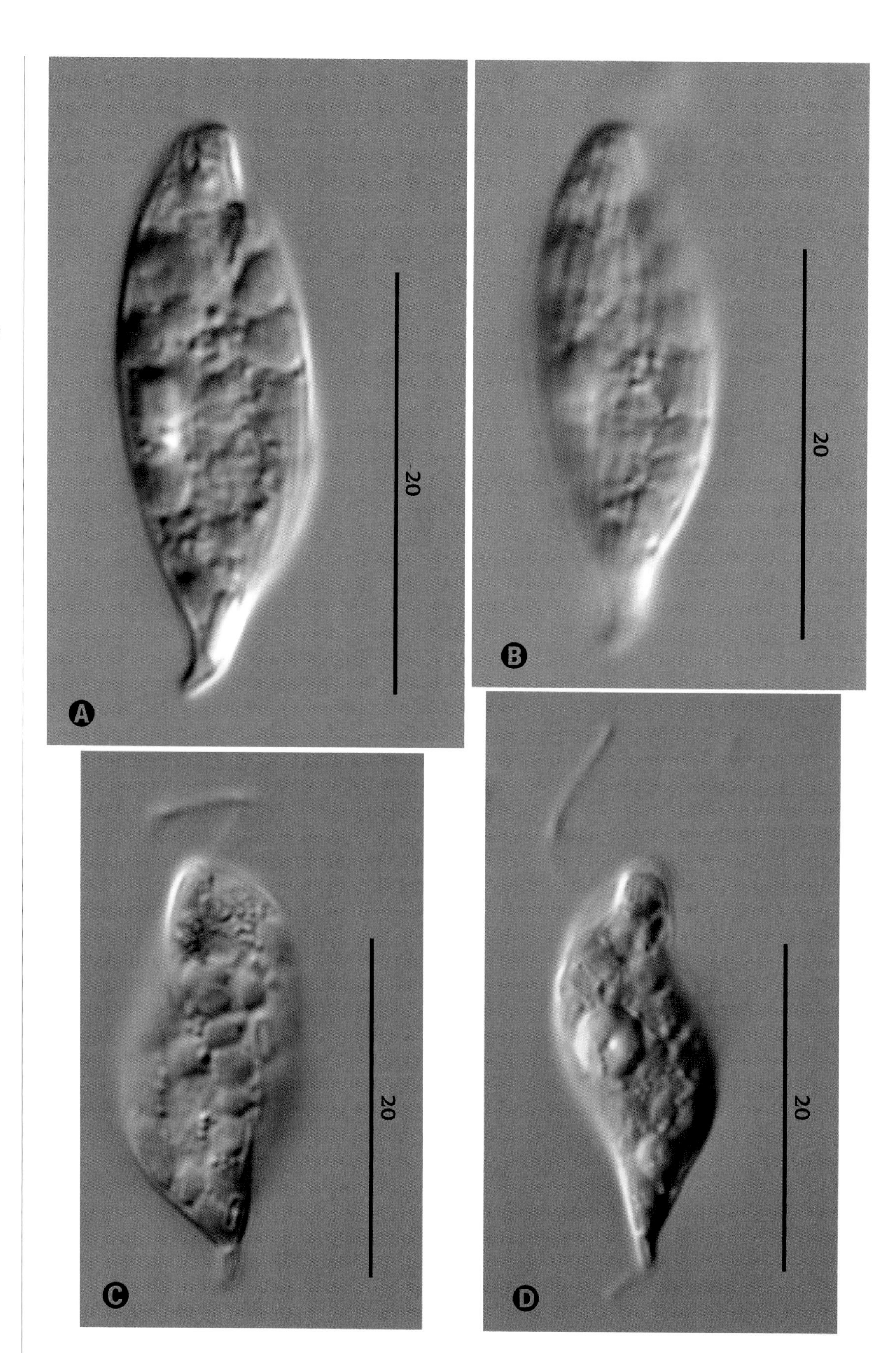

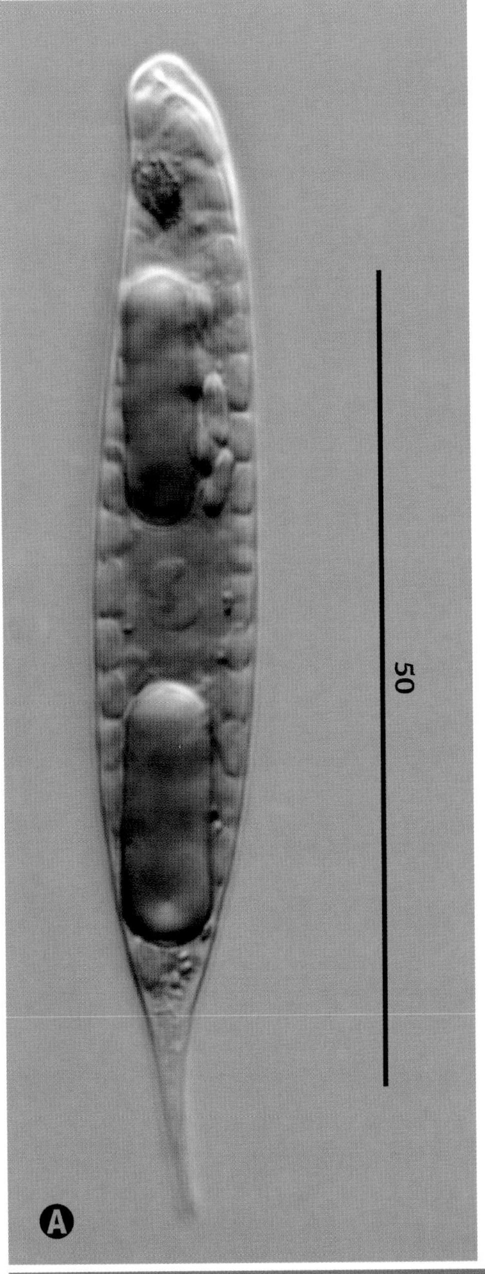

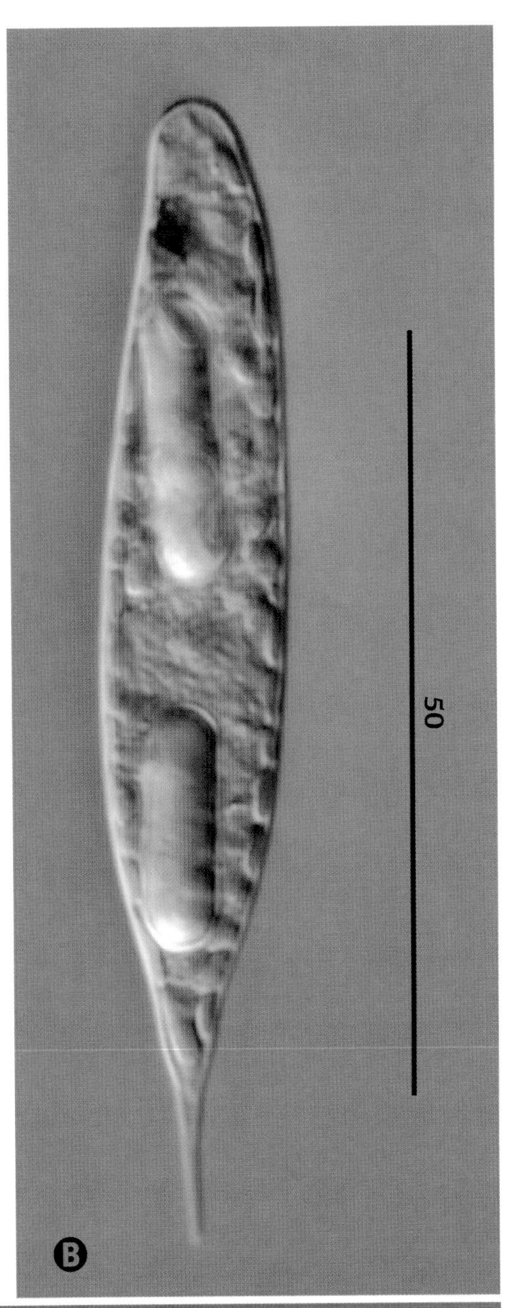

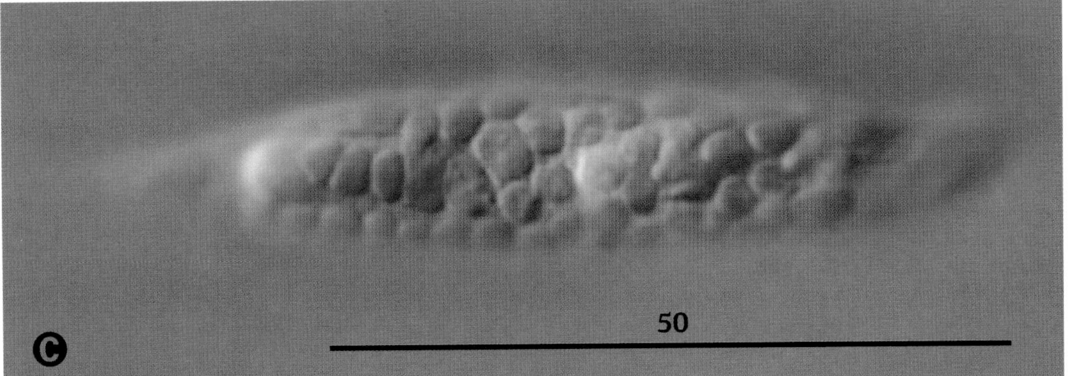

Phacus limnophila

(Lemmermann) Linton et Karnkowska 2010

SIZE: 37–82 µm long × 4–13 µm wide.

Spindle-shaped cells with eyespot, numerous disc-shaped chloroplasts, and two large paramylon bodies on either side of the nucleus (A, B) next to much smaller ones (A). Very fine pellicular striations and choloroplasts lacking pyrenoids (C). Flagellum is one-half to two-thirds body length.

Phacus lismorensis

Playfair 1921

SIZE: 85–100 µm long × 30 µm wide.

Cell showing spatulate body and elongate bent caudus (A). Oblique view showing eyespot and large paramylon grain (B). Pellicular striations curve at the cell anterior and run longitudinally down the body (arrow) (C). Chloroplasts are numerous small discs (C).

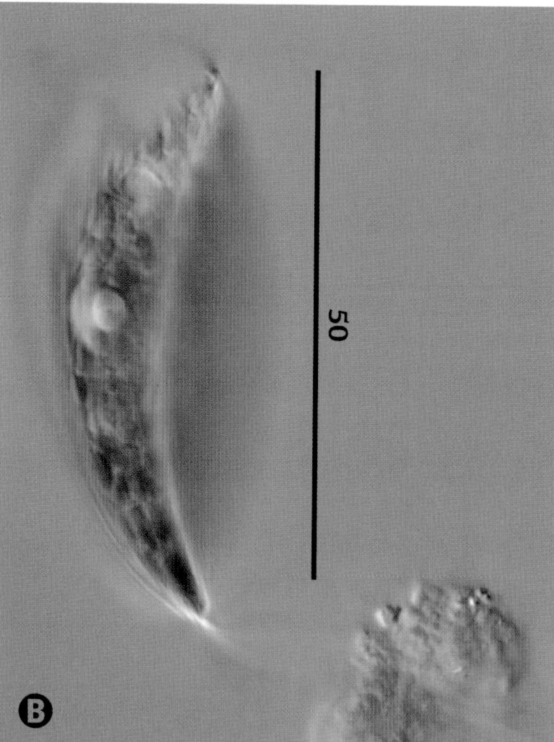

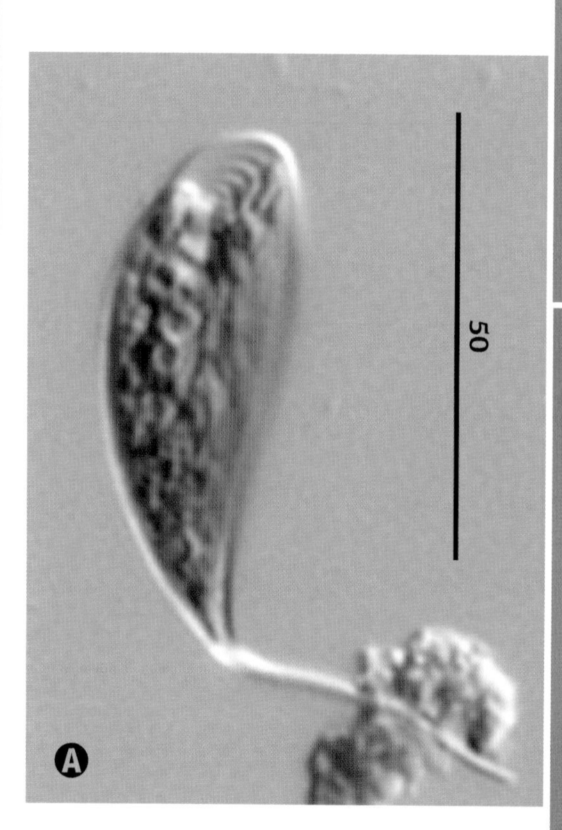

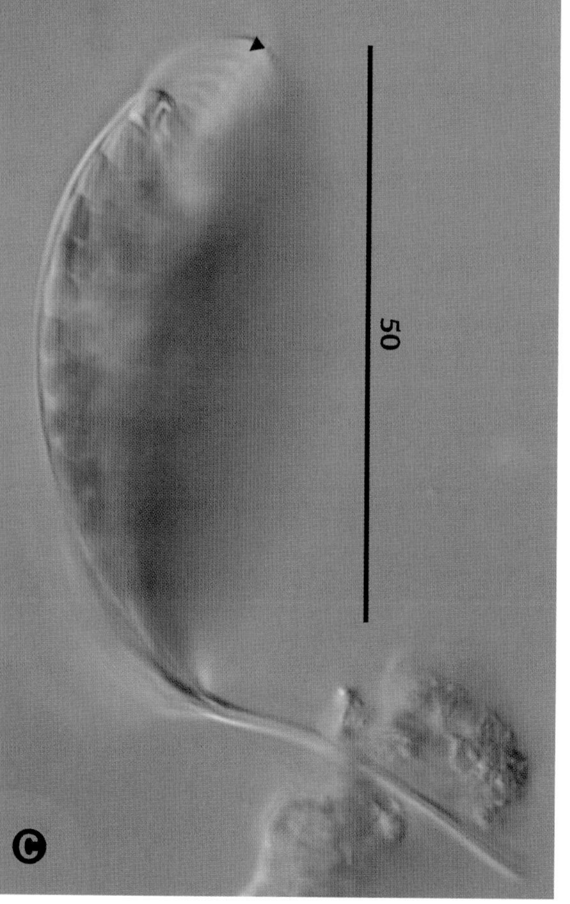

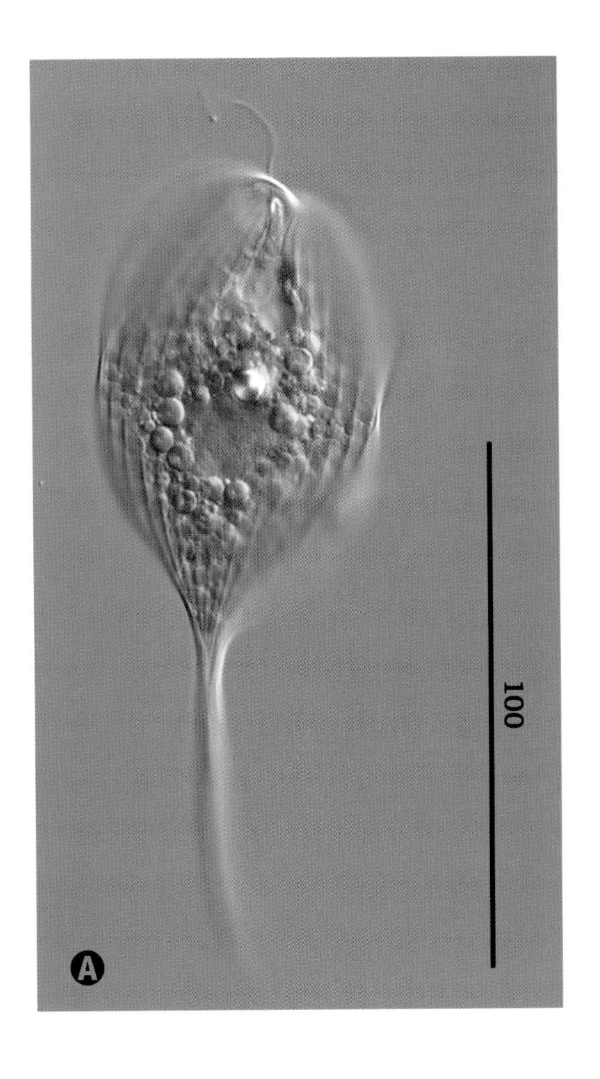

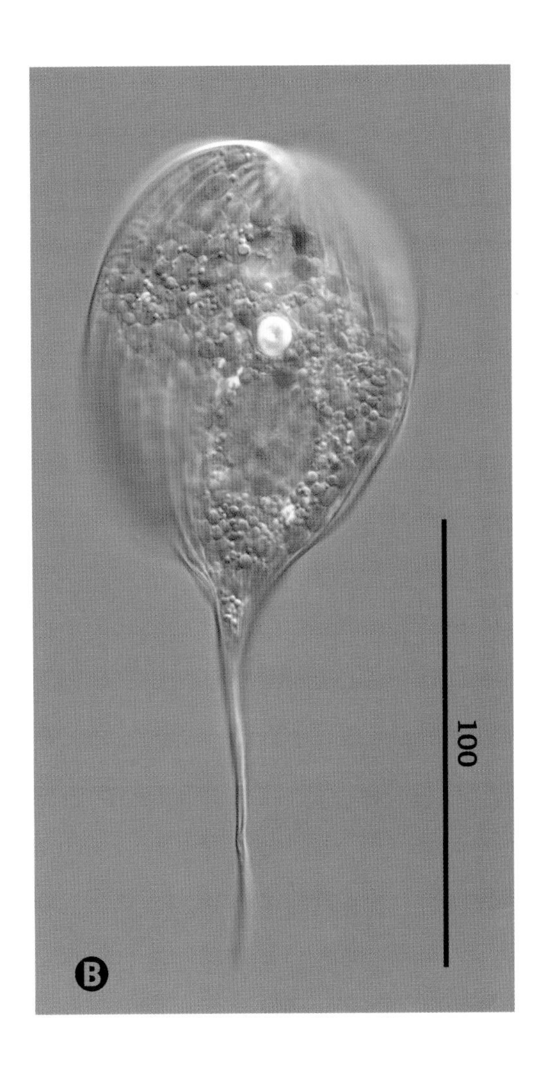

Phacus longicauda

(Ehrenberg) Dujardin 1841

SIZE: 75–181 µm long × 33–76 µm wide.

Ellipsoid to ovoid cells showing flagellum (one-sixth to one-fourth body length) (A), eyespot, disc-shaped chloroplasts, one large and numerous smaller paramylon bodies, longitudinally disposed pellicular striations, and elongate caudus (A, B).

Phacus mangini

Lefèvre 1933

SIZE: 38–51 µm long × 22–29 µm wide.

Ellipsoid cell with hyaline caudus showing long flagellum (body length), red eyespot, disc-shaped chloroplasts, and longitudinal groove (A, B). Detail of cell with two adjacent paramylon discs (C). Pellicle striations are longitudinally disposed (D).

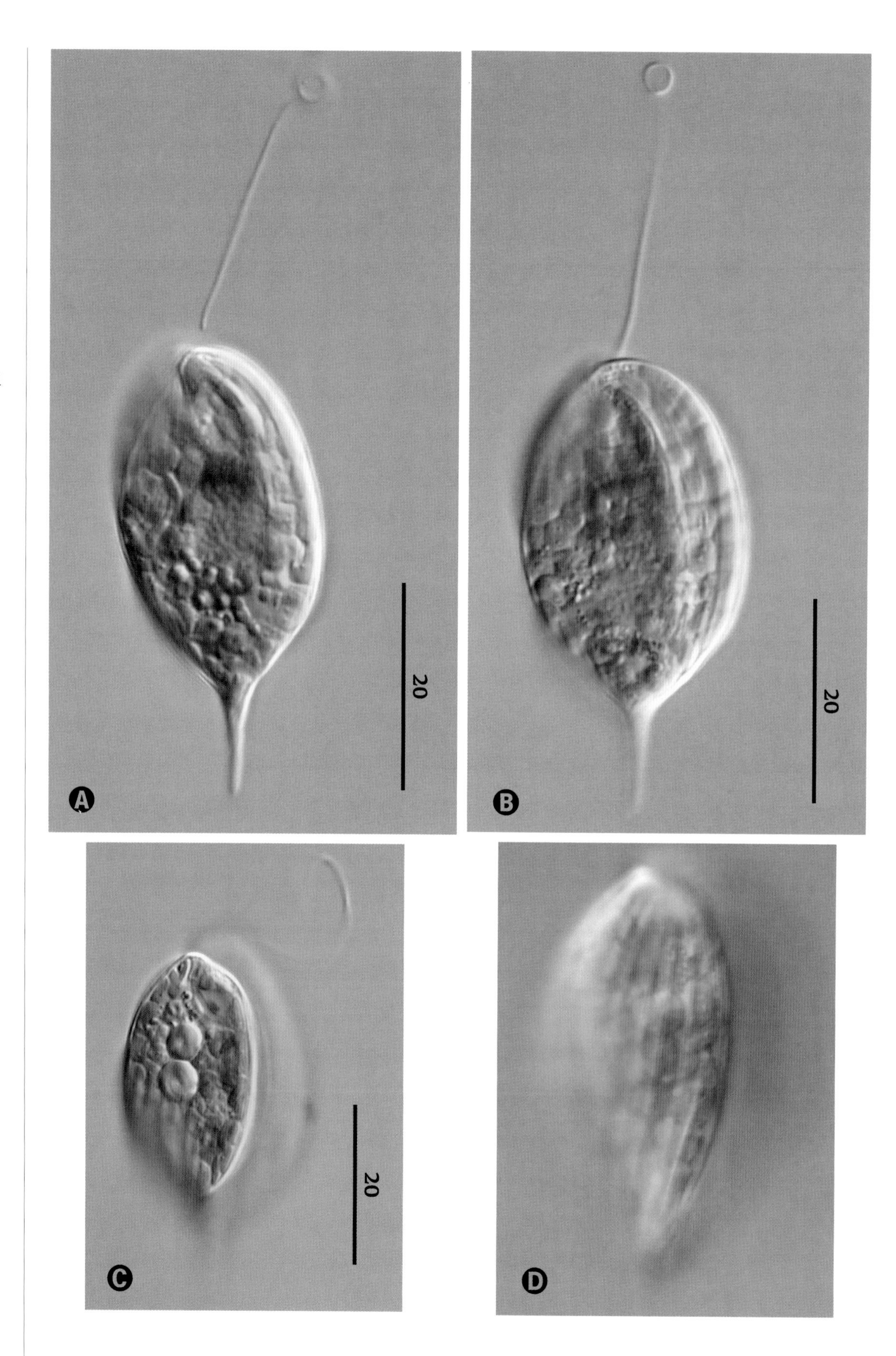

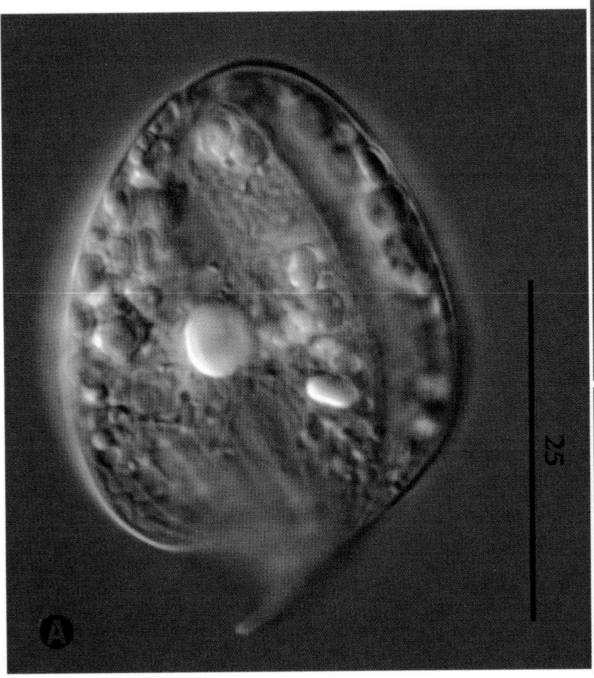

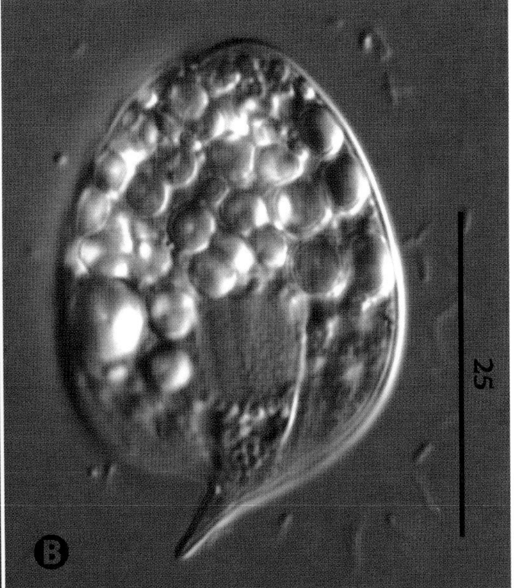

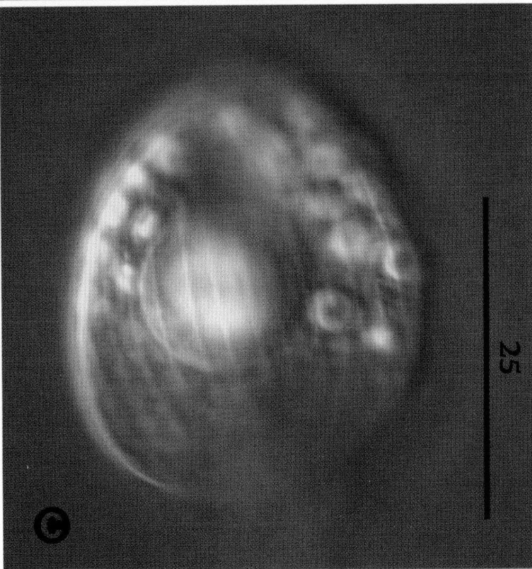

Phacus ocellatus

(Pringsheim) Marin et Melkonian 2003

SIZE: 21–35 μm long × 15–25 μm wide.

Colorless, obovoid cell with bent caudus, red eyespot, longitudinal groove, and paramylon grains (A). Cell with numerous paramylon bodies (B). Detail of longitudinal pellicular strips and large paramylon disc (C). Flagellum is about body length.

Phacus orbicularis

Hübner emend. Zakryś et Kosmala 2007

SIZE: 29–75 µm long × 22–49 µm wide.

Broadly ovoid cell with bent caudus showing eyespot and numerous disc-shaped chloroplasts (Ⓐ). Longitudinal pellicle strips with cross-striations and large paramylum disc (Ⓑ). Two cells showing nucleus, ventral groove, and flagellum (approximately body length) (Ⓒ).

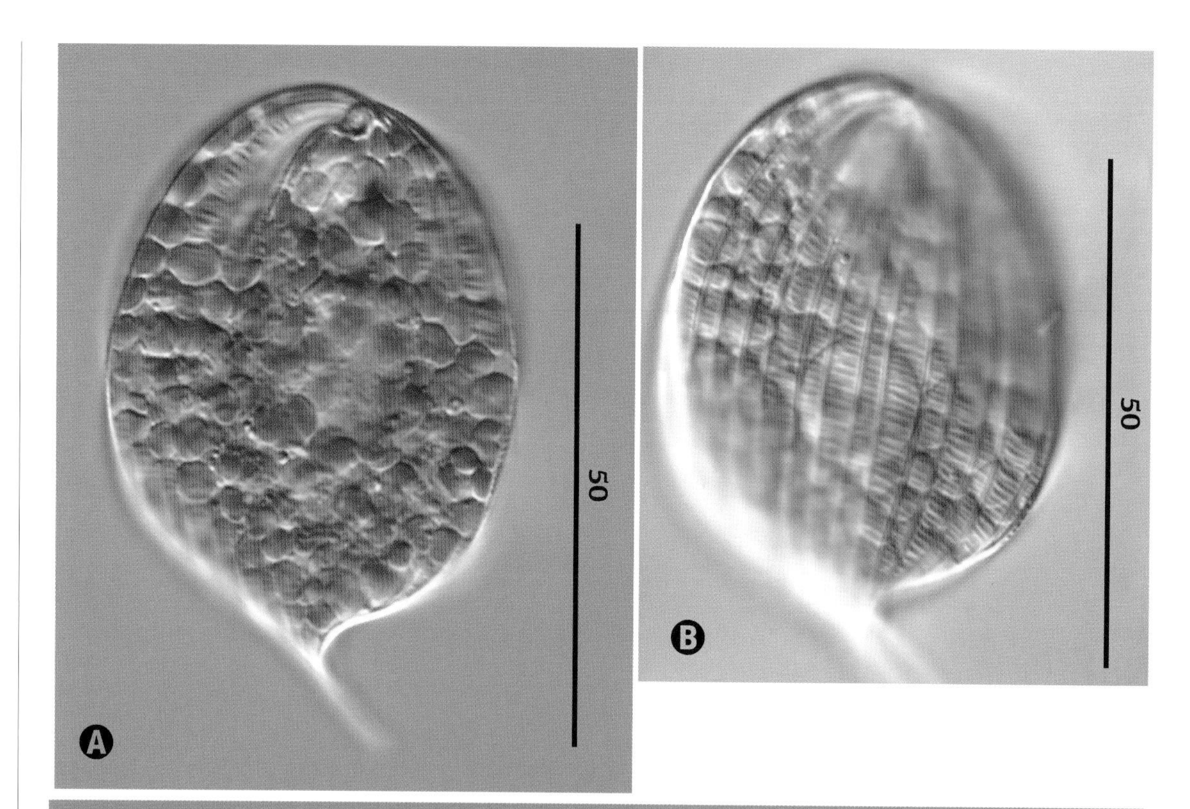

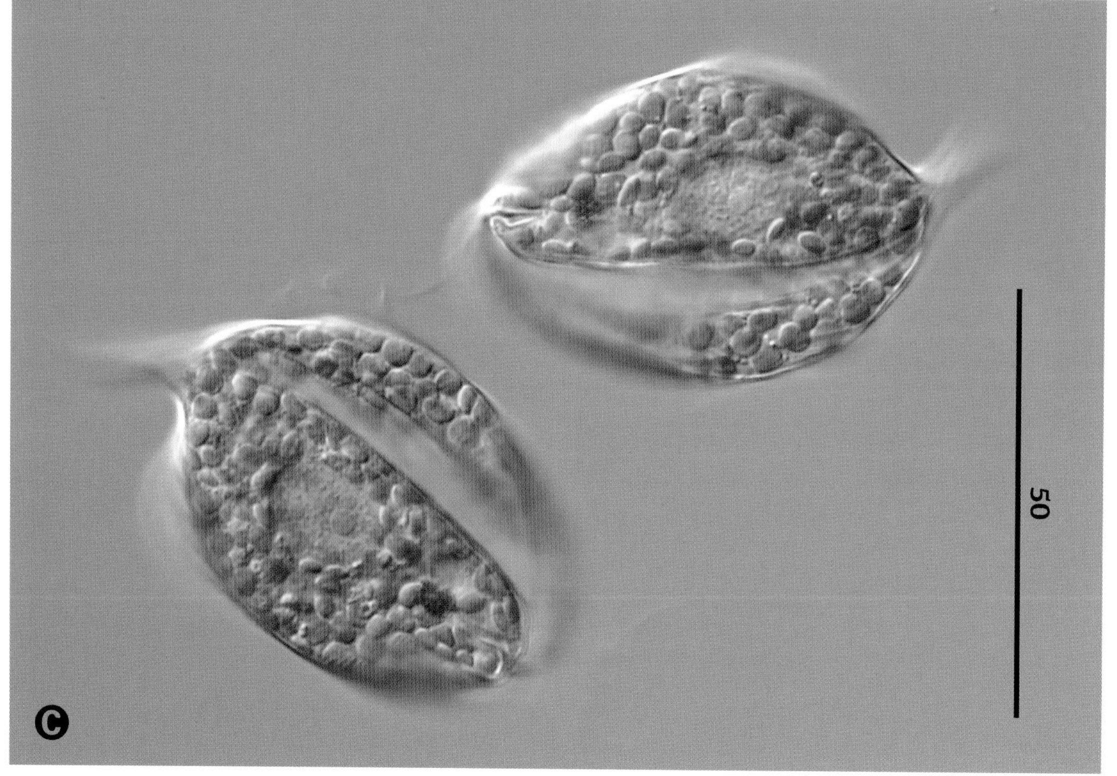

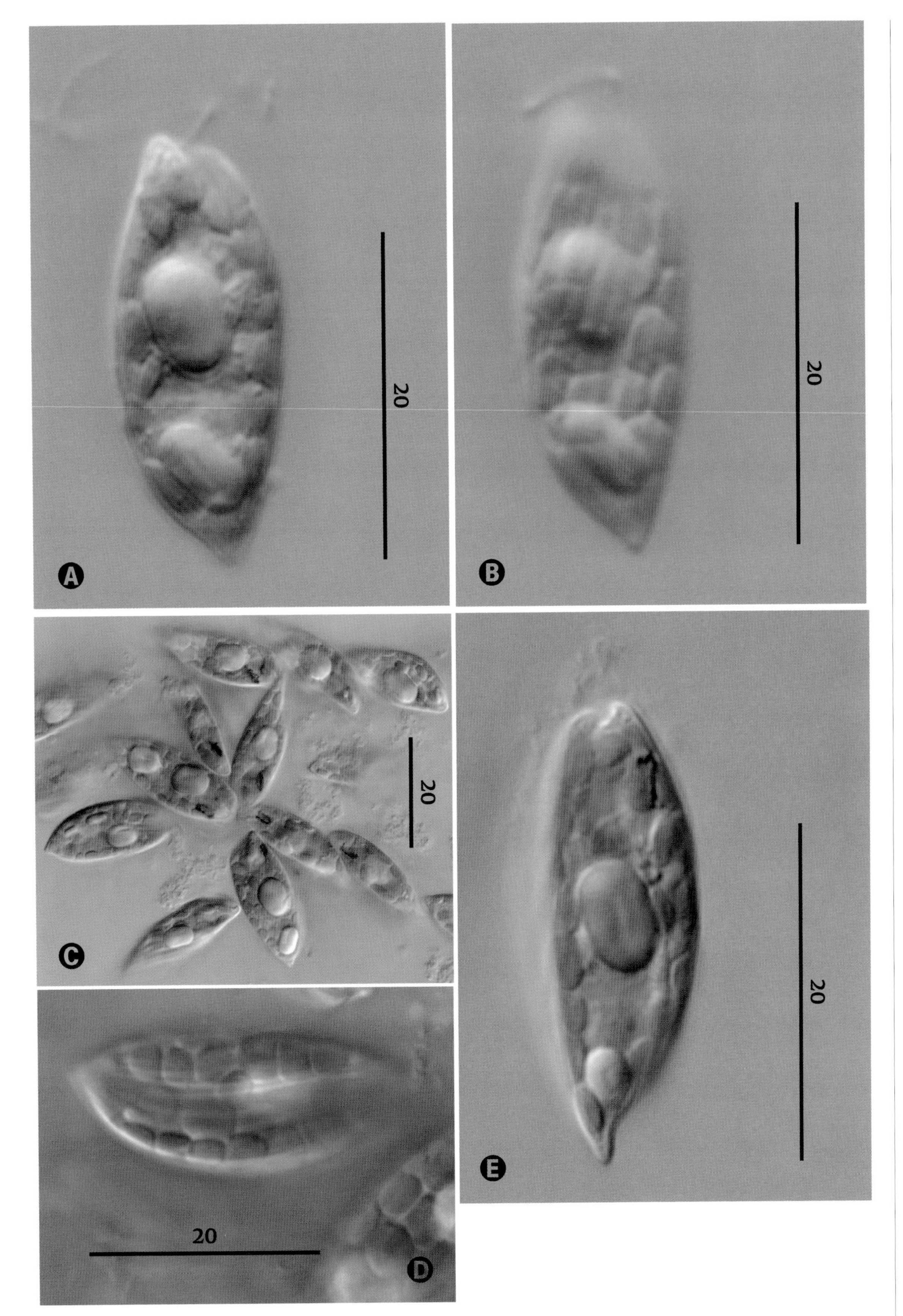

Phacus oscillans

Klebs emend. Zakryś et Karnkowska 2010

SIZE: 15–35 μm long × 5–13 μm wide.

Elongate cell with flagellum (body length), red eyespot, two large paramylon bodies, and disc-shaped chloroplasts (A). Pellicle striations longitudinally arranged (B). Group of cells with prominent eyespots (C). Longitudinal groove and chloroplasts (D). Cell showing eyespot, paramylon grains, and short caudus (E).

Phacus parvulus

Klebs emend. Zakryś et Karnkowska 2010

SIZE: 17–24 µm long × 7–16 µm wide.

Ellipsoid cell lacking furrow; flagellum (about one-half body length), canal, eyespot, numerous small disc-shaped chloroplasts, large paramylon disc (Ⓐ, Ⓑ), and contractile vacuole are visible (arrow) (Ⓐ). Pellicle striations are arranged in a right-handed spiral (Ⓒ).

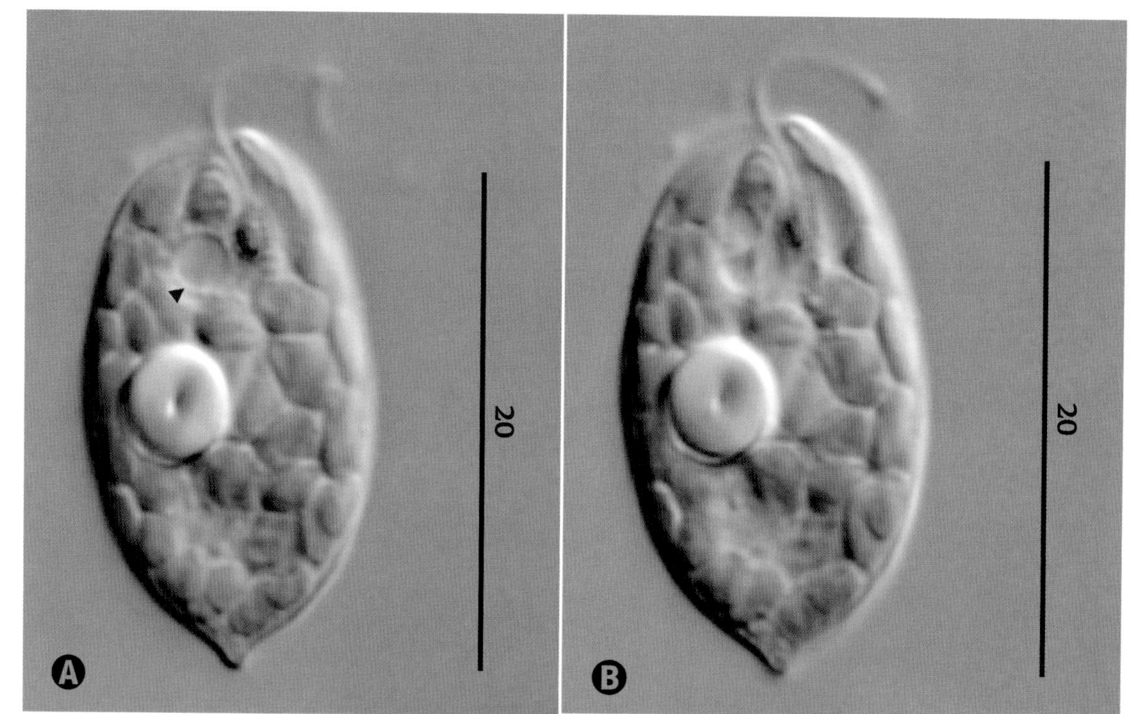

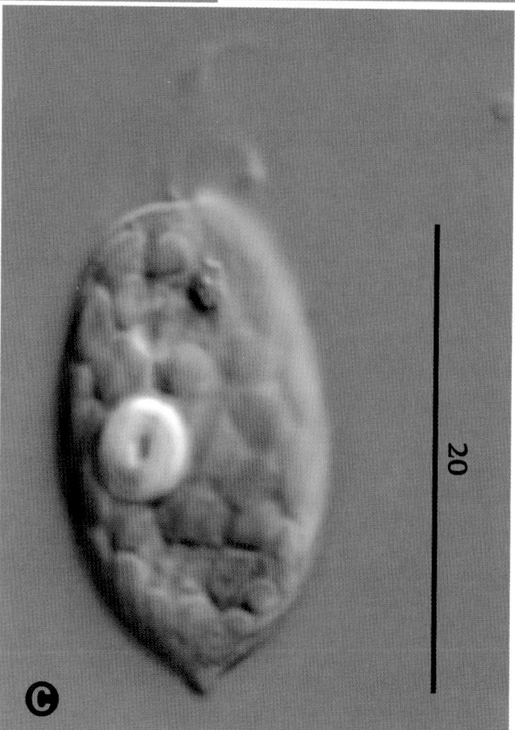

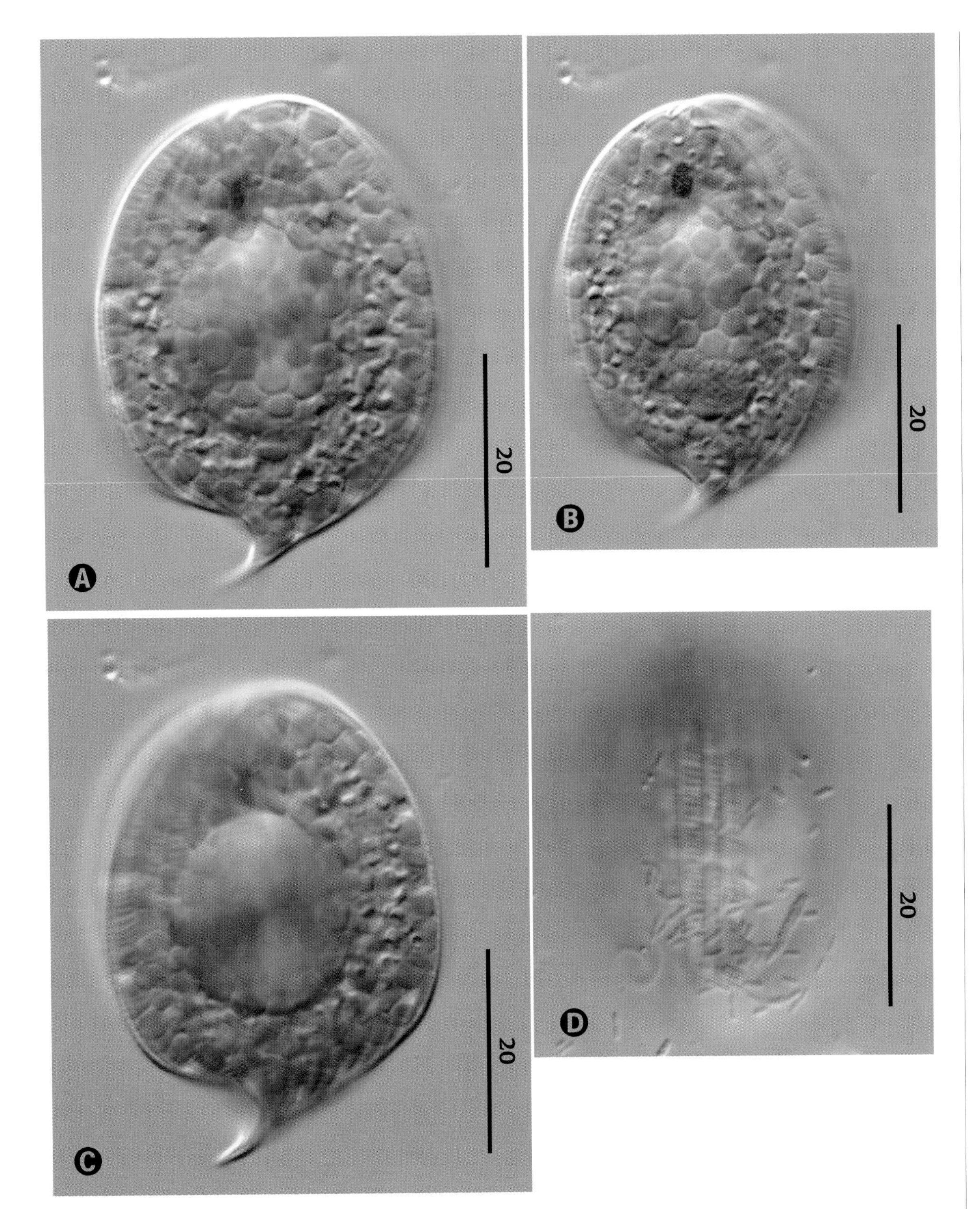

Phacus platalea

Dreżepolski 1925

SIZE: 37–56 µm long × 25–35 µm wide.

Broadly ovoid cell showing red eyespot, large paramylon disc, and numerous small disc-shaped chloroplasts (A, B). Cross-striations between pellicle strips are visible on the right side of the cell (B). Detail of large paramylon disc (C). Longitudinal pellicle strips with distinctive cross-striations (D). Flagellum is body length.

Phacus platyaulax

Pochmann 1942

SIZE: 30–38 µm long × 22–28 µm wide.

Broadly ovoid cell with short bent caudus showing red eyespot, disc-shaped chloroplasts, and two large lateral paramylon discs (A). Surface view of cell showing pellicle striations (B). Longitudinal median groove, chloroplasts, and small paramylon discs (C). Side view of cell with flagellum (body length) and groove (D). Cell with groove, chloroplasts, and large paramylon disc (E). Cell showing groove, chloroplasts, and paramylon grains (F).

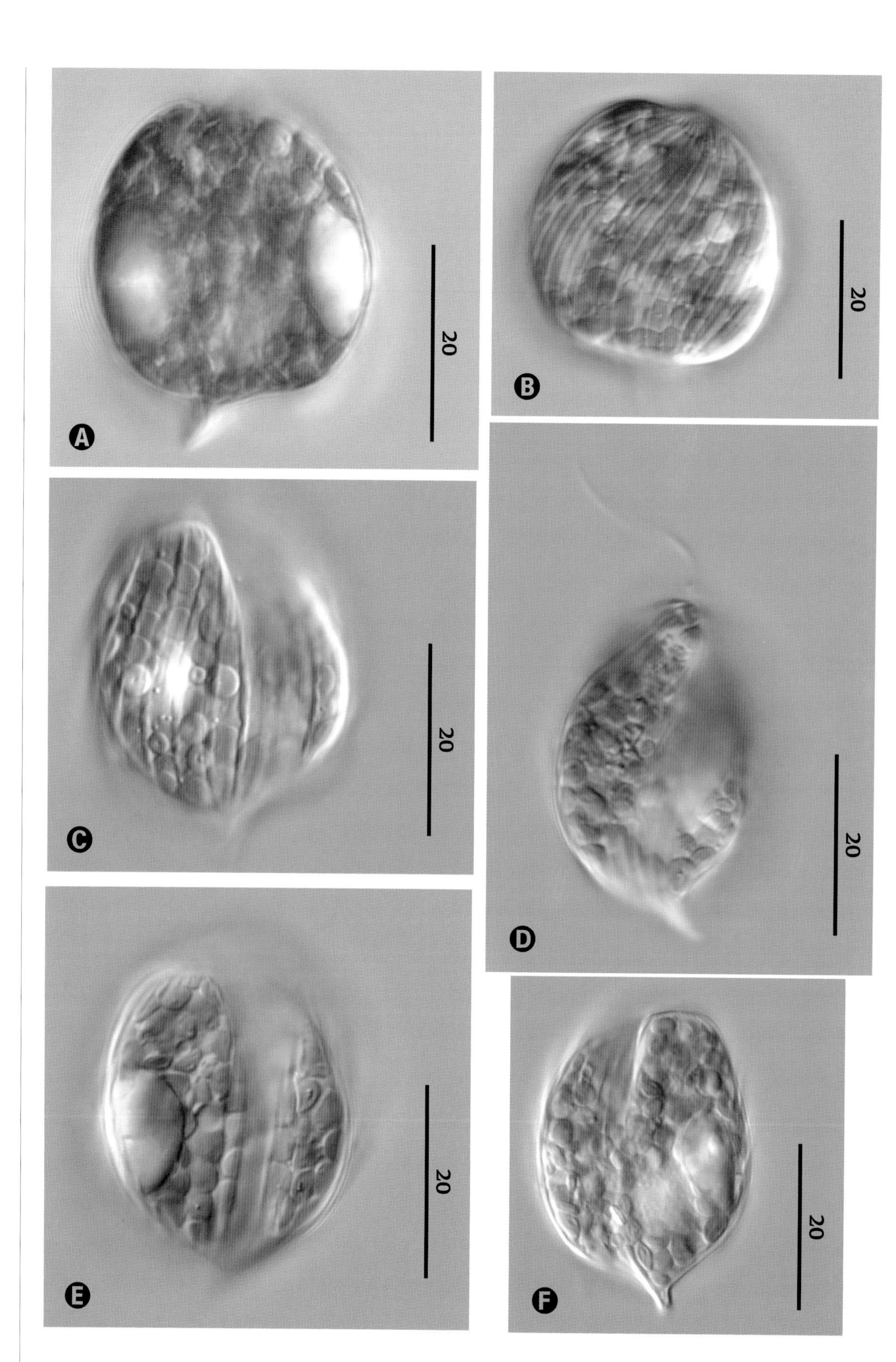

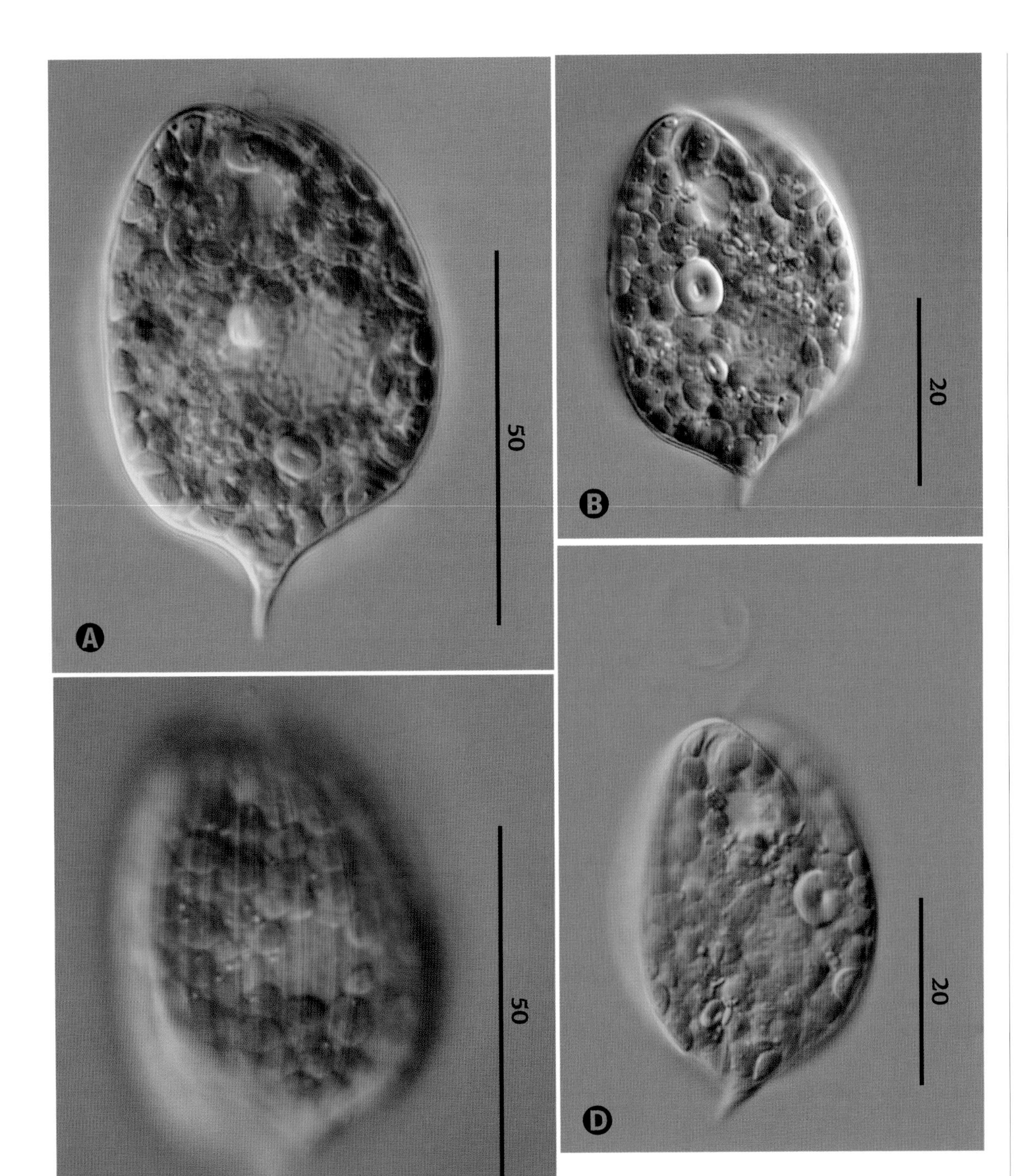

Phacus pleuronectes

(O. F. Müller) Dujardin emend. Zakryś et Kosmala 2007

SIZE: 28–80 µm long × 21–50 µm wide.

Broadly oval cell showing flagellum (body length or longer), numerous small disc-shaped chloroplasts, and paramylon bodies (A). Cells with flagellum (D), eyespot, large paramylon disc, and longitudinal groove (B, D). Detail showing longitudinal pellicular striations (C).

Phacus pusillus

Lemmermann emend. Zakryś et Karnkowska 2010

SIZE: 14–20 µm long × 6–13 µm wide.

Ellipsoid cell showing shallow furrow, disc-shaped chloroplasts, paramylon grain, and nucleus with nucleolus (**A**). Surface view showing shallow furrow (arrow) and pellicle striations arranged in a right-handed spiral (**B**).

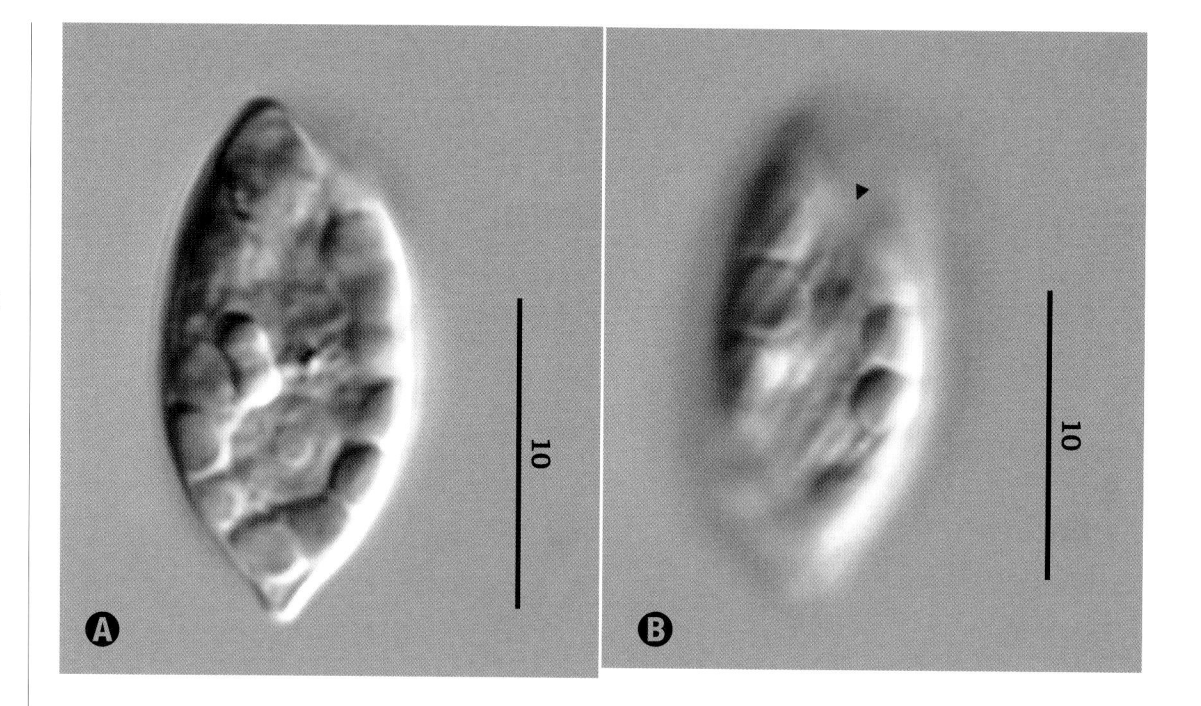

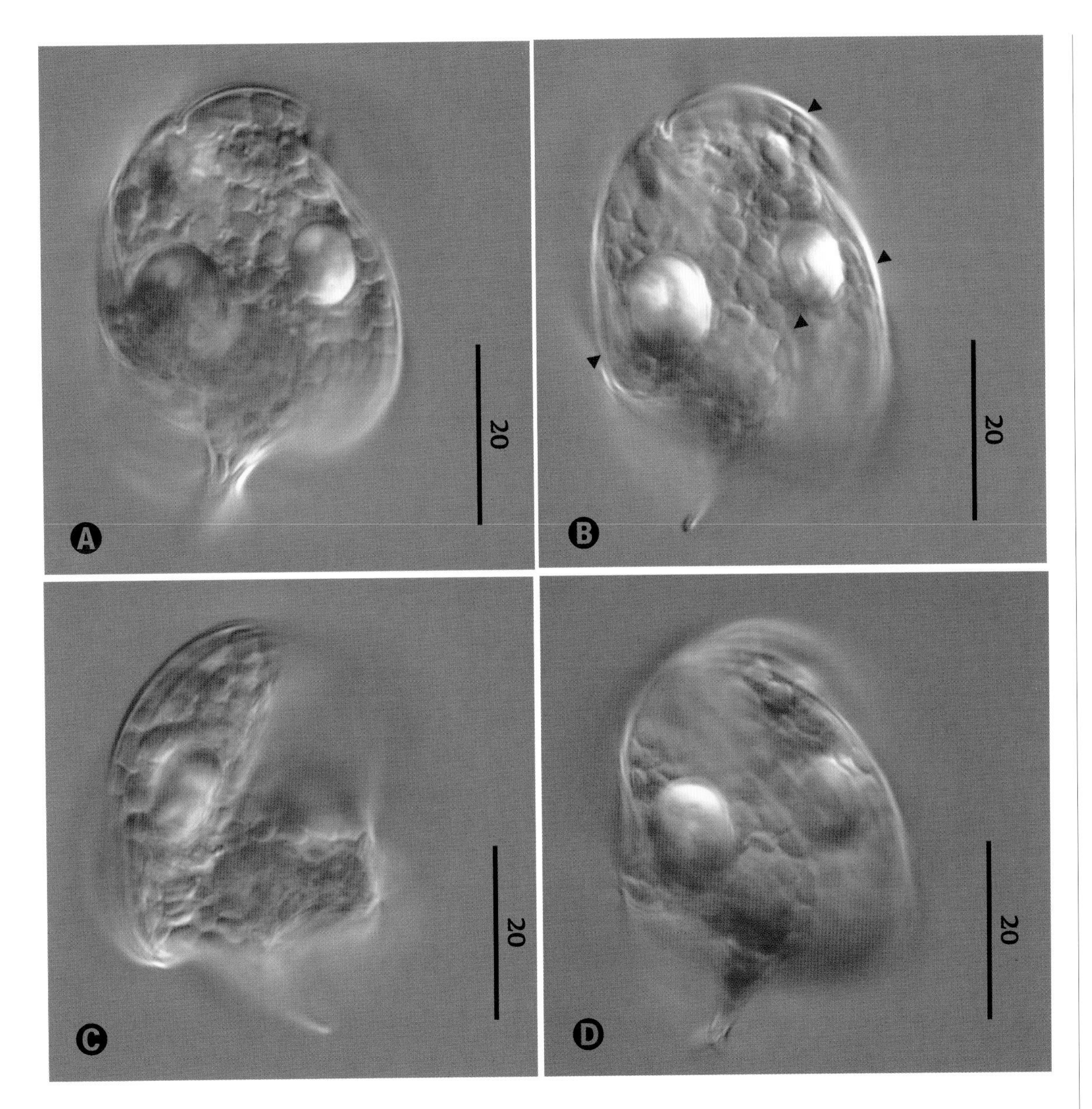

Phacus quinque-marginatus

Jahn et Shawhan 1942

SIZE: 35–52 µm long × 25–40 µm wide.

Ovoid cell with red eyespot, numerous small disc-shaped chloroplasts, and two large paramylon discs (Ⓐ). Same cell in a different focal plane; cell is twisted, four out of five cell margins visible (arrows); red eyespot, numerous chloroplasts, two large paramylon discs, and bent caudus present (Ⓑ, Ⓓ). Detail of twisted cell (Ⓒ). Flagellum is one to one-and-a-half times body length.

Phacus raciborskii

Drežepolski 1925

SIZE: 35 µm long × 10–12 µm wide.

Cell showing eyespot, fine pellicular striations (arrow) (A), and two large paramylon bodies (A, B). Detail showing twisted nature of the cell body (C). Cell with numerous small disc-shaped chloroplasts and two large paramylon bodies (D).

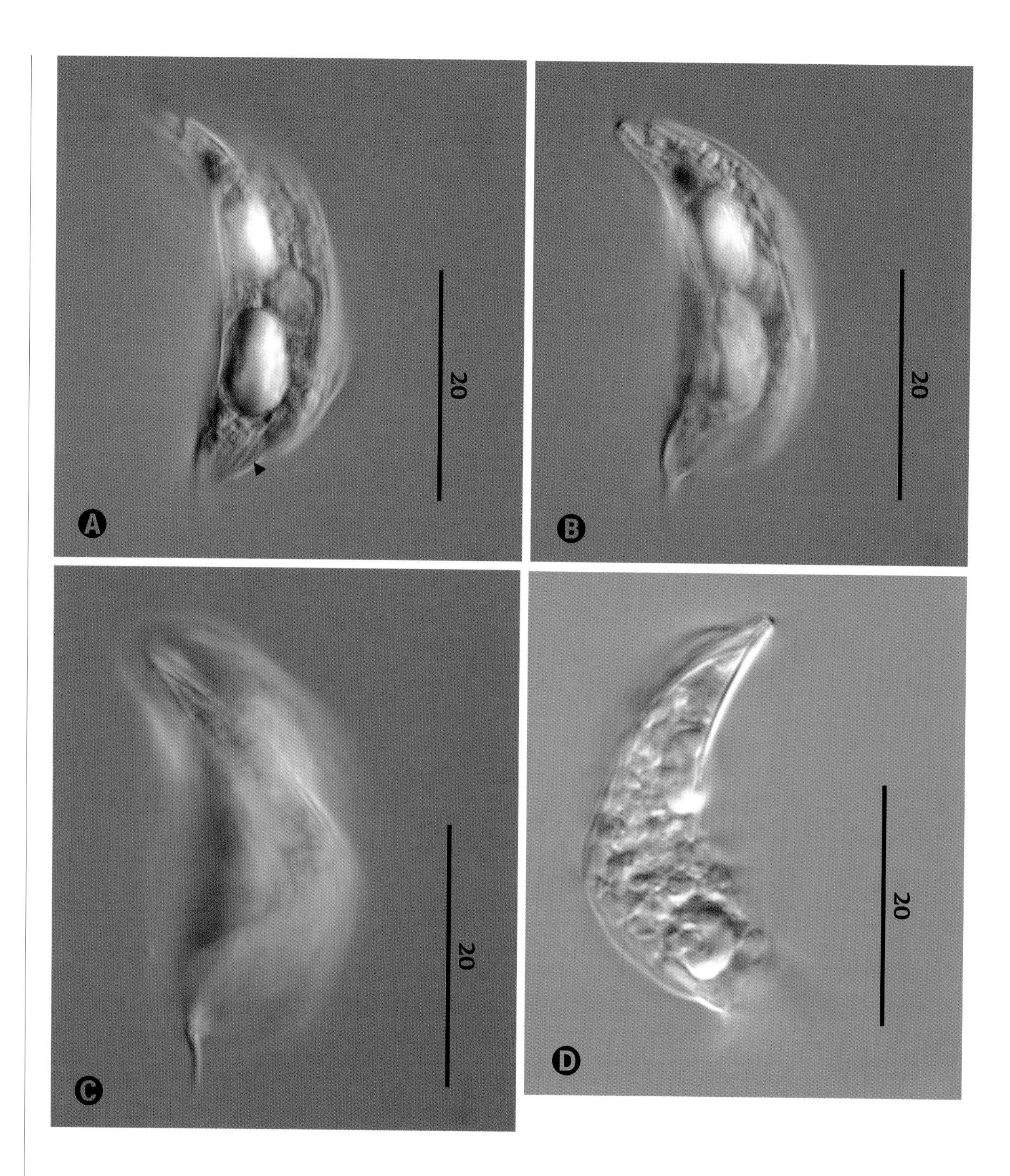

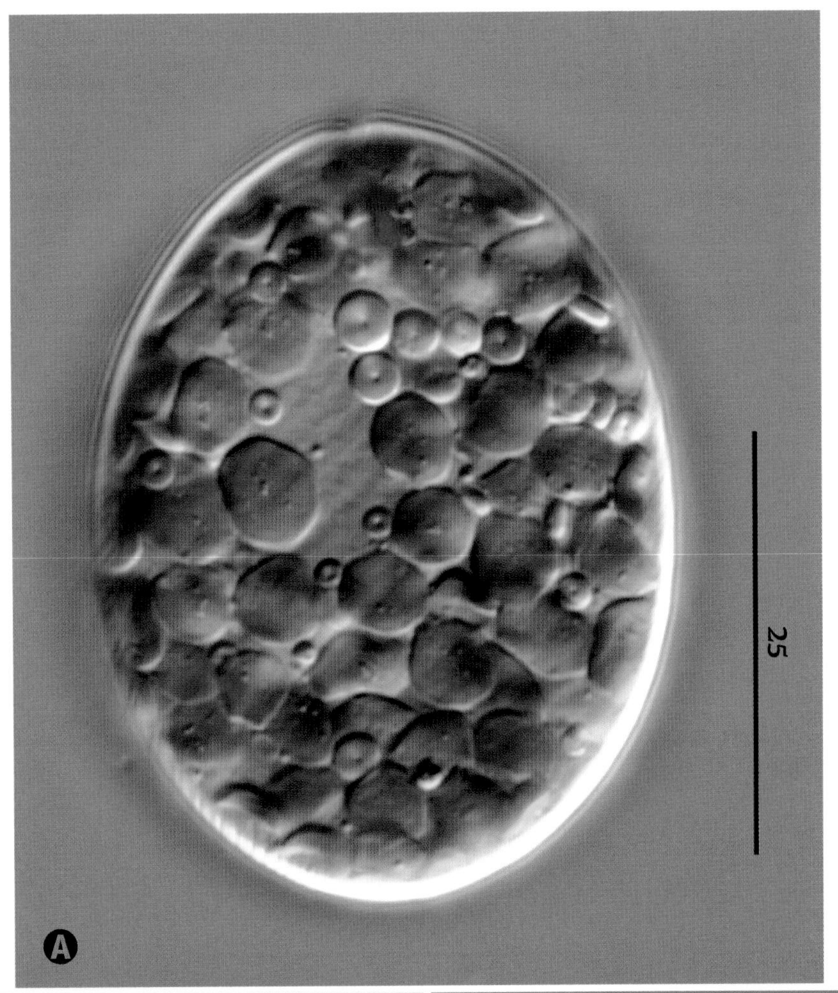

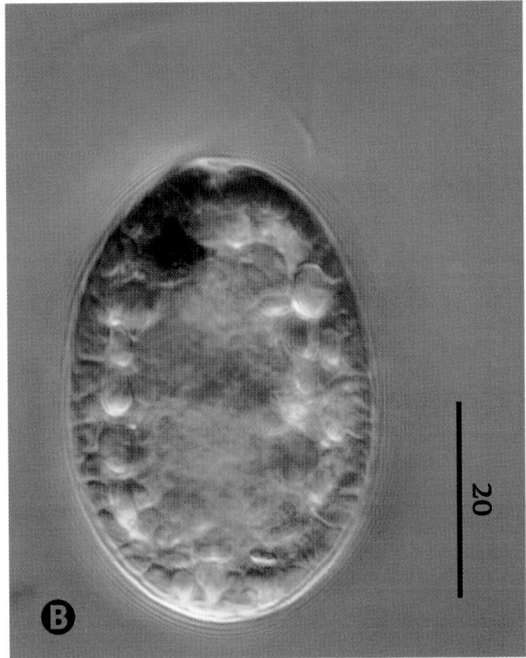

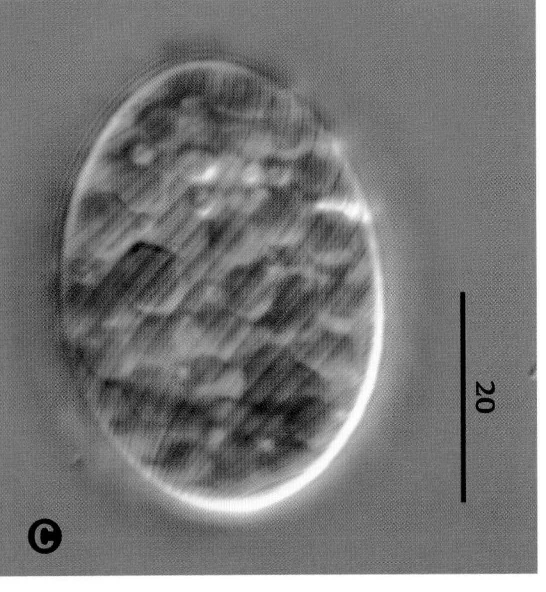

Phacus salina

(Fritsch) Linton et Karnkowska 2010

SIZE: 38–60 µm long × 26–45 µm wide.

Ovoid cells showing numerous disc-shaped chloroplasts, eyespot (A–C), paramylon discs (A), flagellum (twice body length) (B), and right-handed pellicular striations (C).

Phacus salina fa. *minor*

(= ***Lepocinclis salina*** fa. ***minor*** (Huber-Pestalozzi) Conrad 1934)

SIZE: 34–37 µm long × 27–28 µm wide.

Broadly ovoid cell with red eyespot (A). Cell showing discoid to ovoid paramylon bodies (B). Detail of right-handed pellicle striations and disc-shaped chloroplasts (C). Notice the difference in size between *P. salina* var. *minor* (*left*) and *P. salina* (*right*) (D).

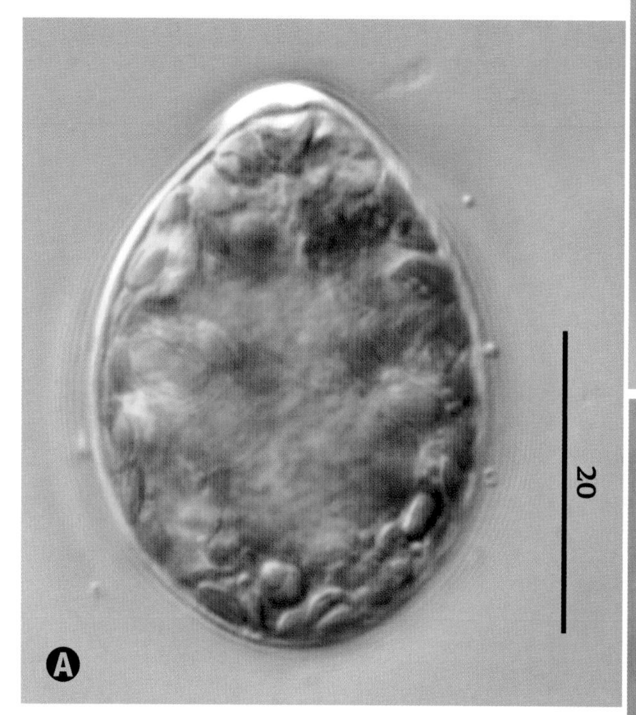

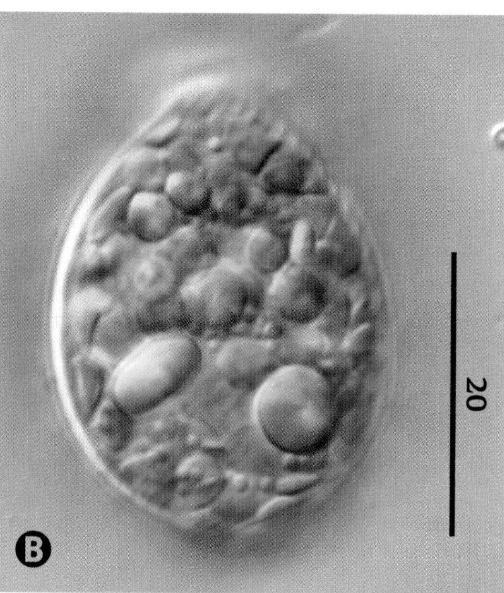

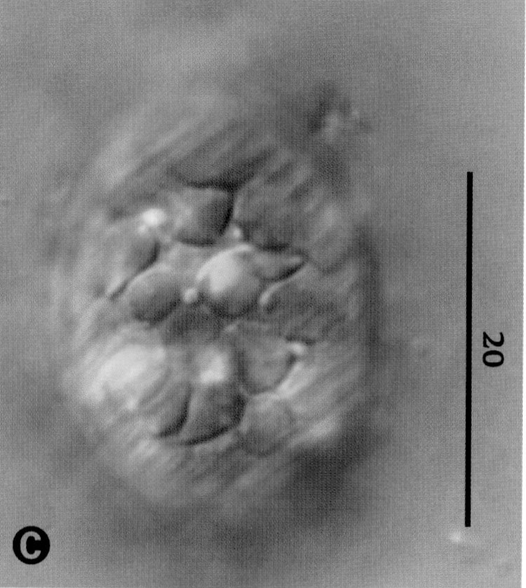

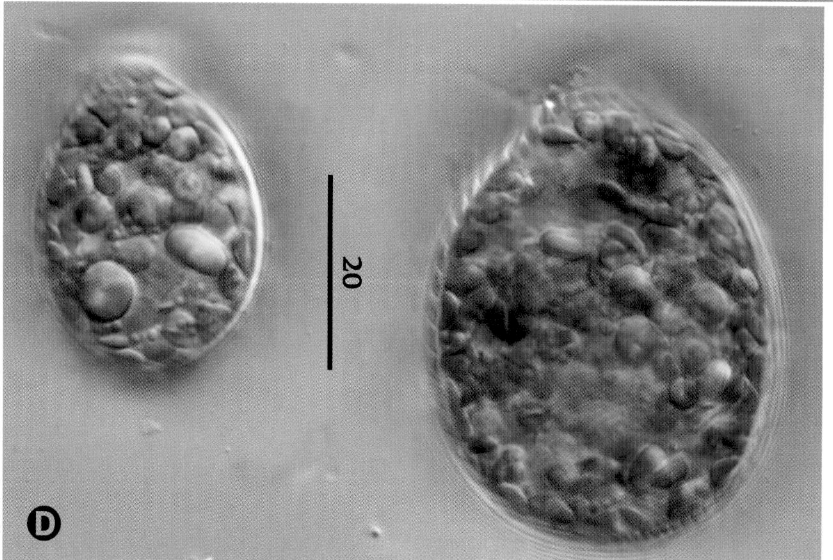

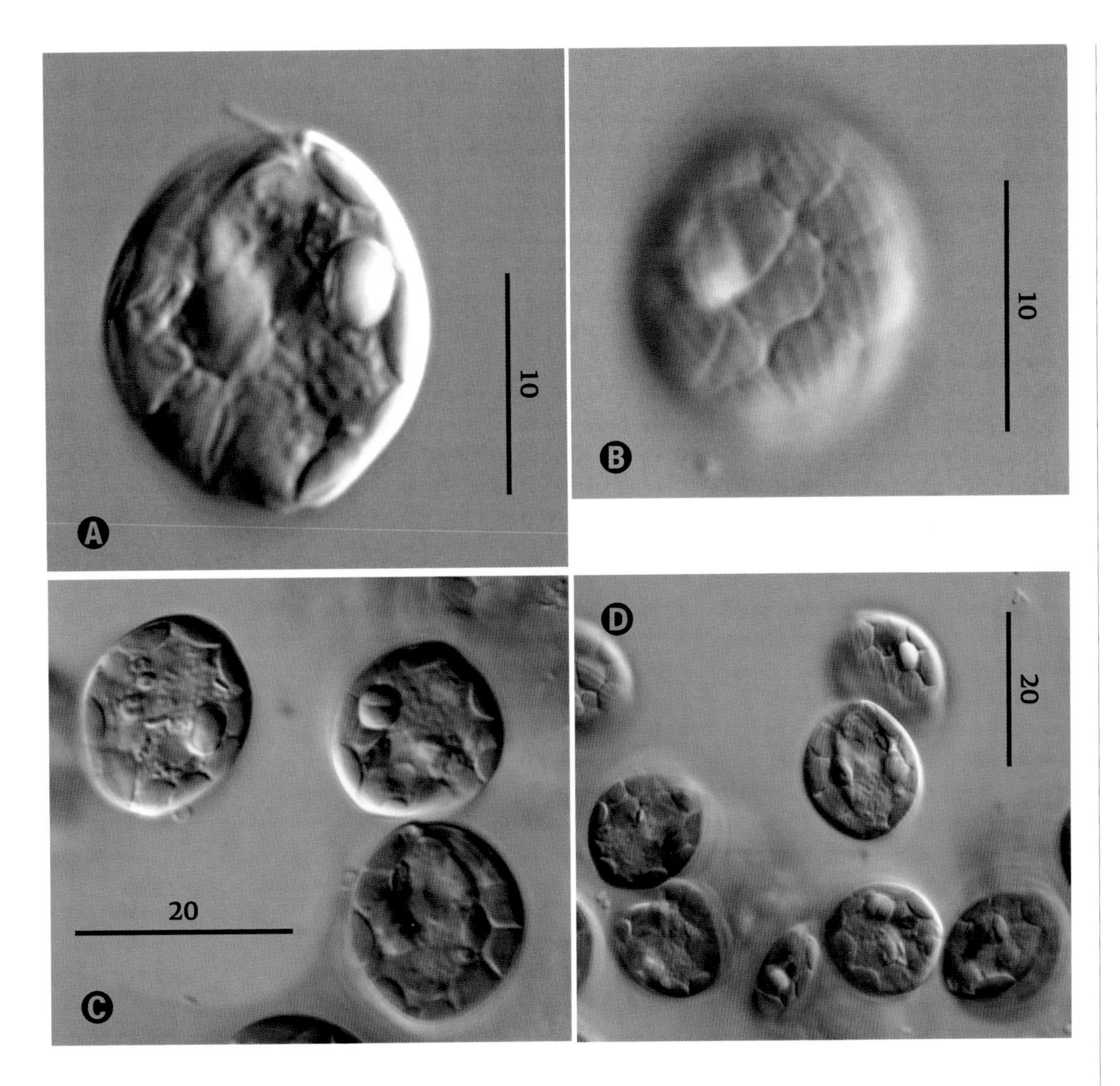

Phacus segretii

Allorge et Lefèvre 1925

SIZE: 17–28 µm long × 14–22 µm wide.

Broadly ovoid, acaudate cell with flagellum, eyespot, small disc-shaped chloroplasts, and one large paramylon body (A). Pellicular striations longitudinally arranged (B). Groups of cells showing typical body shape (C, D).

Phacus skujae

Skvortzov emend. Zakryś et Karnkowska 2010

SIZE: 18–29 µm long × 7–19 µm wide.

Ellipsoidal cell with eyespot, one large paramylon body, and small disc-shaped chloroplasts (A). Pellicular striations longitudinally arranged (B, C). Two dissimilarly sized paramylon grains (D).

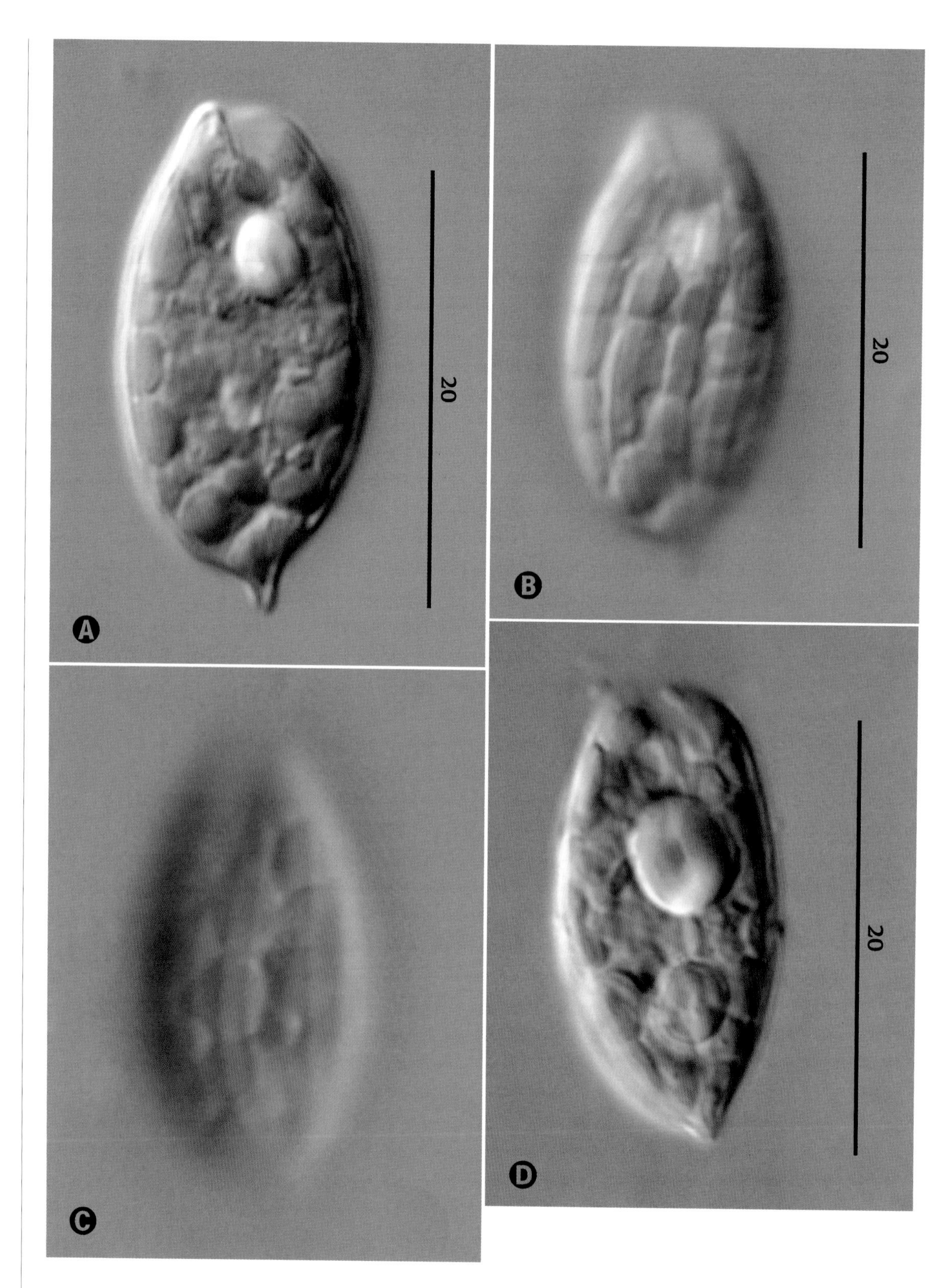

Phacus smulkowskianus

(Zakryś) Kusber emend. Zakryś et Karnkowska 2010

SIZE: 20–41 µm long × 7–16 µm wide.

Twisted cell showing eyespot, disc-shaped chloroplasts, and large paramylon grain (A). Finely striated pellicle (B). Flagellum and two sizes of paramylon grains (C). Several swimming cells showing typical body shape (D).

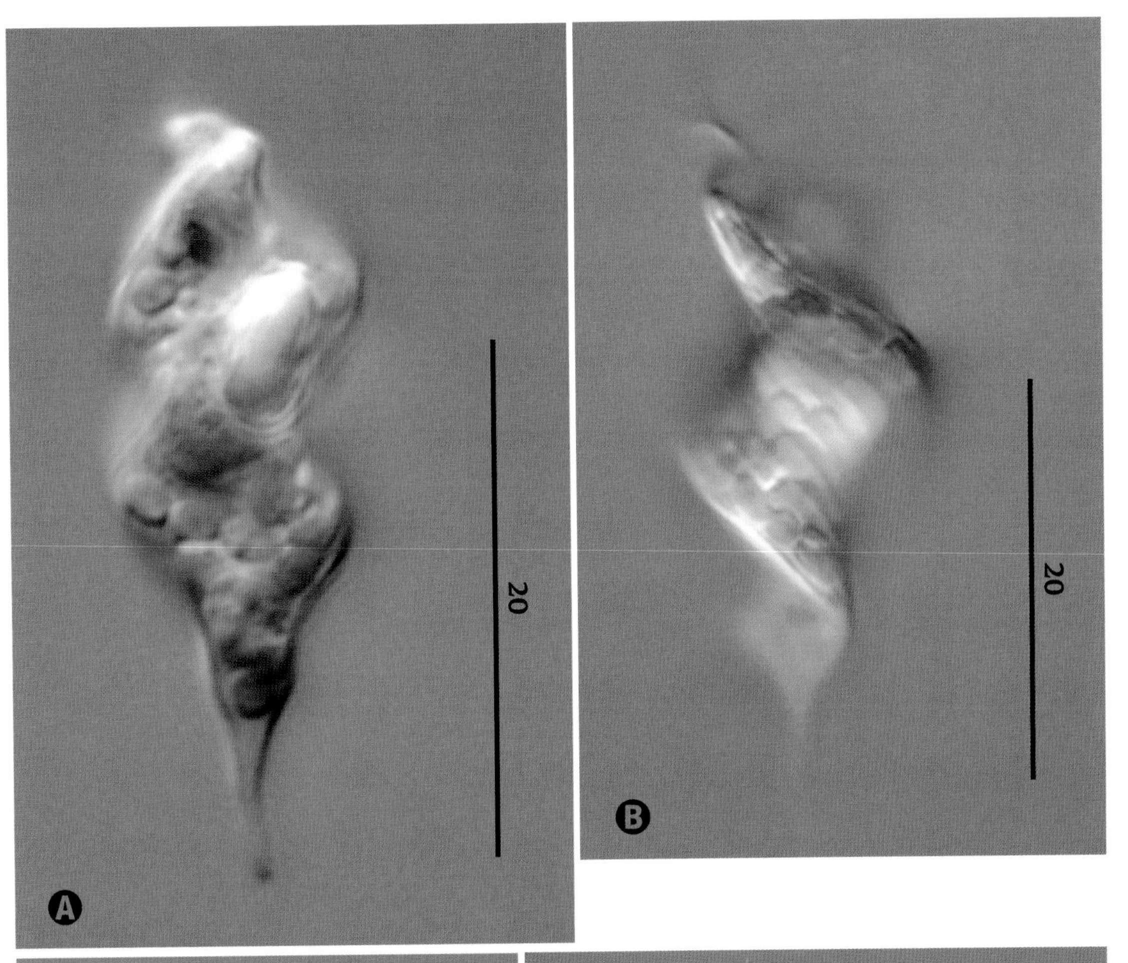

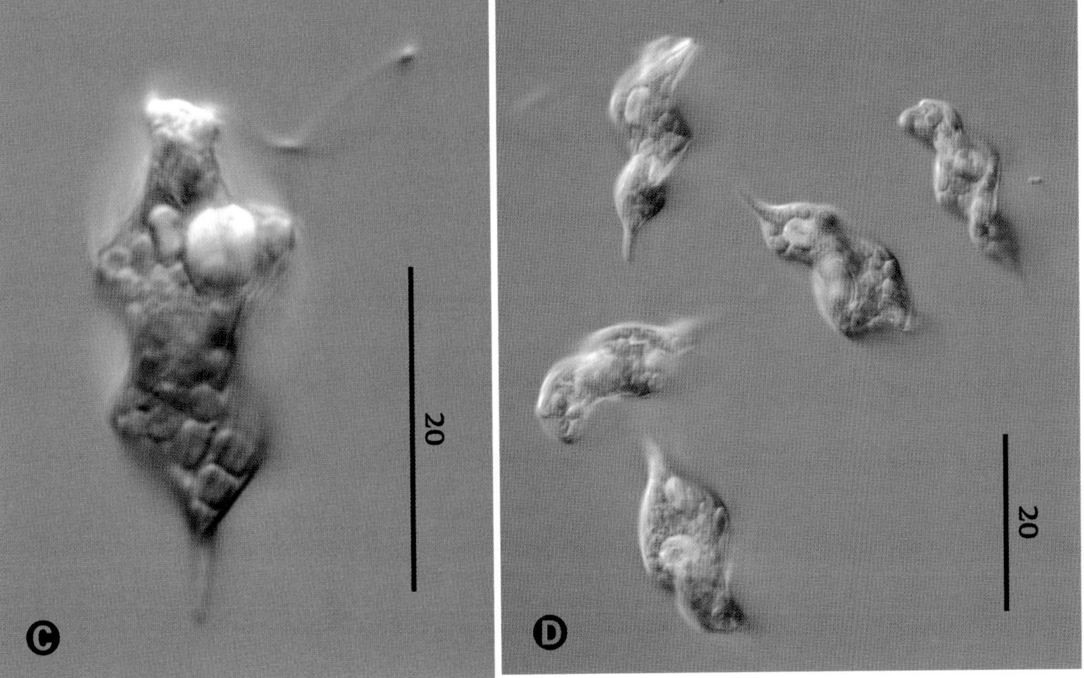

Phacus stokesii

Lemmermann 1901

SIZE: 38–48 µm long × 30–42 µm wide.

Broadly ovoid cell exhibiting eyespot, numerous disc-shaped chloroplasts, prominent paramylon disc, and nucleus (Ⓐ). Cell showing longitudinal groove (Ⓑ). Detail of longitudinally striated pellicle, chloroplasts, and paramylon (Ⓒ). Flagellum is body length.

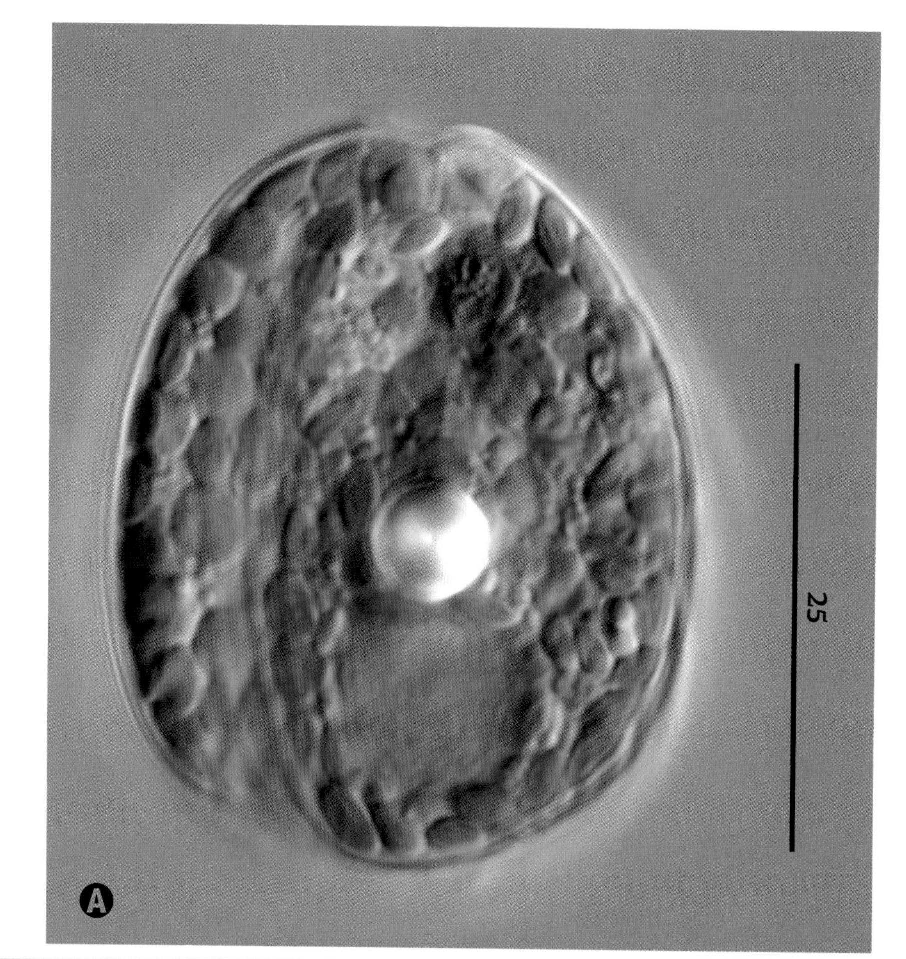

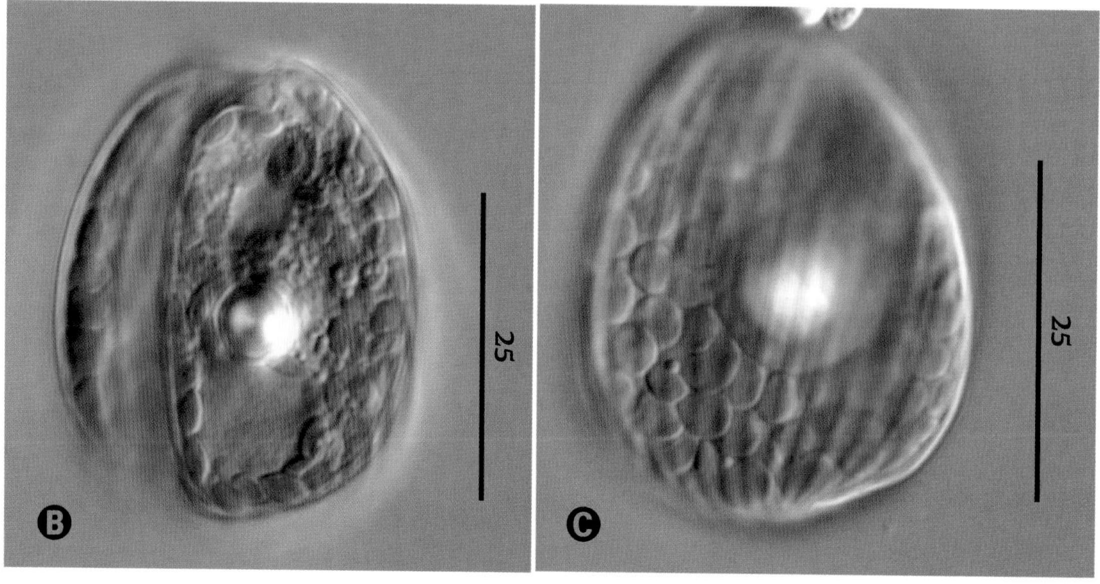

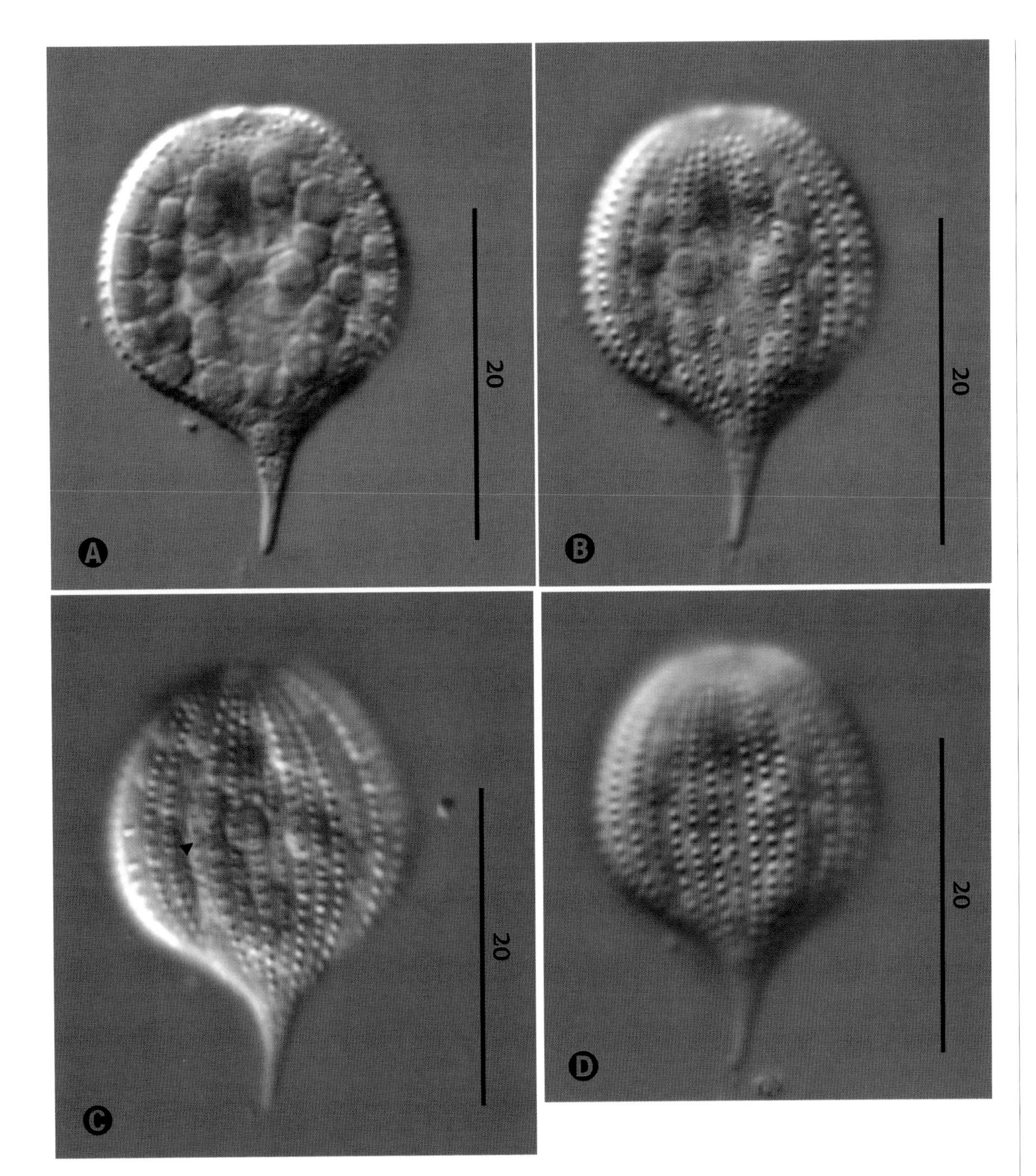

Phacus suecicus

Lemmermann 1913

SIZE: 27–40 μm long × 17–25 μm wide.

Ovate cell with a median protuberance at the anterior end; eyespot and numerous small disc-shaped chloroplasts (A). Different focal planes of cell showing eyespot, chloroplasts, and longitudinally arranged rows of surface protuberances (B–D); large lateral paramylon links (arrow) (C). Flagellum is one-and-a-half times body length.

Phacus tortus

(Lemmermann) Skvortzov 1928

SIZE: 70–150 µm long × 34–48 µm wide.

Broadly ovoid cell in different focal planes showing twisted body with long caudus, eyespot, one large paramylon disc, and numerous small disc-shaped chloroplasts (Ⓐ–Ⓒ). Longitudinally disposed pellicular strips exhibit cross-striations (Ⓒ). Flagellum is about body length.

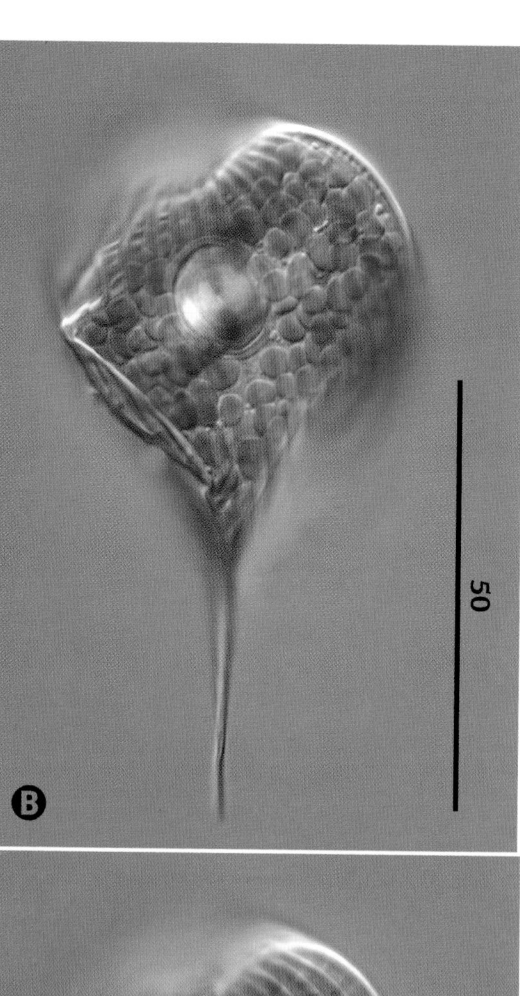

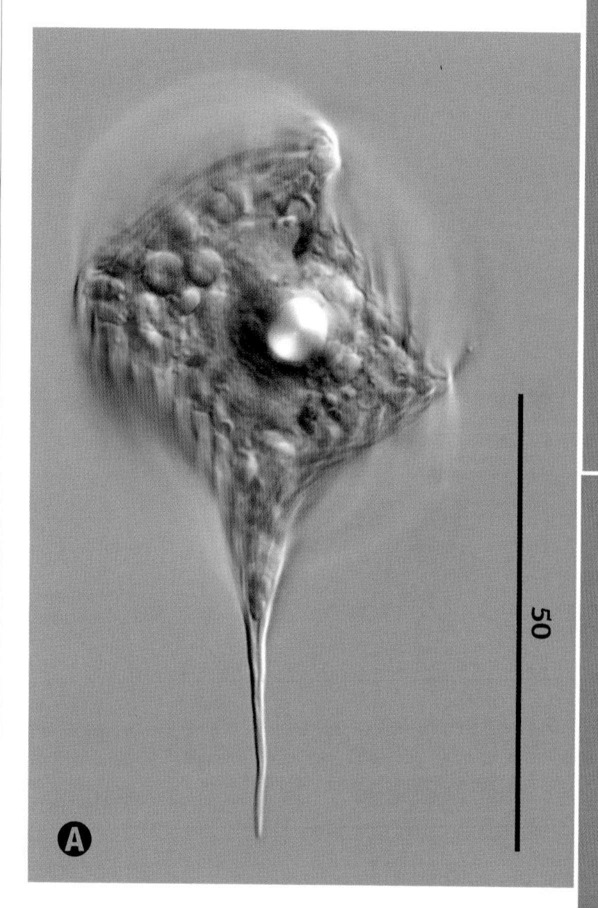

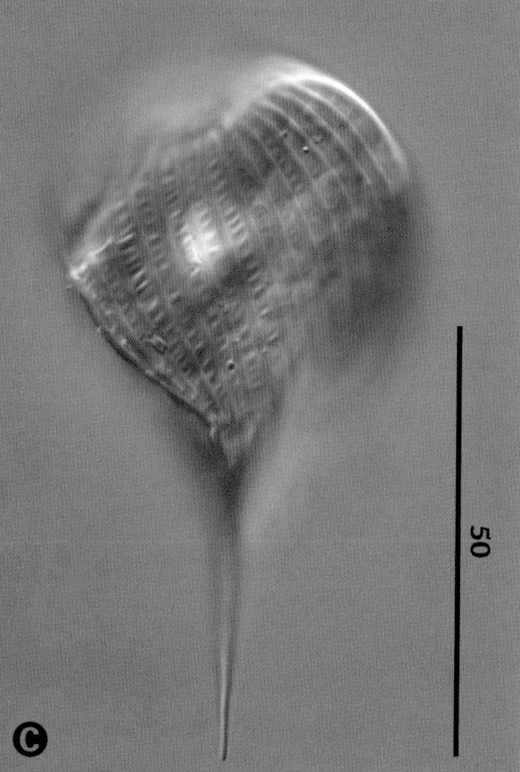

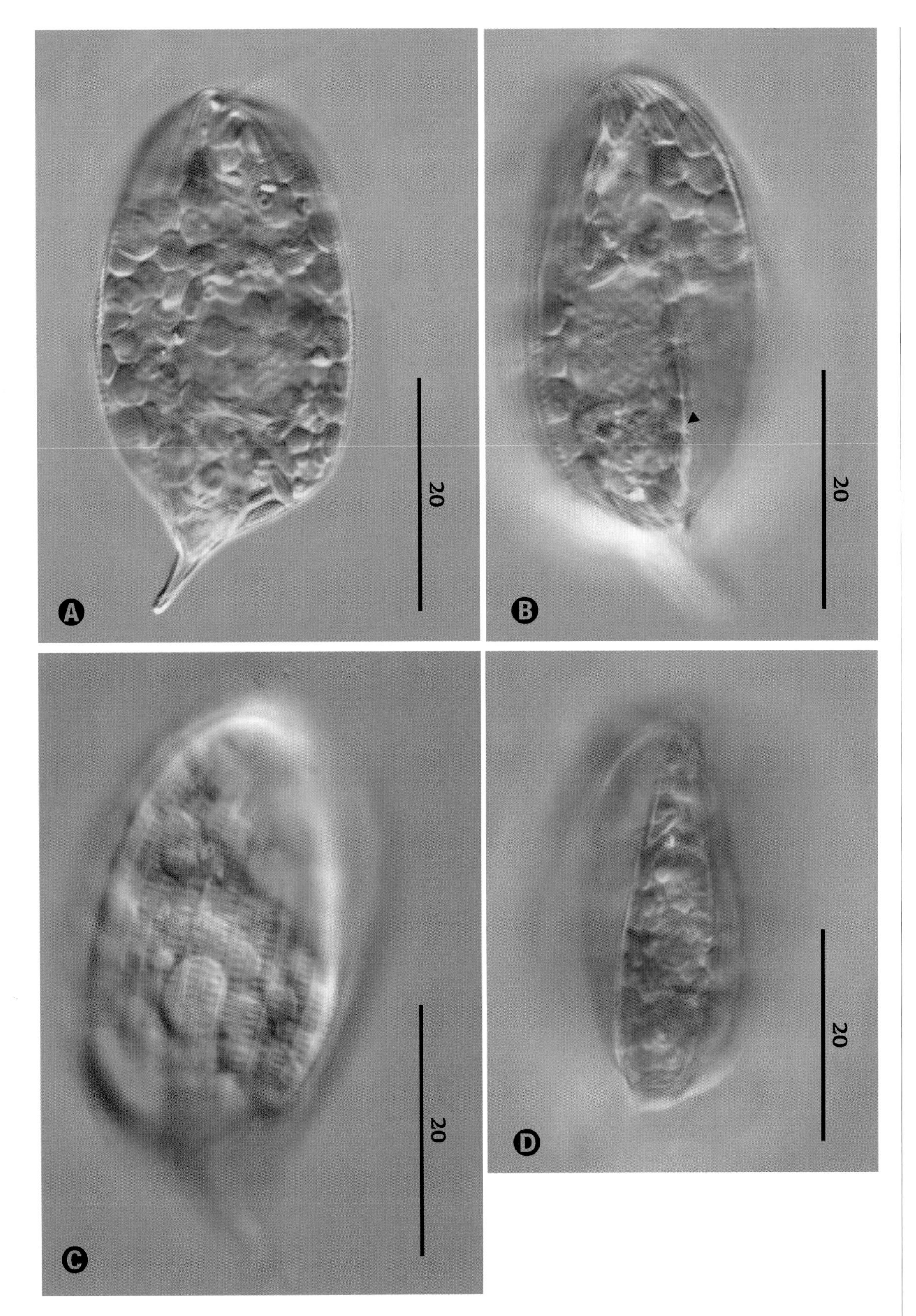

Phacus triqueter

(Ehrenberg) Dujardin 1841

SIZE: 34–78 µm long × 22–39 µm wide.

Ellipsoid cell with short bent caudus showing flagellum (body length), eyespot, and numerous small disc-shaped chloroplasts (A). Same cell revealing the keel (arrow) (B). Detail of longitudinally oriented pellicular strips with cross-striations (C). Cell in side view, slightly twisted showing keel (D).

Phacus warszewiczii

Dreżepolski 1925

SIZE: 55–72 µm long × 35–40 µm wide.

Broadly oval cell with straight caudus; pellicle strips spirally disposed showing cross-striations; numerous small disc-shaped chloroplasts (A). Detail of the thick longitudinal keel; eyespot, nucleus, and chloroplasts visible (B). Cell seen almost from above; eyespot, pellicle strips with cross-striations, and chloroplasts (C). Cell with keel, pellicle strips, and chloroplasts; one large paramylon disc partially visible in the center of the cell (arrow) (D). Flagellum is body length or longer.

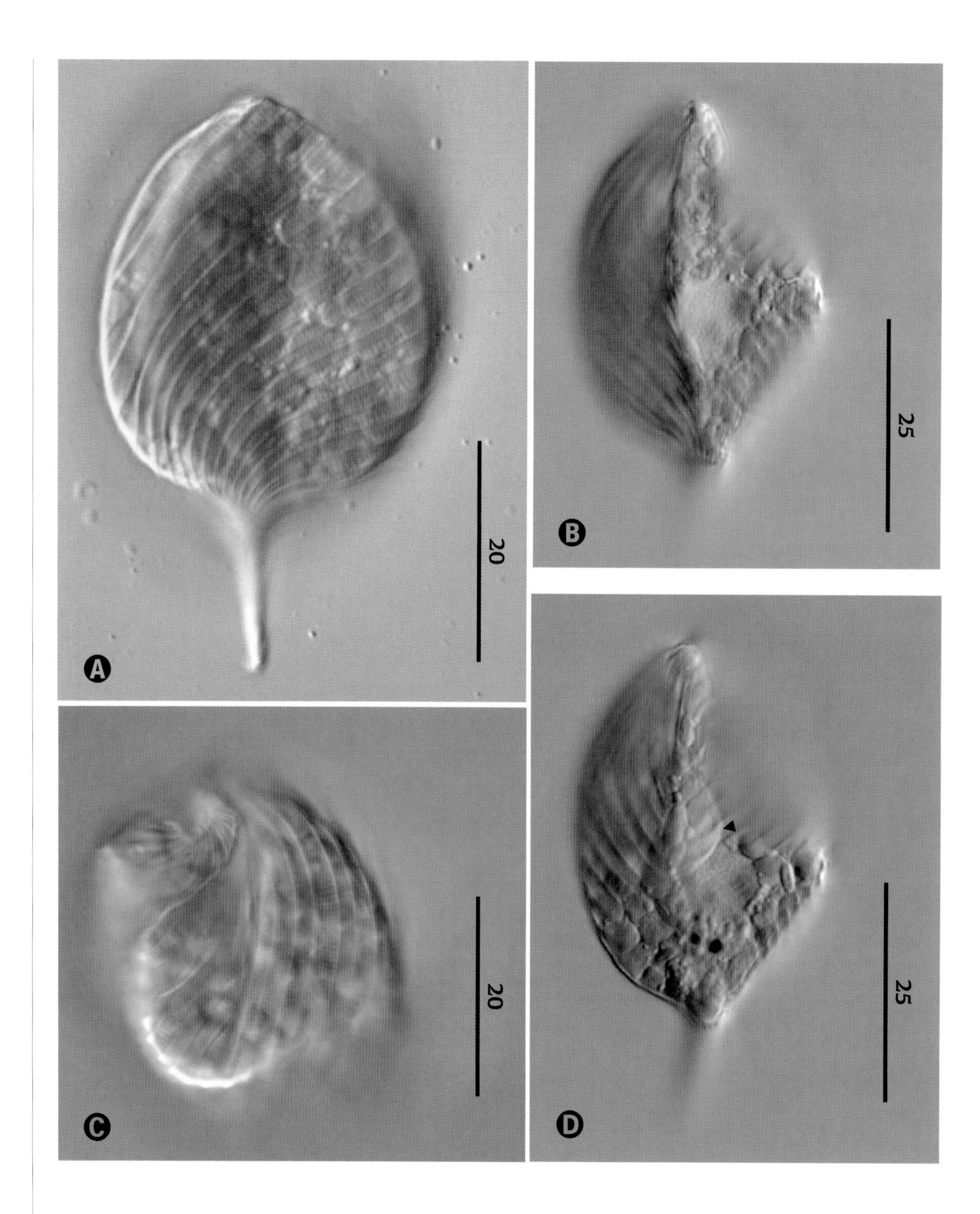

STROMBOMONAS

DEFLANDRE 1930

Cells enclosed in variously shaped (ovoid, ellipsoidal, trapezoidal, spindle, or irregular) mineralized envelope (lorica); lorica often with a long collar at the anterior end, usually with a conical posterior end; lorica surface smooth with no pores, warts, or spines; commonly aggregated particles/debris on the exterior of the loricas; a single emergent flagellum; numerous parietal chloroplasts, plate-like or polygonal in shape, with or without pyrenoids; paramylon bodies small; large eyespot at the anterior end; cells metabolic inside the lorica; cysts not known; found in fresh or brackish water.

Strombomonas borysthenienesis

(Roll) Popova 1955

SIZE: 22–32 µm long × 18–22 µm wide.

Ovoid lorica with short neck and very short tail; small particles attached to the lorica surface; protoplast with red eyespot and disc-shaped chloroplast (A). Loricate cell with more rounded base (B). Detail of lorica surface (C). Protoplast showing chloroplasts and small paramylon grains (D). Cell exiting lorica (E).

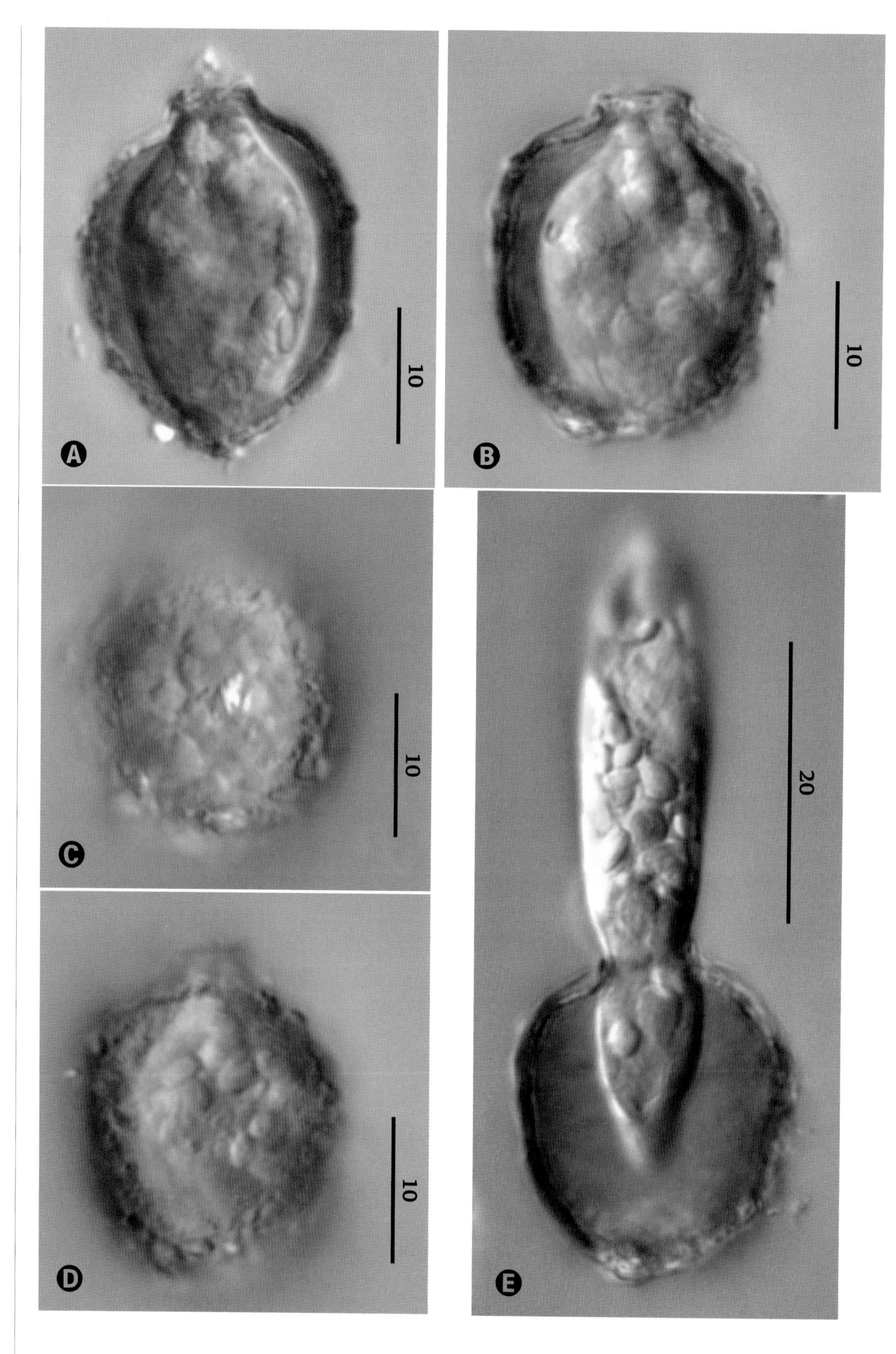

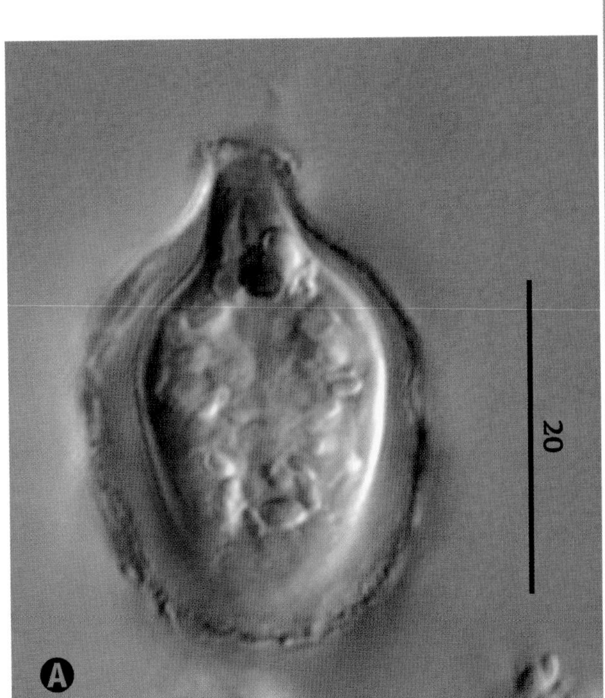

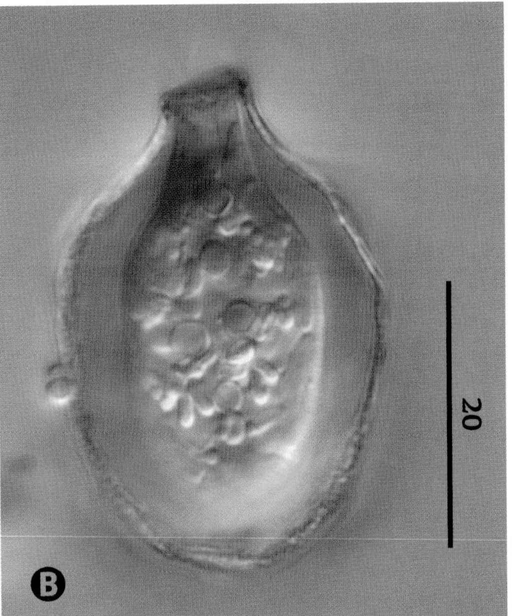

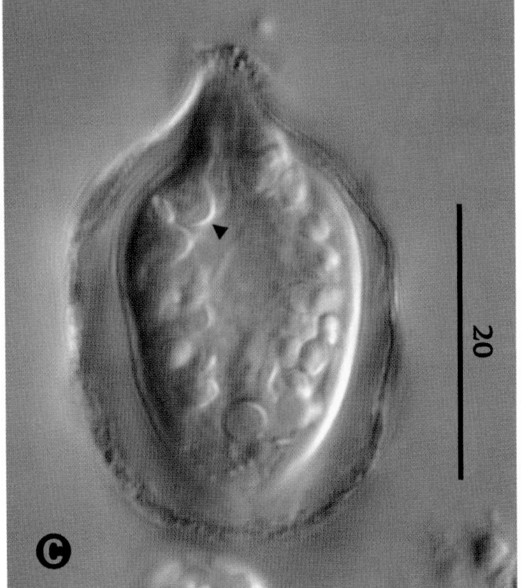

Strombomonas conspersa

(Pascher) Tell et Conforti 1984

SIZE: 25–35 µm long × 10–25 µm wide.

Ellipsoidal lorica with well-defined neck; cell with flagellum, eyespot, disc-shaped chloroplasts, and paramylon grains (A, B). Chloroplasts with haplopyrenoids (arrow) (C).

Strombomonas deflandrei

(Roll) Deflandre 1930

SIZE: 31–44 µm long × 19–27 µm wide.

Broadly ovoid lorica with short tail; protoplast with red eyespot, disc-shaped chloroplasts, and small ovoid paramylon bodies (Ⓐ). Chloroplast with haplopyrenoid (arrow) (Ⓑ). Cell with chloroplasts and paramylon bodies of various shapes and sizes (Ⓒ). Flagellum is one-and-a-half times body length.

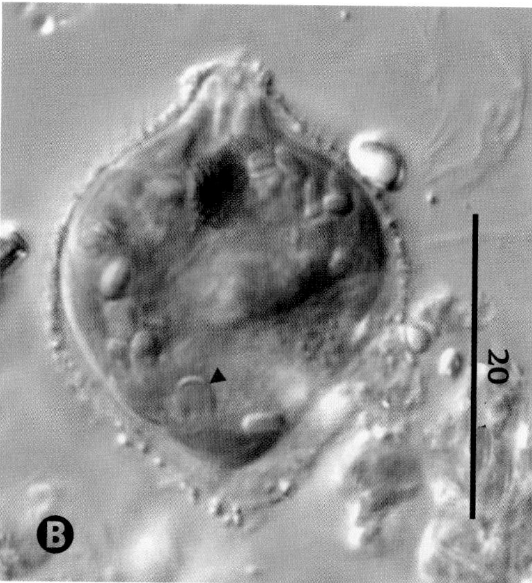

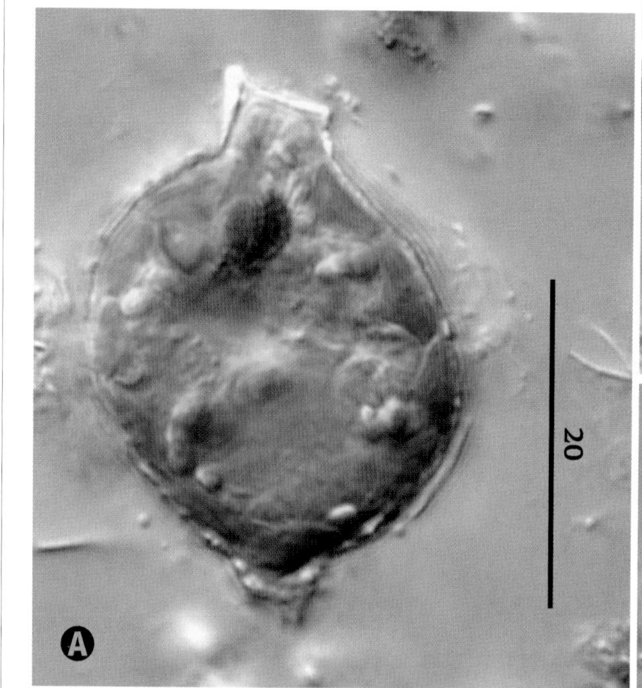

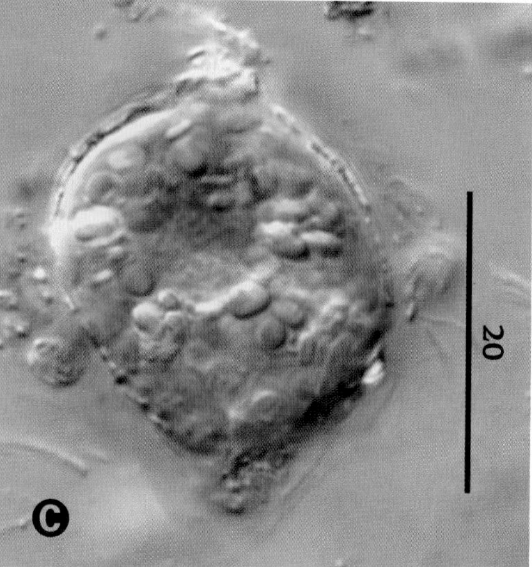

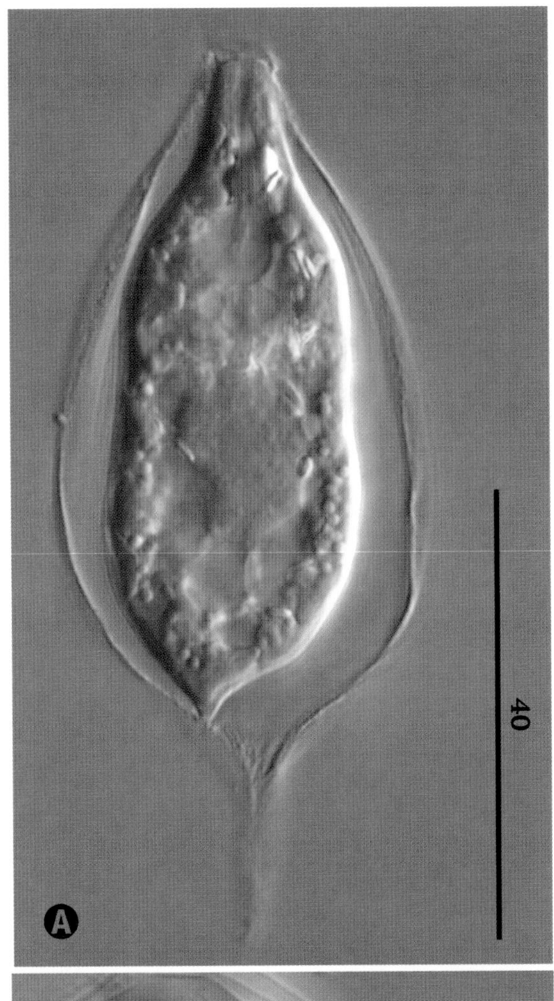

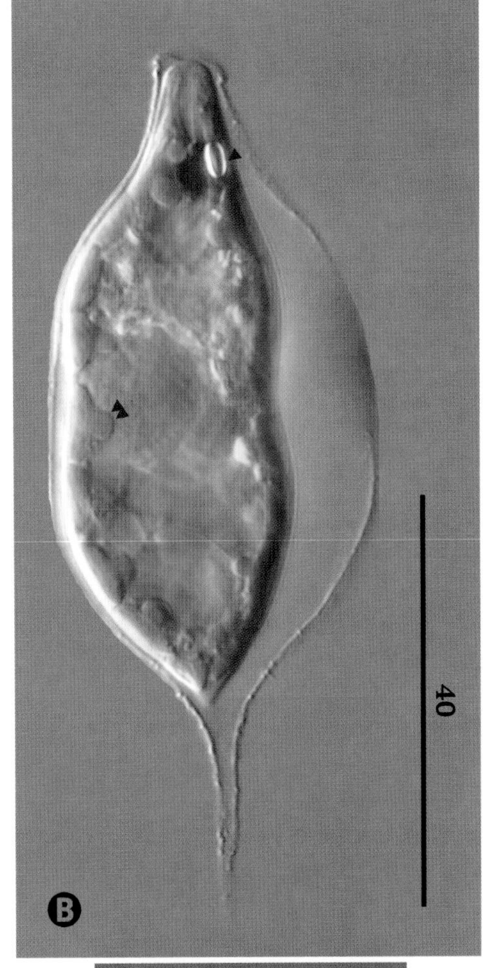

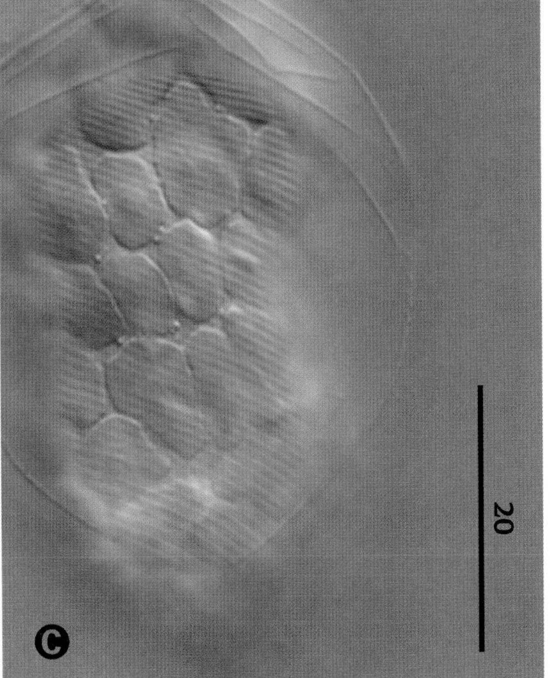

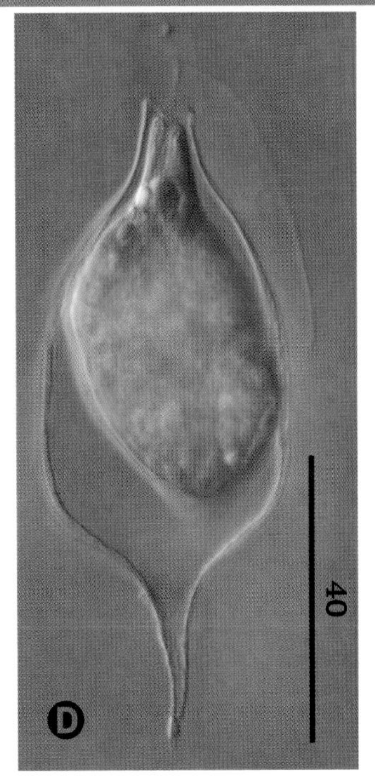

Strombomonas maxima

(Skvortzov) Deflandre 1930

SIZE: 58–107 µm long × 22–44 µm wide.

Large urn-like lorica with long caudus; cell with red eyespot, reservoir, chloroplasts, and central nucleus (**A**). Smooth lorica and protoplast with eyespot, a paramylon grain opposed to eyespot (arrow), and chloroplasts with haplopyrenoids (double arrow) (**B**). Detail of pellicle striations and disc-shaped chloroplasts (**C**). Urn-like lorica and protoplast with flagellum (**D**).

Strombomonas napiformis var. *brevicollis*

(Playfair) Deflandre 1930

SIZE: 36–53 µm long × 19–25 µm wide.

Ellipsoid lorica with short neck; cell with red eyespot, disc-shaped chloroplasts, and nucleus with nucleolus (A). Cell with small paramylon grains (B); lorica surface rough with attached particles (B, C).

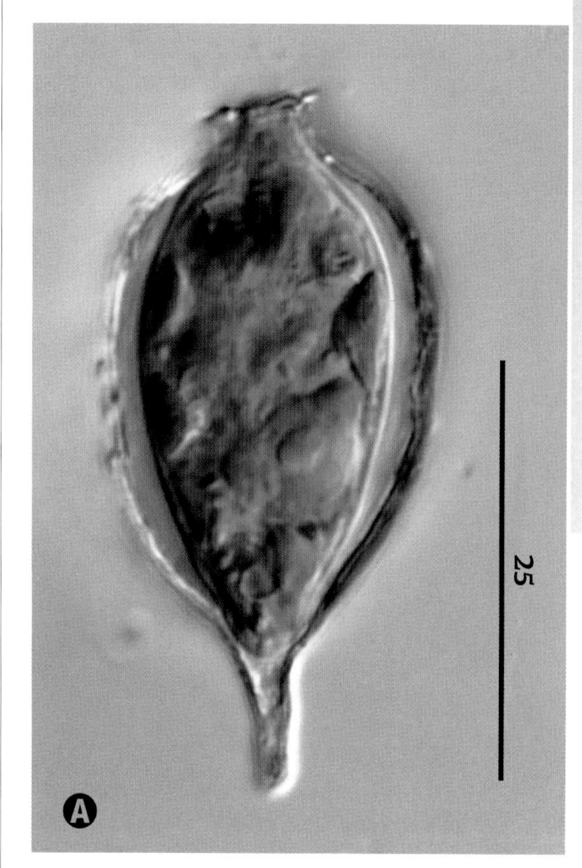

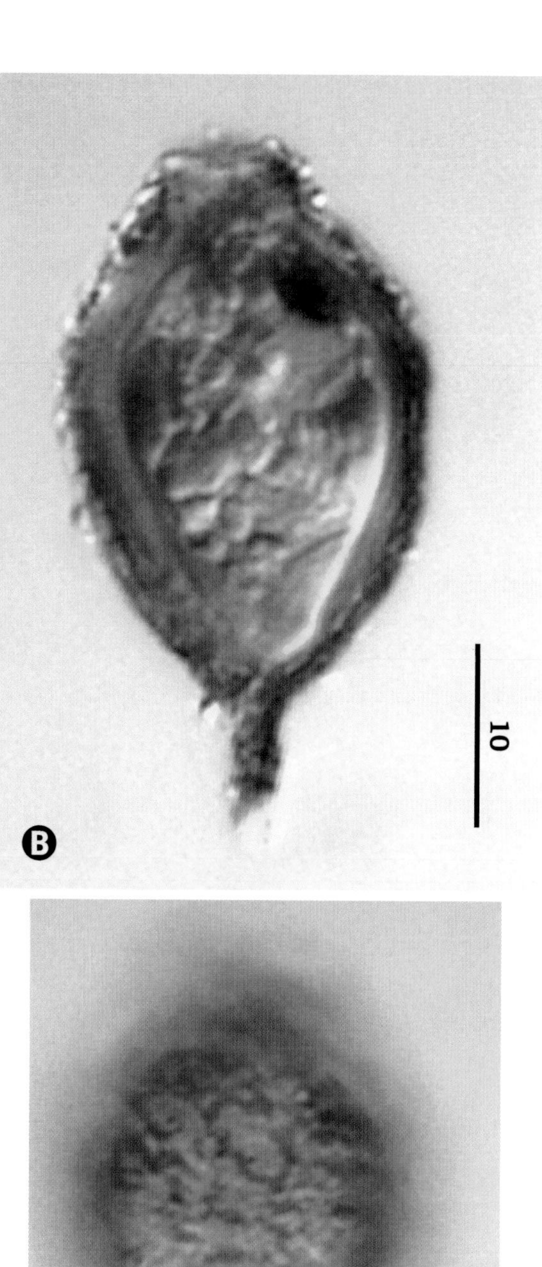

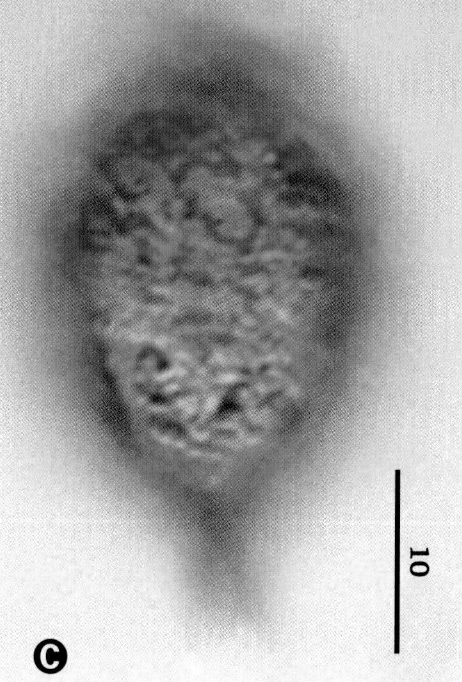

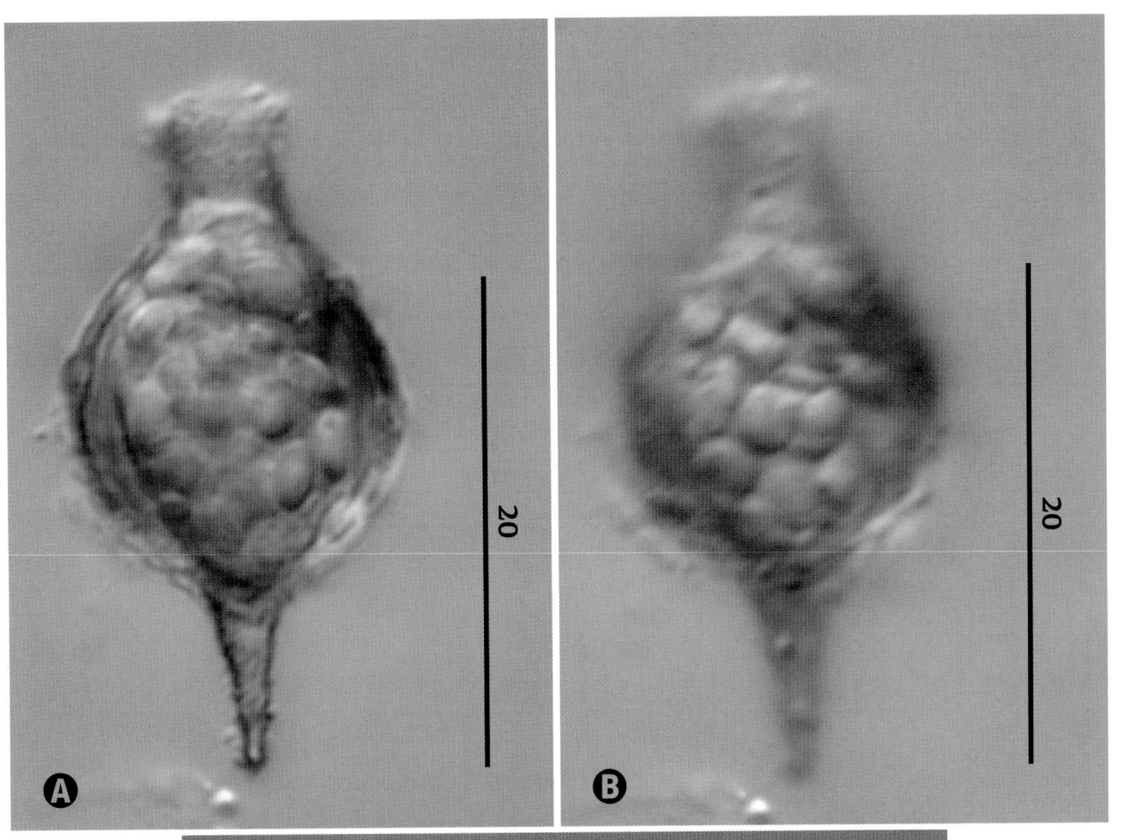

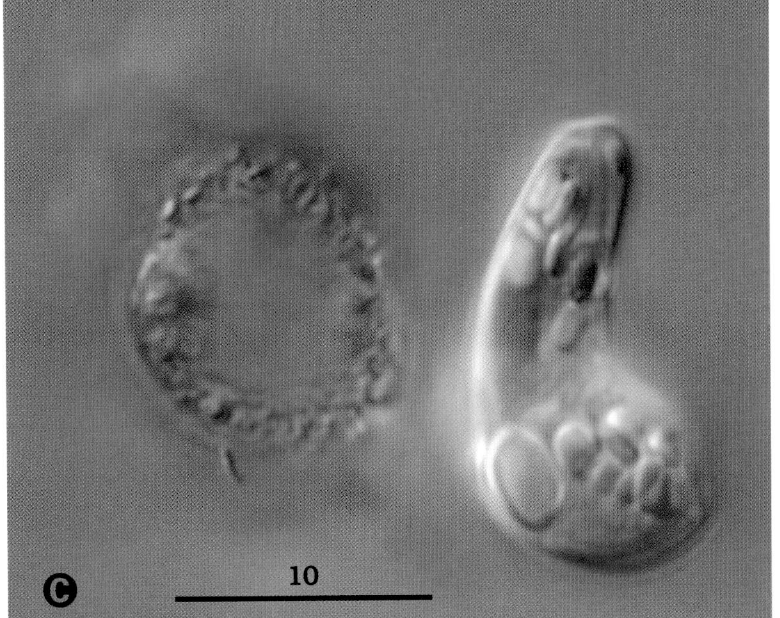

Strombomonas rotunda

(Playfair) Deflandre 1930

SIZE: 25–33 μm long × 15–19 μm wide.

Ovoid lorica with wide neck and straight caudus; protoplast showing chloroplasts and paramylon grains (A). Same cell in surface view (B). Empty lorica and next to it the protoplast showing red eyespot, chloroplast with pyrenoid, and paramylon bodies (C).

Strombomonas triquetra

(Playfair) Deflandre 1930

SIZE: 40 µm long × 20 µm wide.

Rectangular lorica with cylindrical neck; cell showing eyespot and disc-shaped chloroplasts with haplopyrenoids (arrow) (Ⓐ). Posterior end of lorica is attenuated into a conical caudus (Ⓑ, Ⓒ); cell shows flagellum (Ⓑ), eyespot, chloroplasts (Ⓑ, Ⓒ) with haplopyrenoids (Ⓑ).

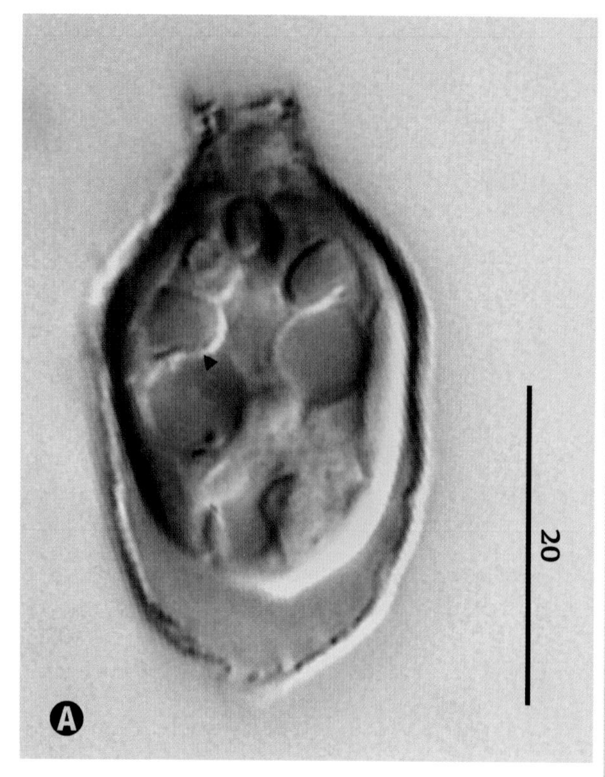

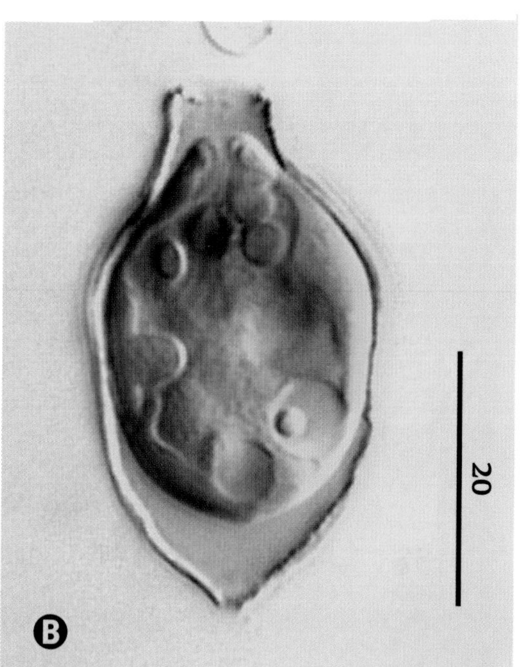

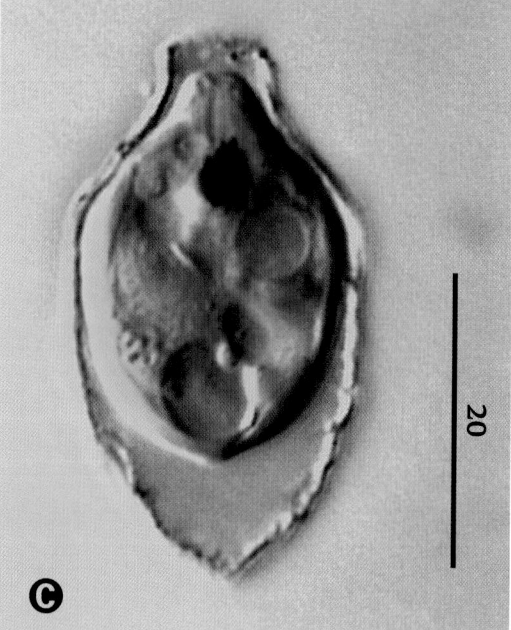

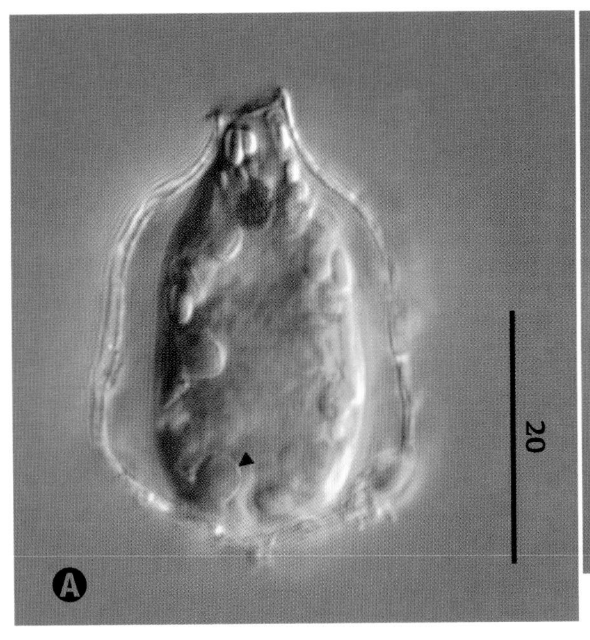

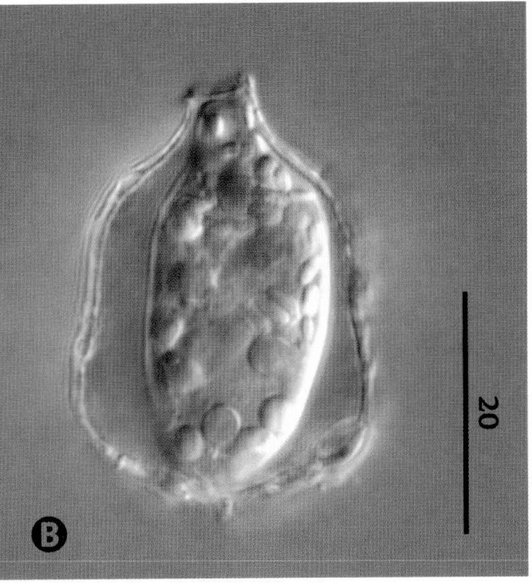

Strombomonas verrucosa

(Von Daday) Deflandre 1930

SIZE: 20–50 µm long × 10–29 µm wide.

Trapezoidal lorica with oblique neck and very short caudus; cell with red eyespot, disc-shaped chloroplasts with haplopyrenoids (arrow) (Ⓐ), and small paramylon grains (Ⓐ, Ⓑ).

Strombomonas verrucosa var. *zmiewika*

(Swirenko) Deflandre 1930

SIZE: 37–53 µm long × 20–31 µm wide.

Trapezoidal lorica with long caudus, flagellum, red eyespot, and disc-shaped chloroplasts (A). Same cell in different focal plane showing haplopyrenoids (arrow) and numerous ovoid paramylon bodies (B).

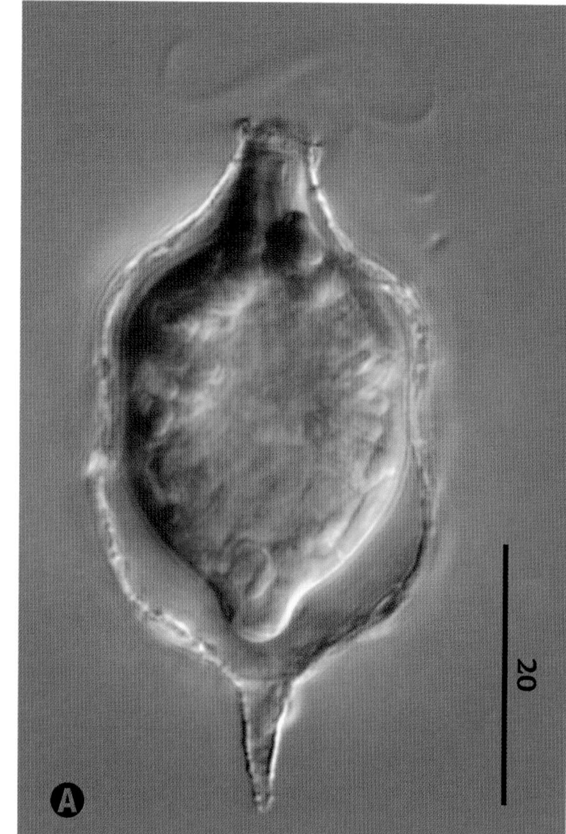

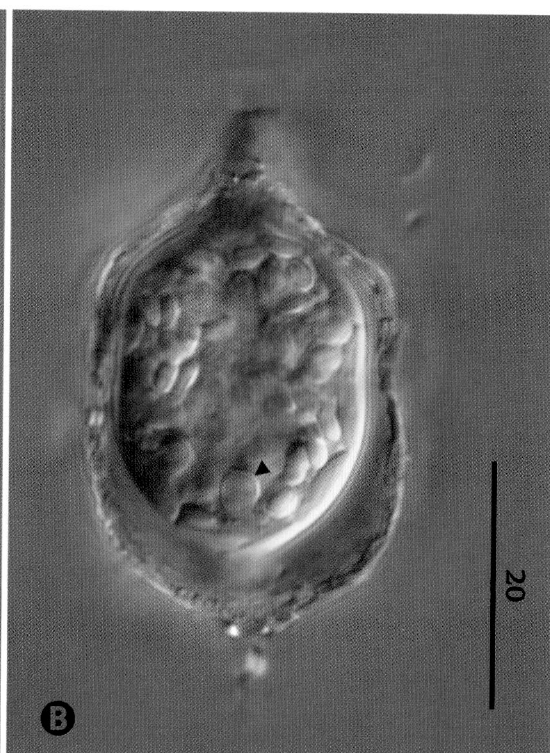

TRACHELOMONAS

EHRENBERG 1833

Cells enclosed in a spherical or ovoid mineralized envelope (lorica), sometimes with a defined neck or collar surrounding an apical pore through which the locomotory flagellum emerges; lorica tends to be porous and ornamented with appendages of different sizes and shapes; a single emergent flagellum, usually longer than the lorica; chloroplast one to many, small or large discs, with or without pyrenoids; when present, pyrenoids can be haplo- (most often) or diplopyrenoids; paramylon bodies small, scattered throughout the cell; some forms colorless (e.g., *Trachelomonas reticulata*); eyespot present at the anterior end; cells exhibit metaboly; cysts known; freshwater.

Trachelomonas abrupta

Swirenko emend. Deflandre 1926

SIZE: 22–34 µm long × 12–21 µm wide.

Elongate, cylindrical lorica ornamented with short spines; protoplast showing flagellum, red eyespot, and disc-shaped chloroplasts with haplopyrenoids (**A**). Same cell in different focal plane; lorica with flagellar pore and protoplast with chloroplasts and haplopyrenoids (arrow) (**B**). Cell with numerous ovoidal paramylon bodies (**C**). Lorica surface showing densely and irregularly arranged short spines (**D**).

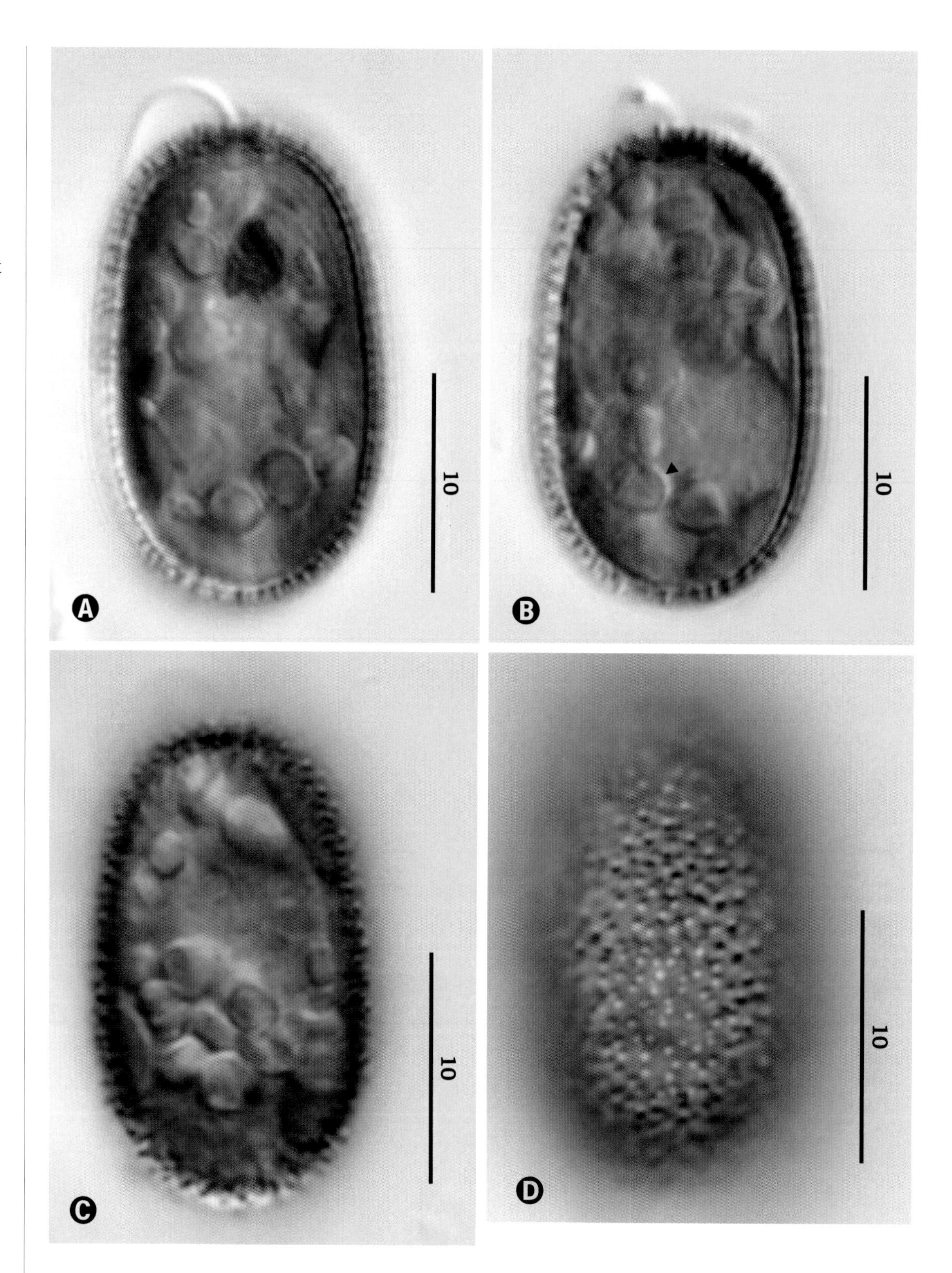

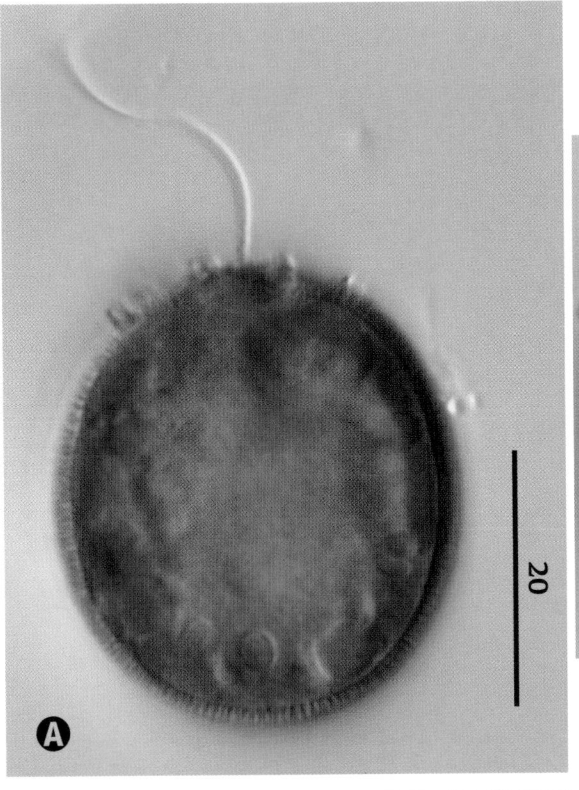

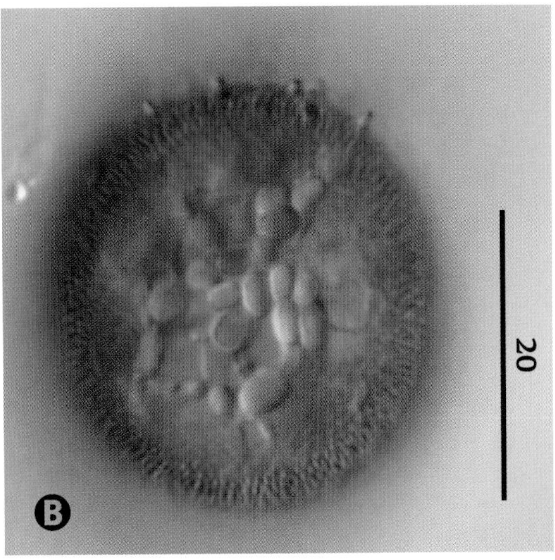

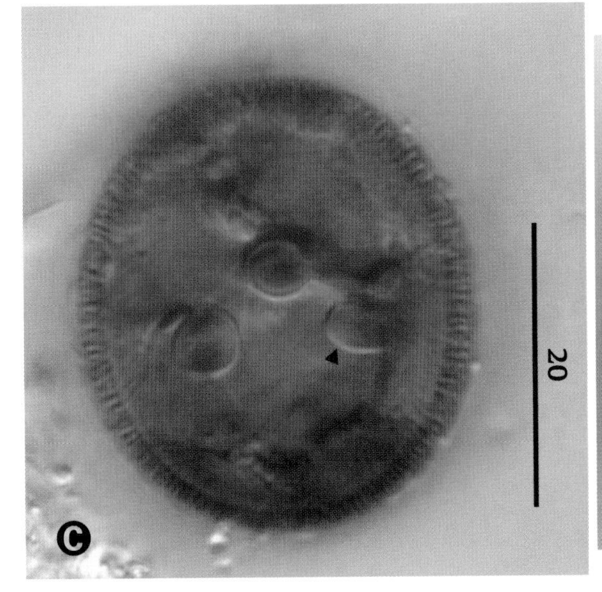

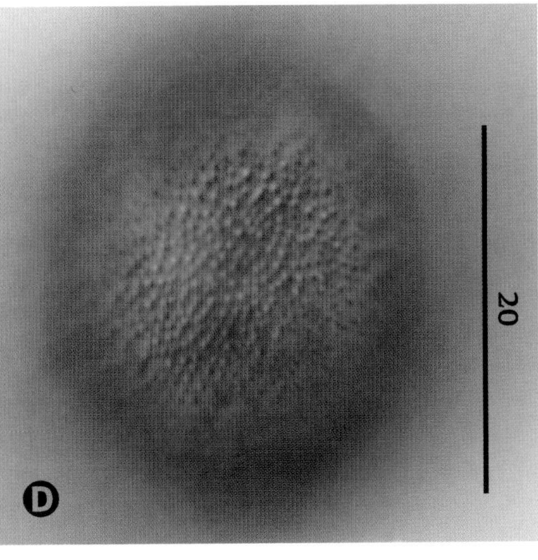

Trachelomonas acanthostoma

Stokes emend. Deflandre 1926

SIZE: 26–36 µm long × 22–28 µm wide.

Broadly ellipsoid lorica with short spines around the flagellar pore; cell showing flagellum, red eyespot, and haplopyrenoids toward the posterior end (A). Detail of protoplast with ovoid paramylon bodies and lorica showing short anterior spines (B). Cell with disc-shaped chloroplasts and haplopyrenoids (arrow) (C). Lorica surface is finely granulated (D).

Trachelomonas allorgei

Deflandre 1926

SIZE: 52–61 µm long × 20–22 µm wide.

Fusiform lorica with long denticulate collar and few short spines; protoplast with flagellum, eyespot, and disc-shaped chloroplasts (A). Detail of lorica surface; collar showing few spines on the margin (B). Cell with chloroplasts and paramylon grains (C).

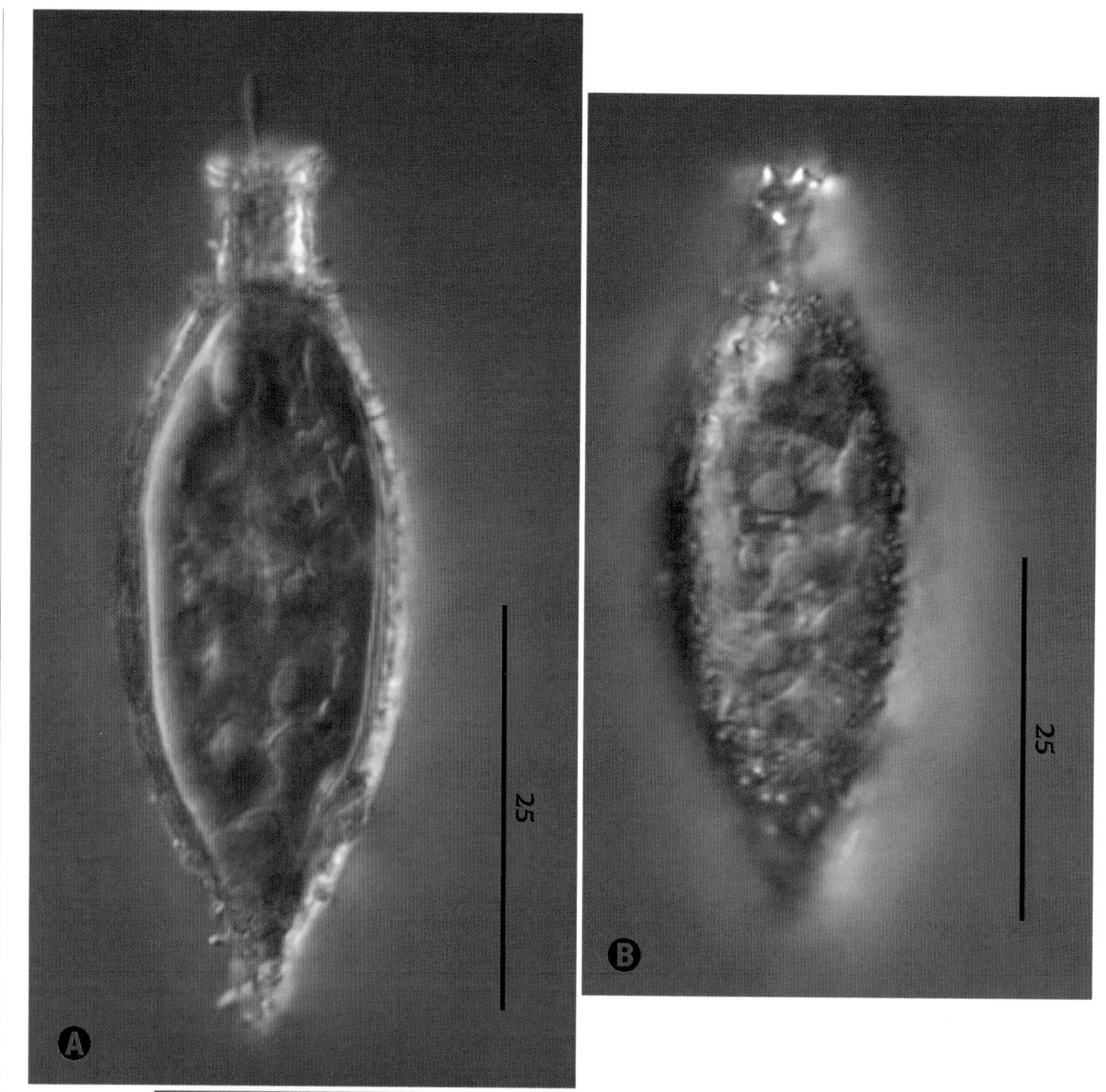

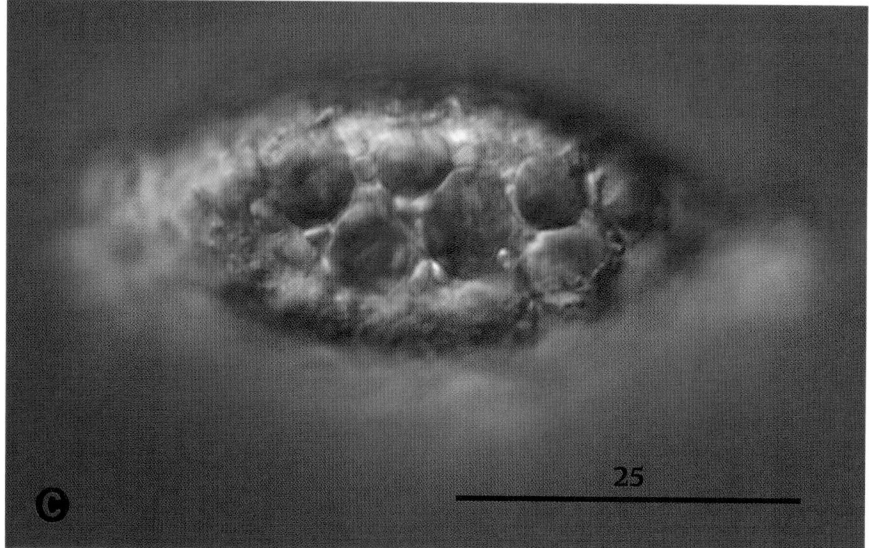

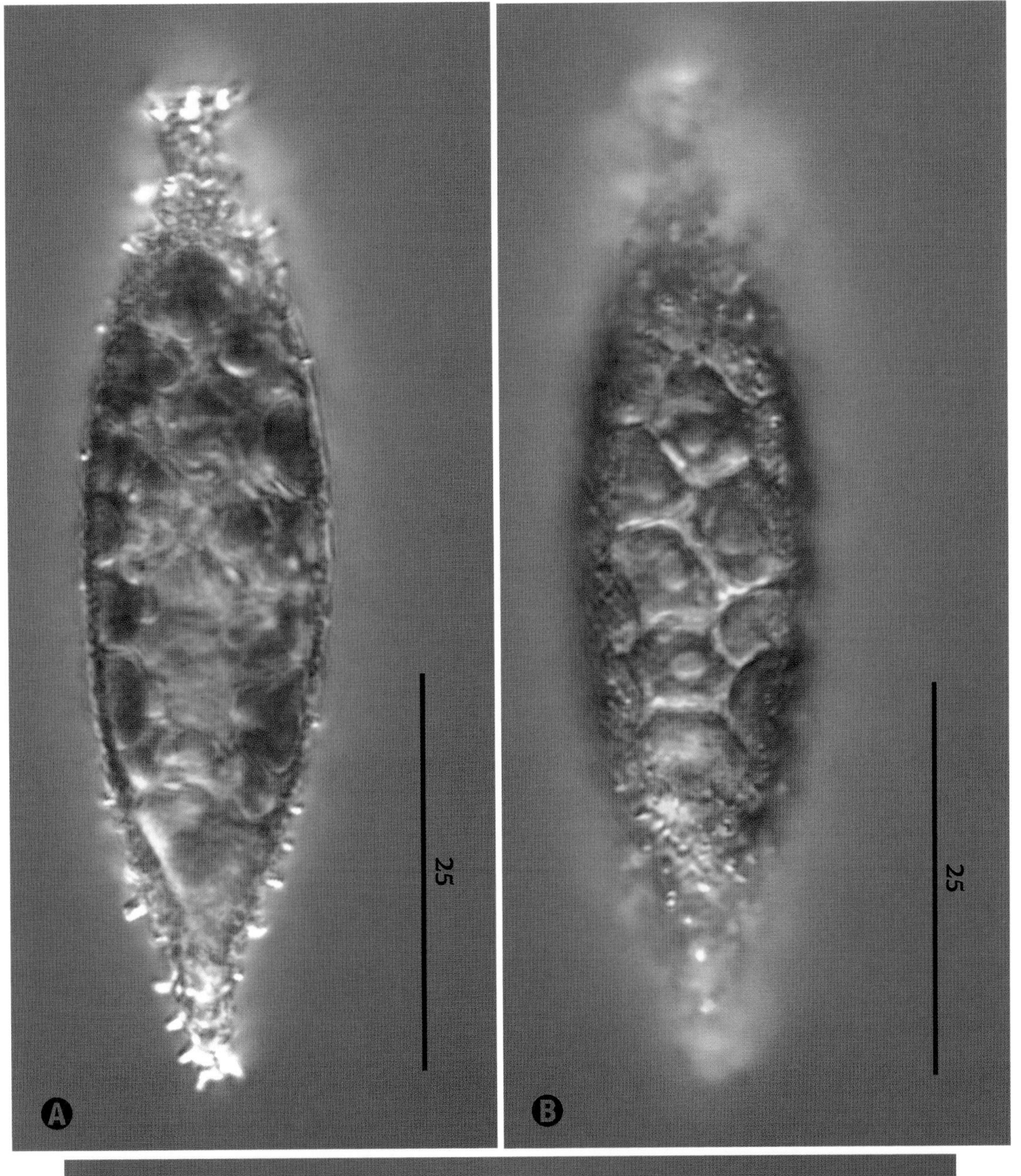

Trachelomonas amphoriformis

Osorio-Tafall 1942

SIZE: 60–63 µm long × 15–18 µm wide.

Fusiform lorica with cylindrical neck; cell with eyespot and chloroplasts (A). Detail of cell showing disc-shaped chloroplasts with haplopyrenoids in face view (B). Protoplast with flagellum, eyespot, and chloroplasts (C).

Trachelomonas armata

(Ehrenberg) Stein 1878

SIZE: 29–54 μm long × 29–39 μm wide.

Ovoid lorica and protoplast with flagellum (about twice body length), red eyespot, and disc-shaped chloroplasts (A). Lorica with long posterior spines and cell with chloroplasts and haplopyrenoids in face view (arrow) (B). Detail of cell showing five large haplopyrenoids (arrow) (C). Lorica with finely punctate surface (D). Cell showing numerous parietal chloroplasts and small paramylon bodies; lorica with dentate flagellar pore (E).

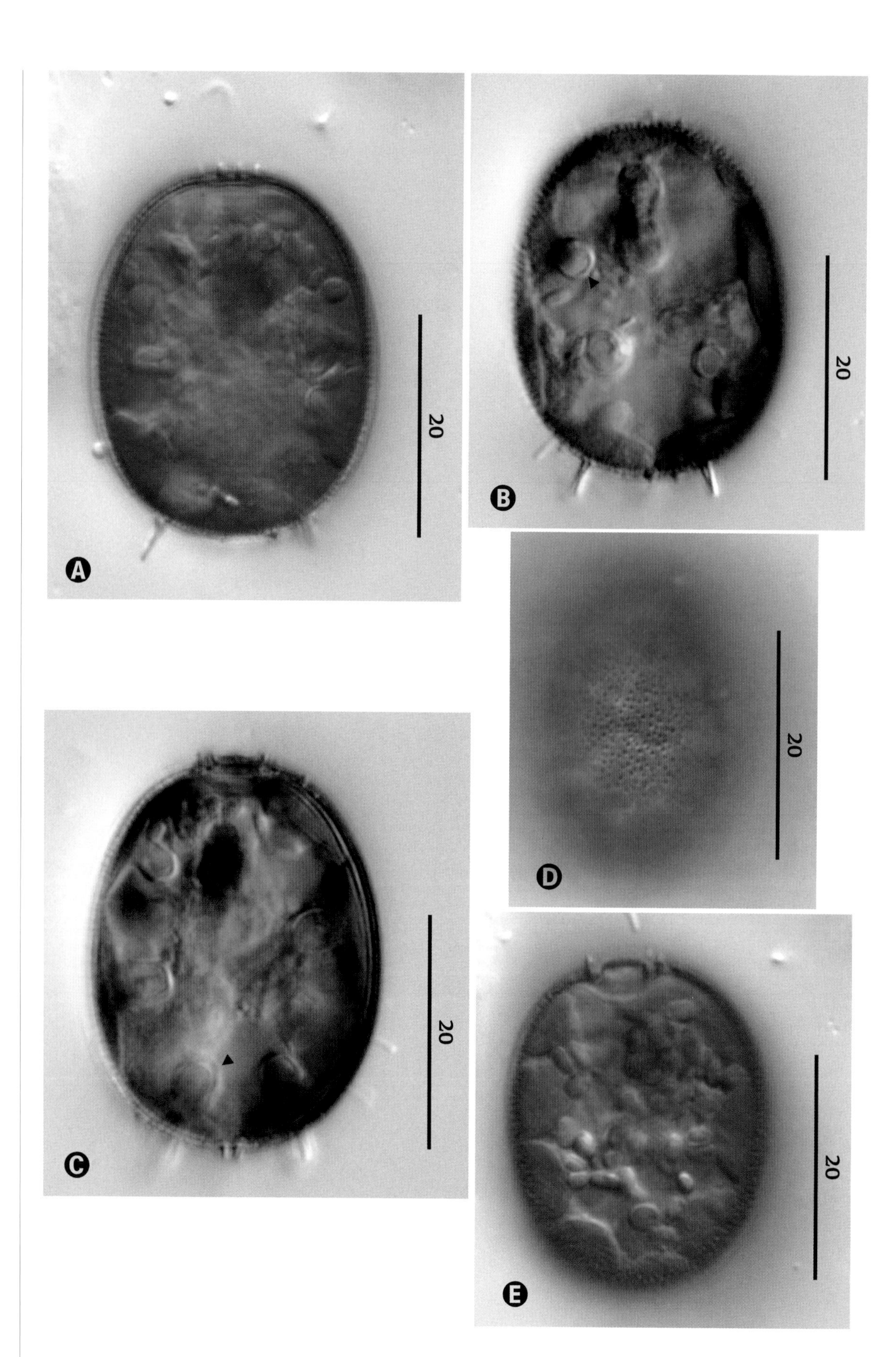

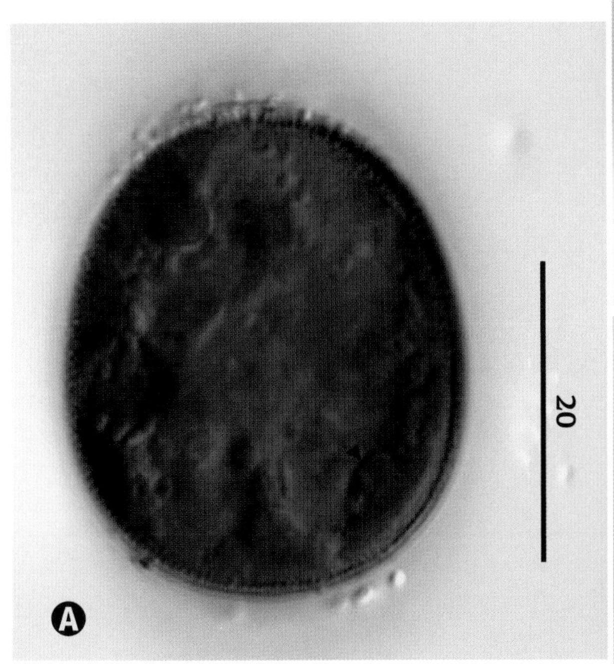

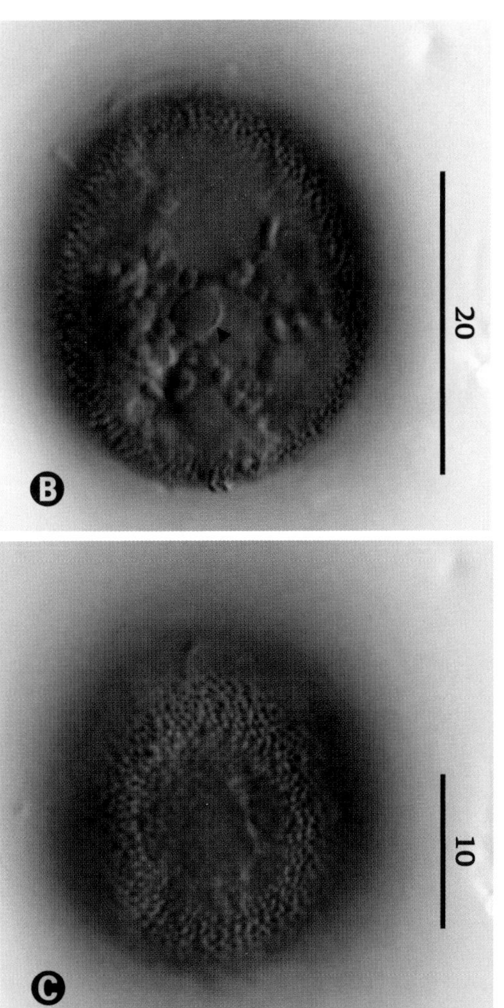

Trachelomonas armata fa. *inevoluta*

Deflandre 1926

SIZE: 29–38 µm long × 26–31 µm wide.

Broadly ovoid lorica; protoplast with red eyespot and haplopyrenoids (arrow) (A). Cell with numerous small paramylon grains, disc-shaped chloroplasts, and haplopyrenoid in face view (arrow) (B). Detail of lorica ornamentation; surface finely punctate (C).

Trachelomonas armata var. *longa* fa. *pseudolongispina*
Deflandre 1926

SIZE: 41–43 µm long × 29–31 µm wide.

Large ovoid lorica bearing short spines around the annular ring and longer spines at the posterior pole; protoplast with red eyespot and disc-shaped chloroplasts (A). Cell with chloroplasts, each with a haplopyrenoid (arrows); several small ovoid paramylon bodies are present (B, C). The region of lorica not bearing spines is densely granulated (D).

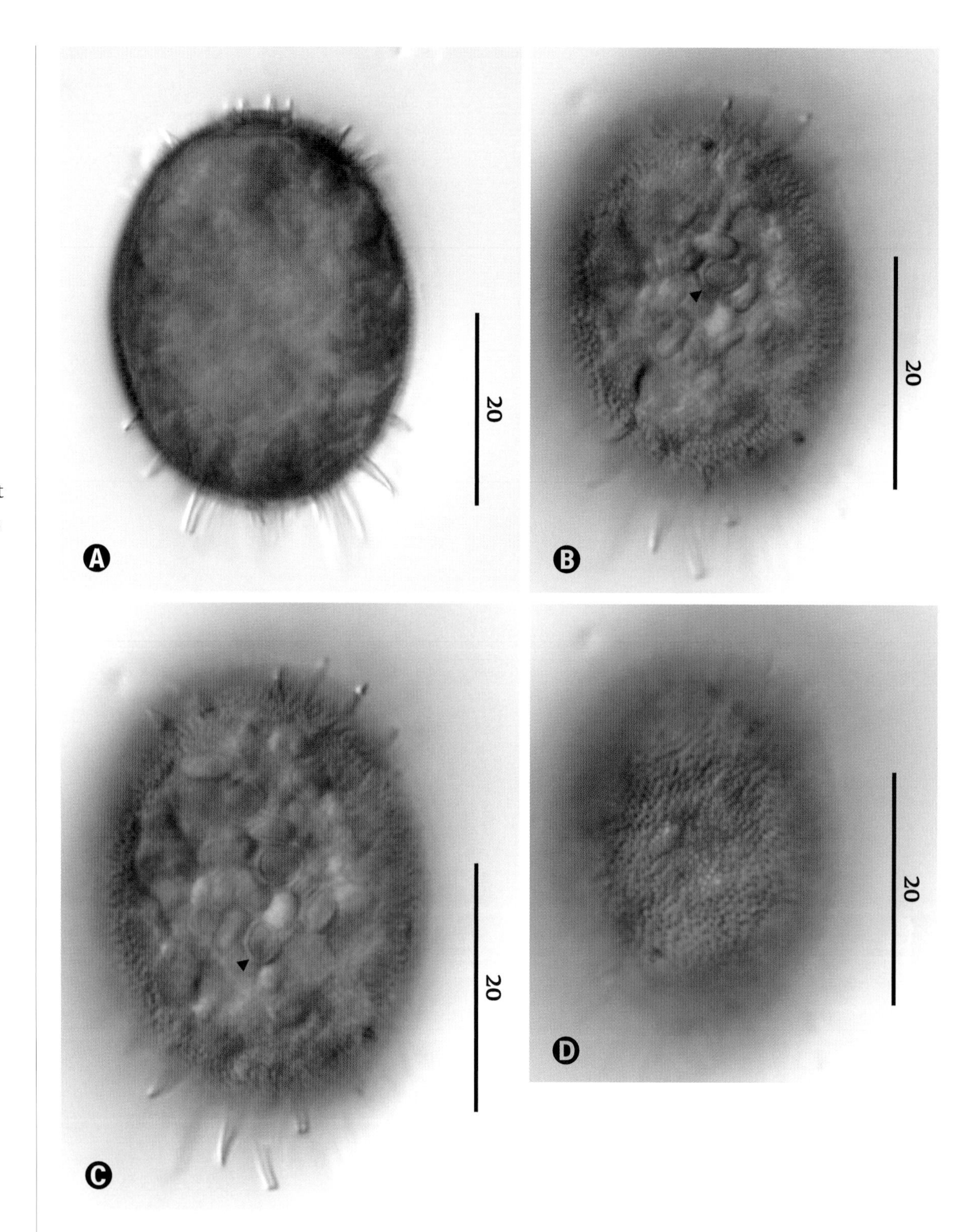

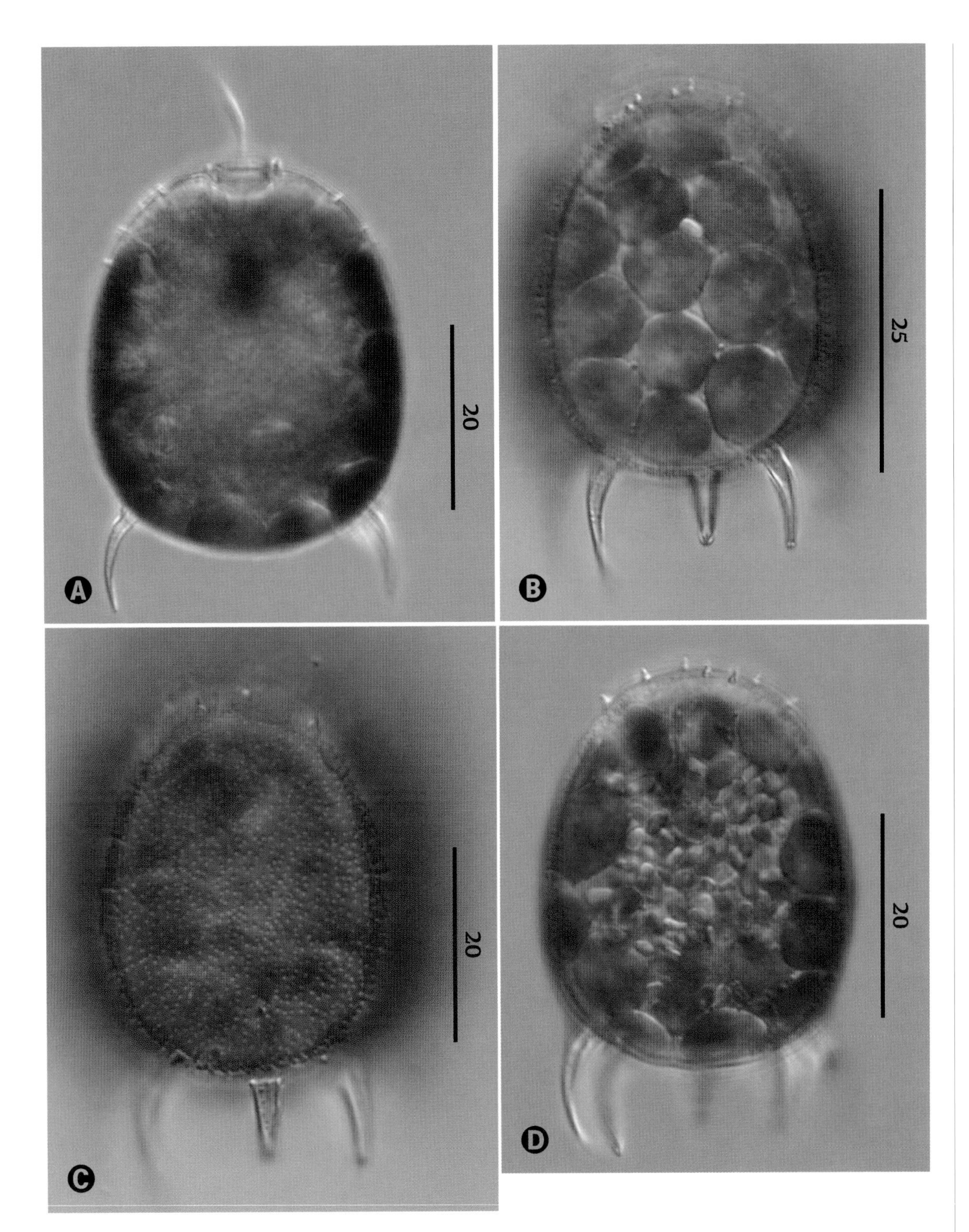

Trachelomonas armata var. *steinii*

Lemmermann 1905

SIZE: 33–45 µm long × 28–38 µm wide.

Cell showing flagellum, flagellar pore, eyespot, and numerous disc-shaped chloroplasts; broadly ovoid lorica with short spines at the anterior pole and long thick spines at the posterior (A). Cell with numerous chloroplasts, lorica with very short anterior spines, and long thick posterior spines (B). Detail of lorica surface with pores (C). Lorica and protoplast with chloroplasts and numerous small paramylon grains (D).

Trachelomonas bulla

Stein ex Deflandre 1926

SIZE: 36–48 µm long × 20–24 µm wide.

Lorica is ellipsoidal with a long collar bearing serrate margins (A). A broken lorica showing collar; cell anterior is positioned at the base of the lorica with red eyespot and disc-shaped chloroplasts (B). Detail of lorica surface and protoplast showing eyespot and paramylon bodies (C). View of cell surface showing fine pellicle strips, more than a dozen large chloroplasts without pyrenoids, and several paramylon bodies (D). Lorica surface with irregular short spines and pellicle with spirally disposed striations (E). Flagellum is longer than lorica.

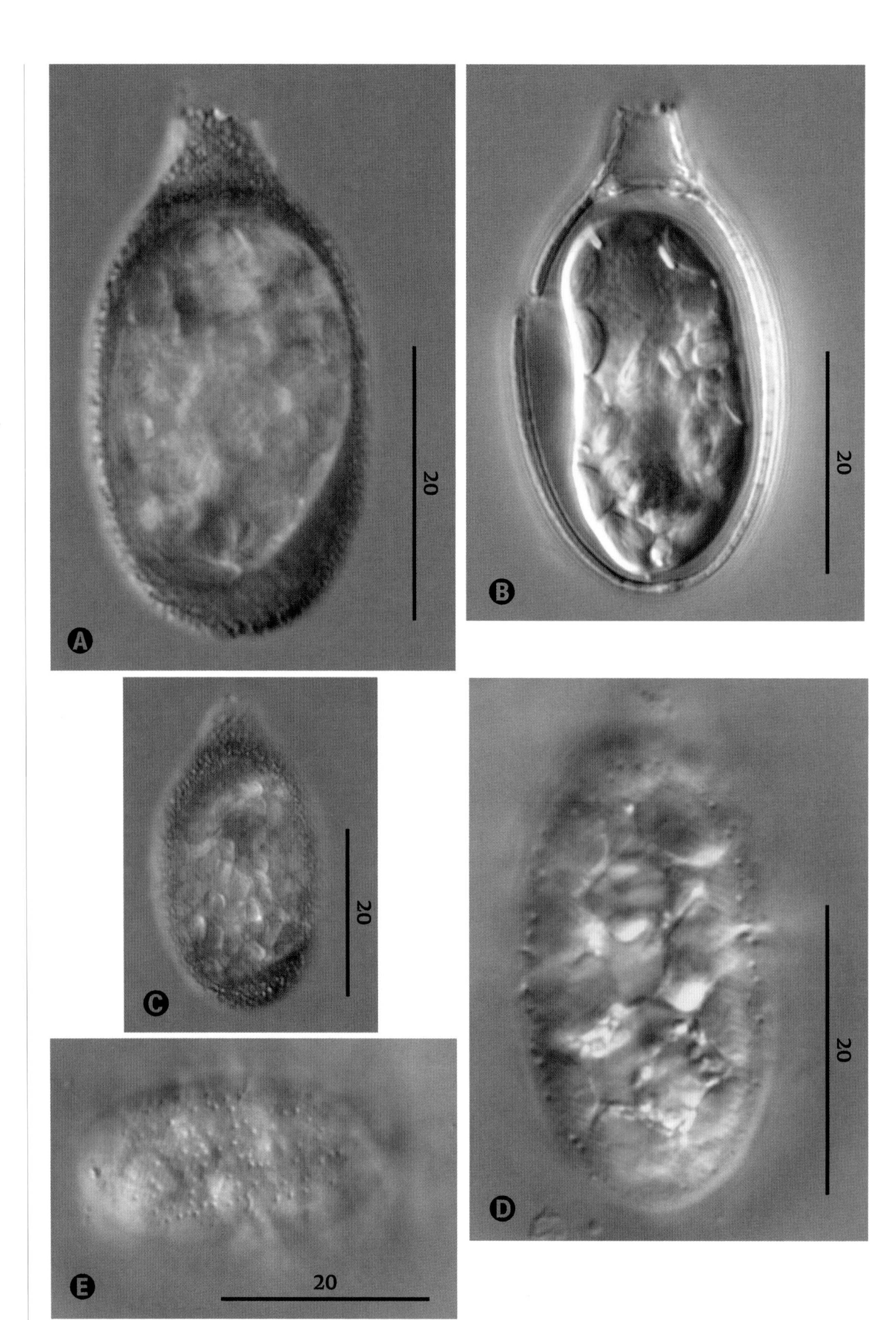

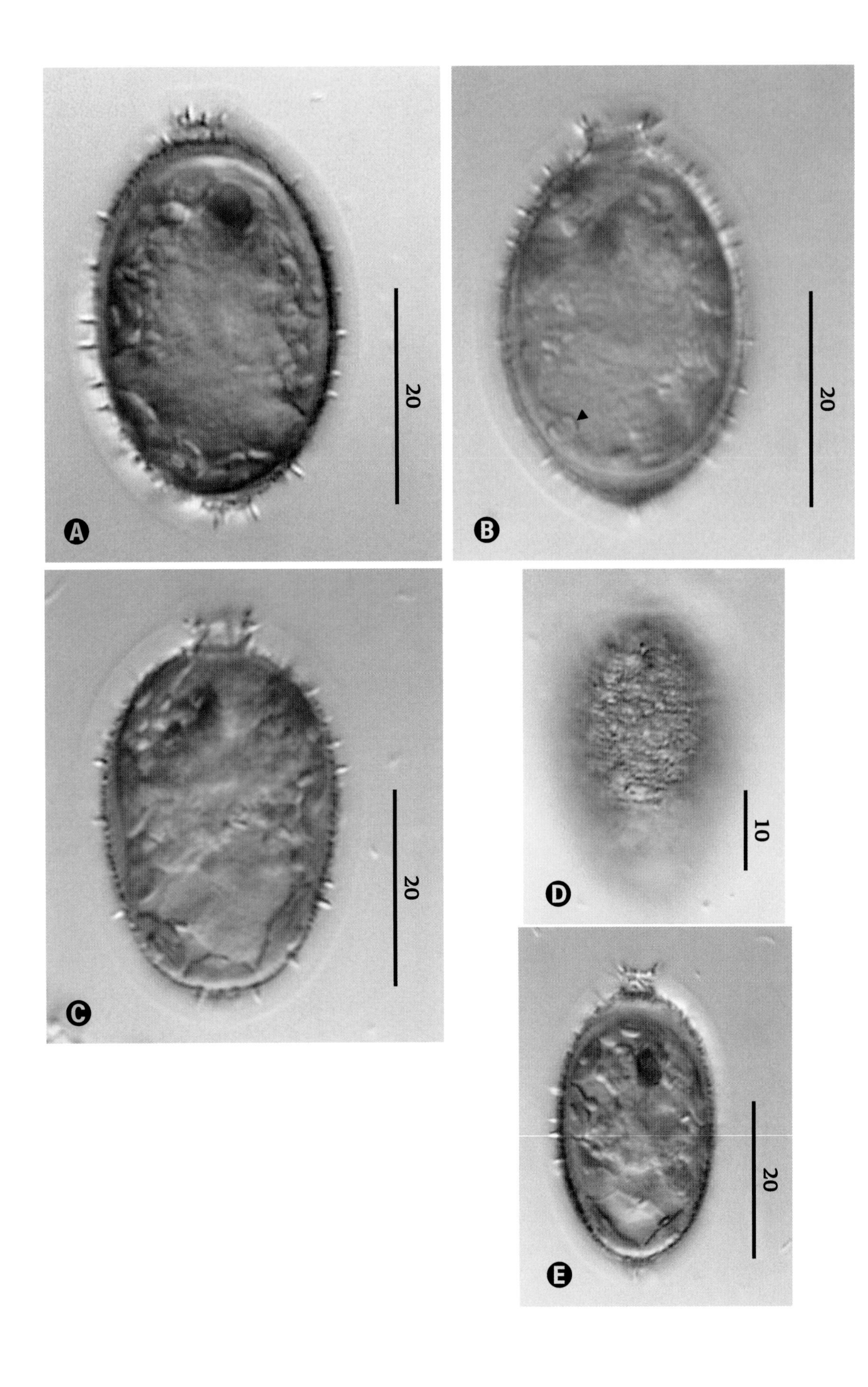

Trachelomonas cingulata

Singh 1956

SIZE: 32–36 µm long × 21–23 µm wide.

Ellipsoidal lorica with denticulate collar and button-like thickening at the posterior end; short sparsely disposed spines adorn the lorica; protoplast shows red eyespot, parietal disc-shaped chloroplasts, and small paramylon bodies (A, C, E). Chloroplast with diplopyrenoid (arrow) (B). Lorica surface is irregularly granulated (D). Flagellum is three to four times body length.

Trachelomonas conradi

Skvortzov 1925

SIZE: 15–23 µm long × 14–18 µm wide.

Broadly ellipsoid lorica; cell with red eyespot (A) and plate-like chloroplasts with haplopyrenoids (arrow) (A, B). Lorica surface showing spirally arranged ridges (C). Flagellum is two to three times body length.

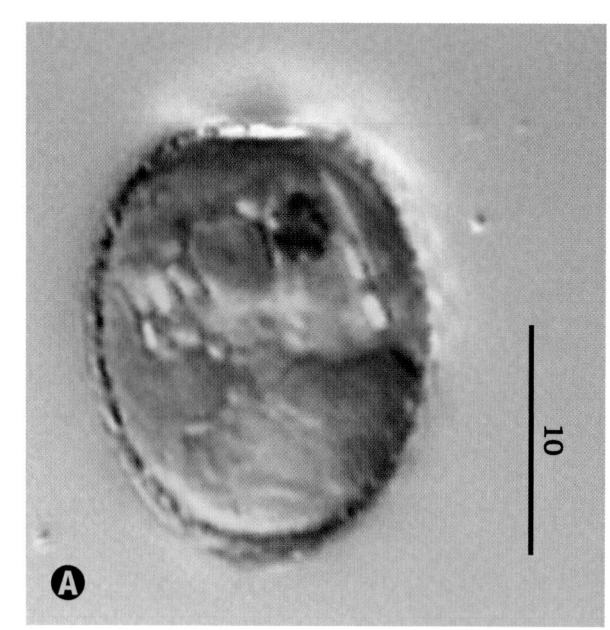

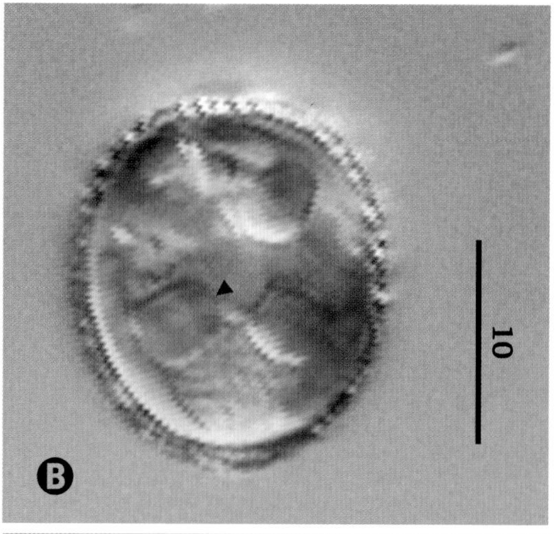

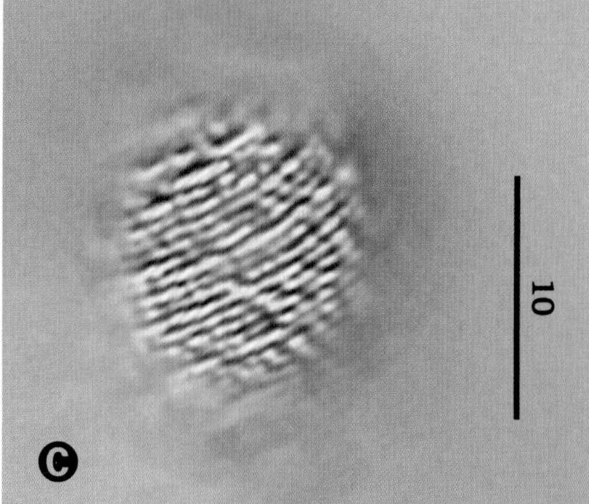

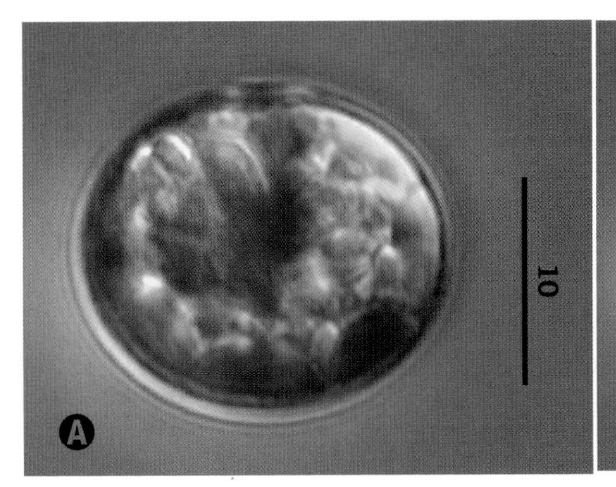

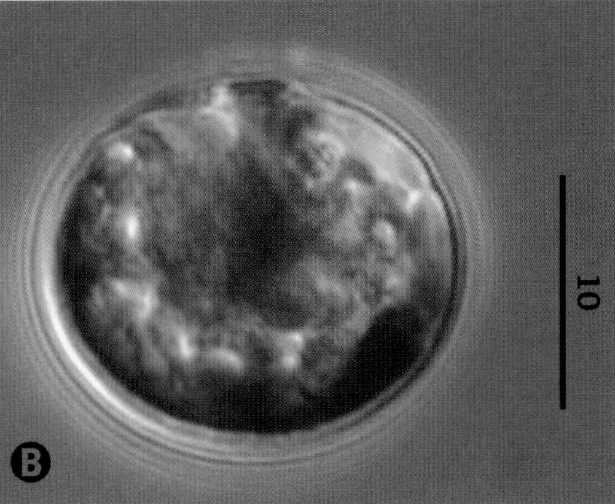

Trachelomonas curta

Da Cunha emend. Deflandre 1927

SIZE: 20–29 µm long × 23–31 µm wide;
smaller forms 17–18 µm long × 20–21 µm wide.

Subspherical lorica, wider than long, disc-shaped chloroplasts and paramylon bodies (A). Same cell in a different focal plane; flagellar opening with slim annular thickening, red eyespot, and chloroplasts (B). Flagellum is about twice body length.

Trachelomonas ellipsoidalis

Singh 1956

SIZE: 22–25 µm long × 15–18 µm wide.

Ellipsoid lorica, protoplast with flagellum (two to two-and-a-half times body length), eyespot, parietal plate-like chloroplasts with haplopyrenoids, and small paramylon bodies (A). Cell showing red eyespot and chloroplasts with haplopyrenoids (arrow) (B). Lorica surface, finely granulated (C).

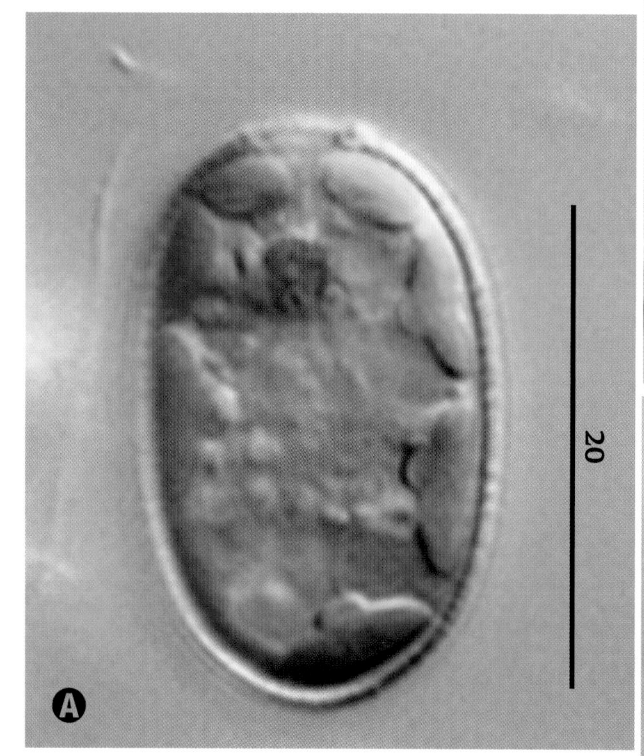

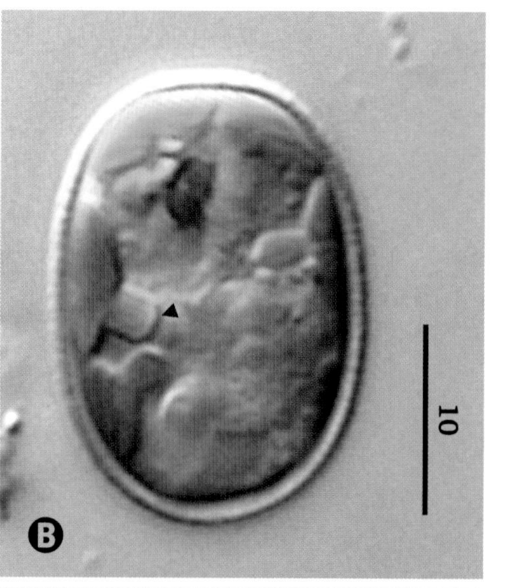

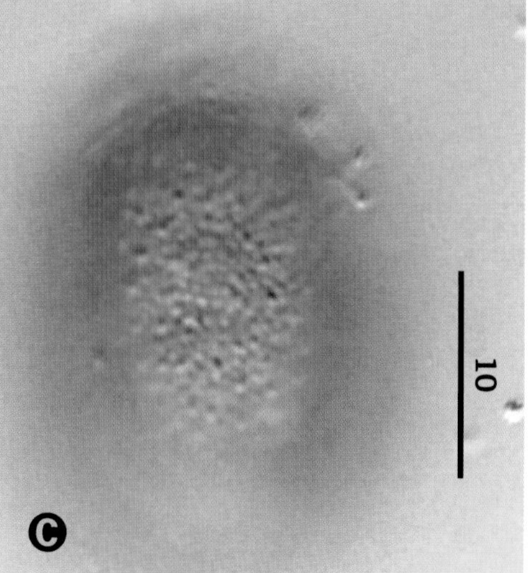

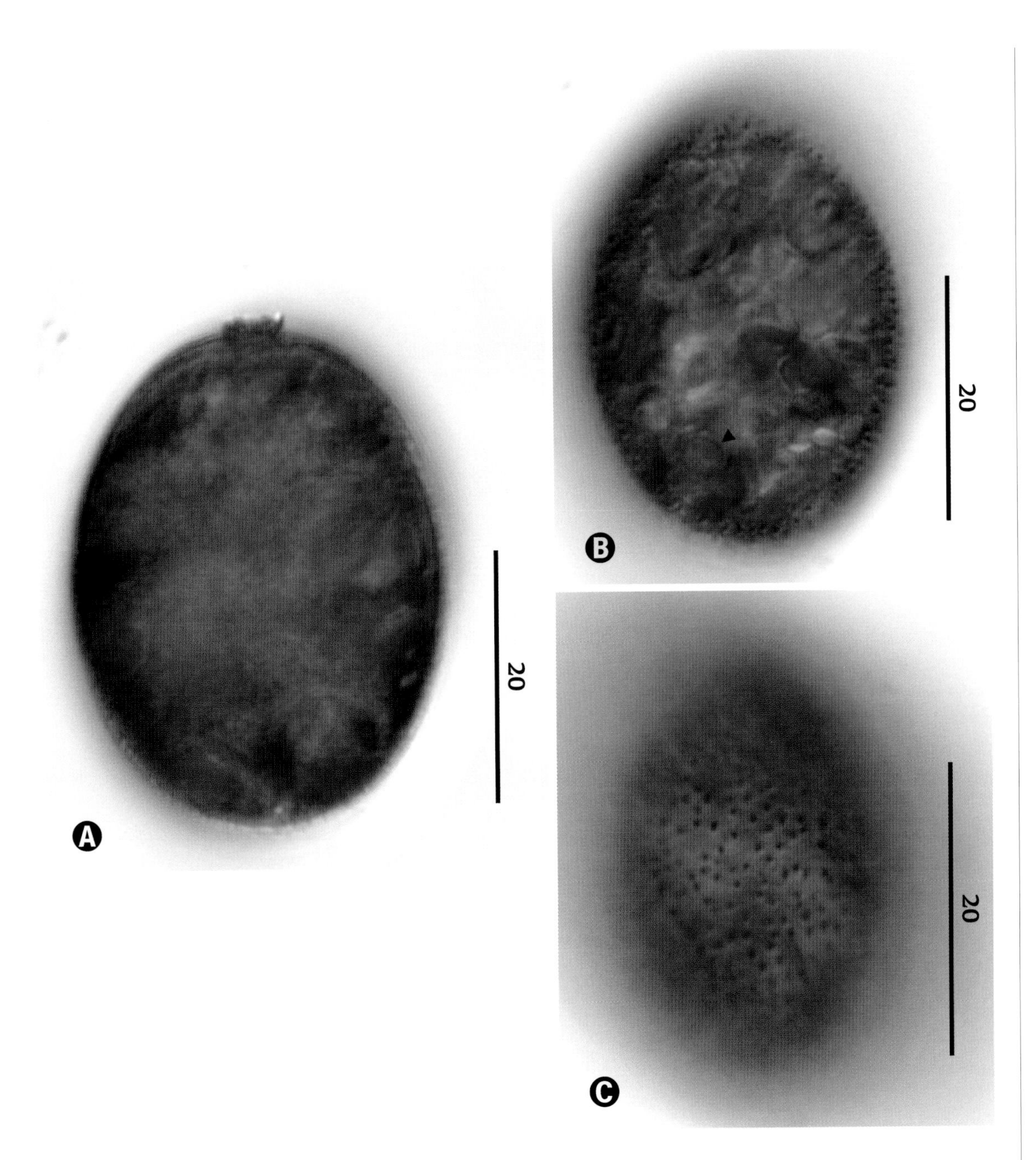

Trachelomonas grandis

Singh 1956

SIZE: 30–43 μm long × 25–32 μm wide.

Large ellipsoidal lorica with short narrow collar (A). Cell showing chloroplasts with haplopyrenoids (arrow) and several small ovoid paramylon bodies (B). Lorica surface showing short irregularly arranged granules (C). Flagellum is four to five times body length.

Trachelomonas granulosa

Playfair 1915

SIZE: 17–26 µm long × 13–22 µm wide.

Ellipsoid lorica; cell showing eyespot and disc-shaped chloroplasts with haplopyrenoids (A, B). Chloroplast and haplopyrenoid in face view (C). Lorica surface showing dense, fine granules (D).

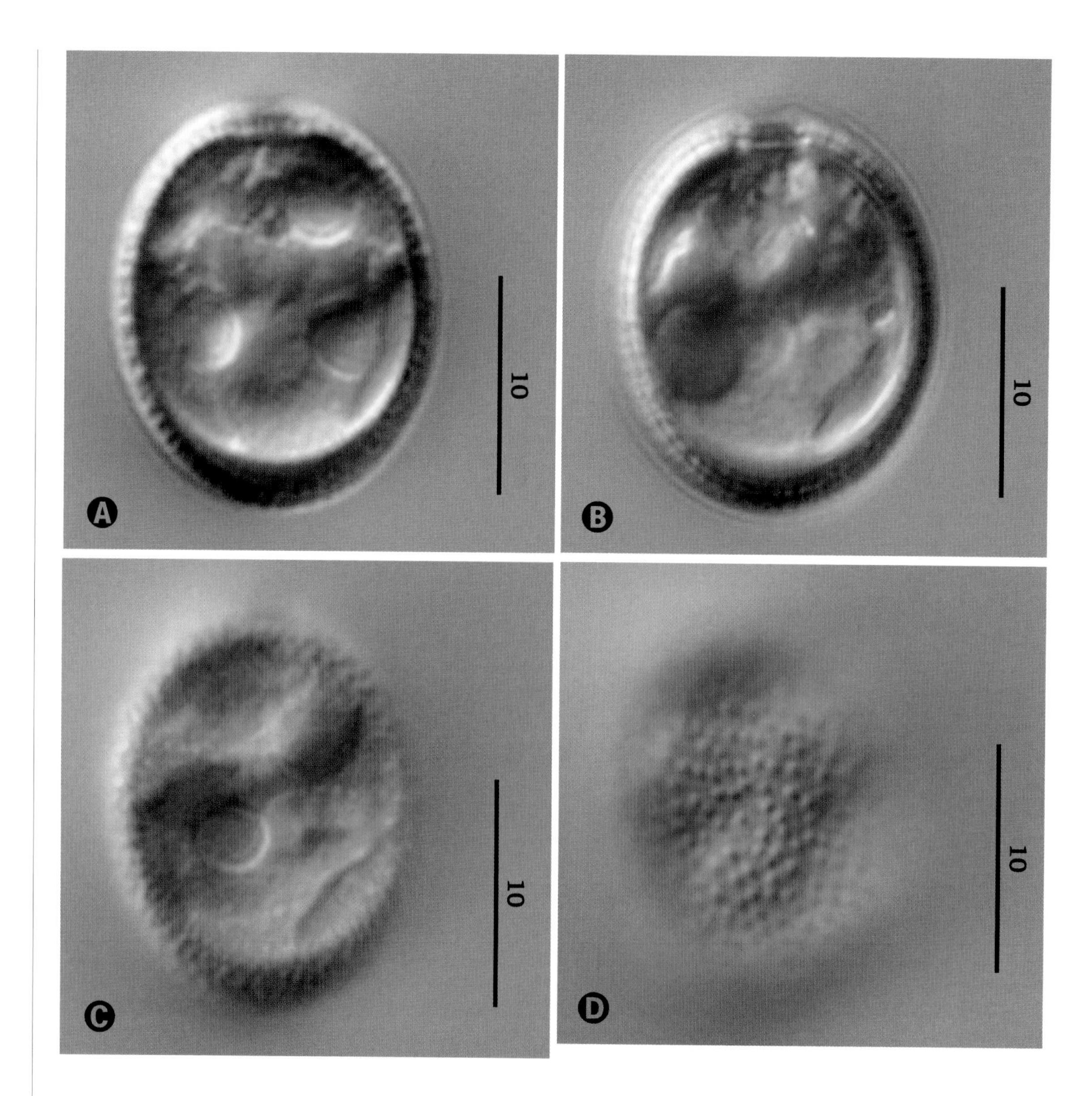

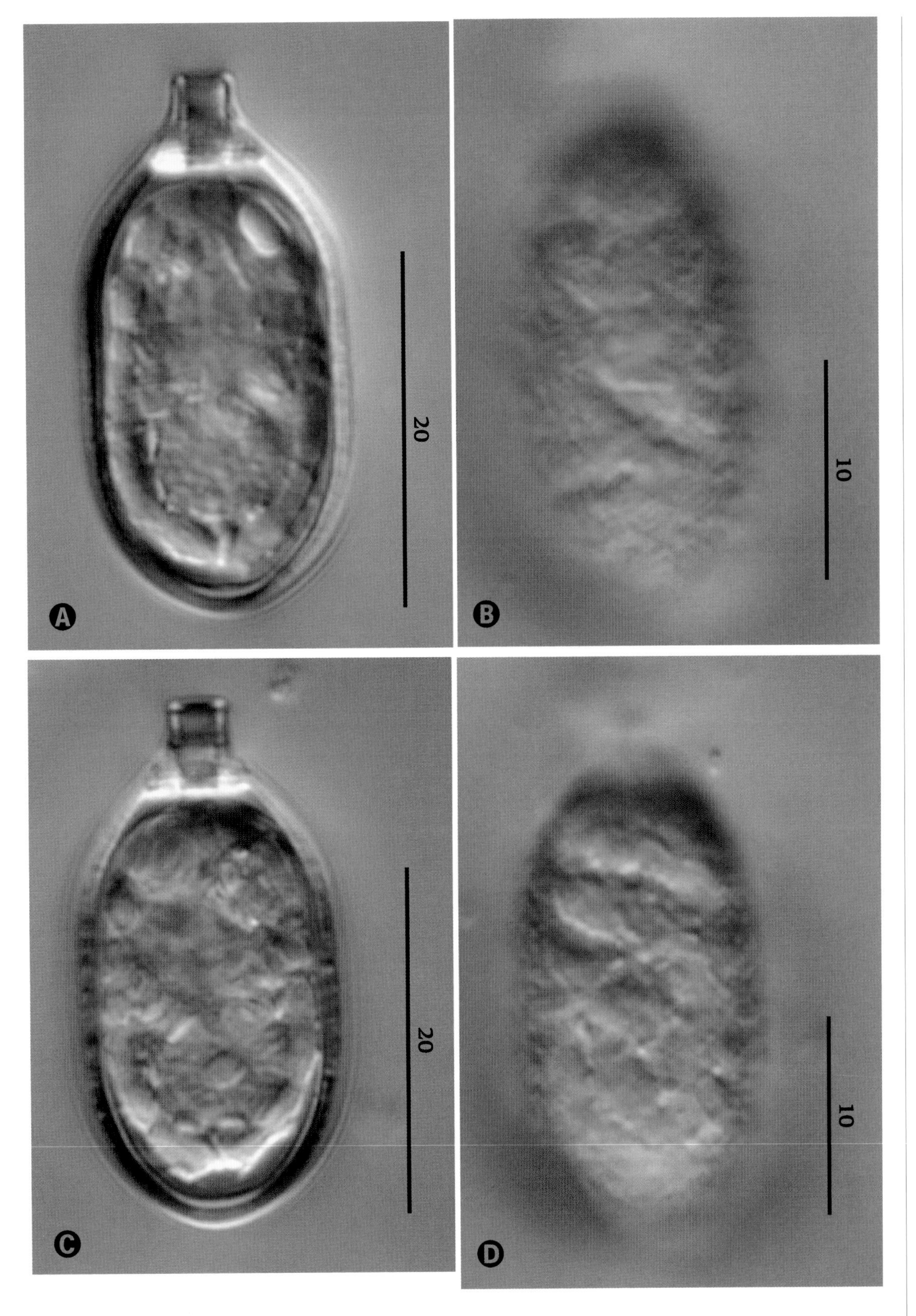

Trachelomonas hexangulata

(Swirenko) Playfair 1915

SIZE: 24–34 µm long × 10–16 µm wide.

Elongate lorica with prominent collar; cell with disc-shaped chloroplasts and paramylon grains (A, C). Detail of smooth lorica and underlying chloroplasts (B, D).

Trachelomonas hispida

(Perty) Stein emend. Deflandre 1926

SIZE: 20–42 µm long × 15–26 µm wide.

Ellipsoid lorica, red eyespot, and disc-shaped chloroplasts (A). Chloroplasts, each with a large diplopyrenoid (arrow) (B). Lorica surface, densely ornamented (C). Flagellum is one-and-a-half to two times body length.

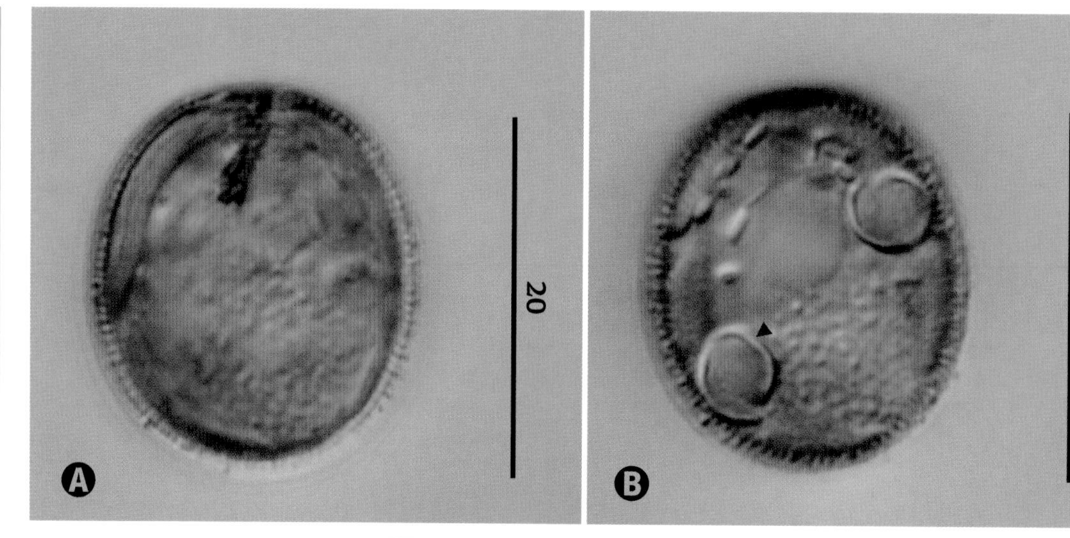

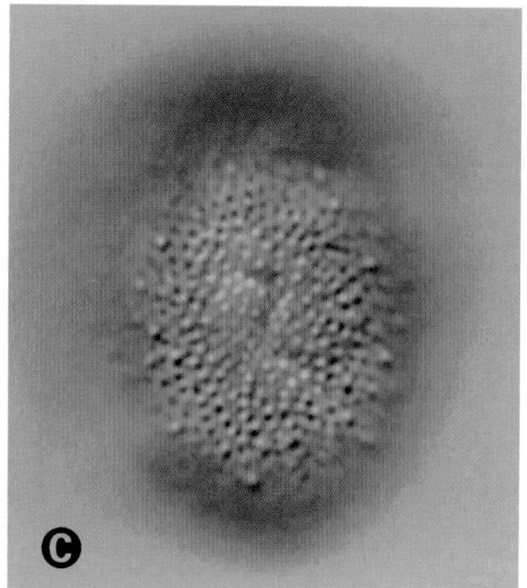

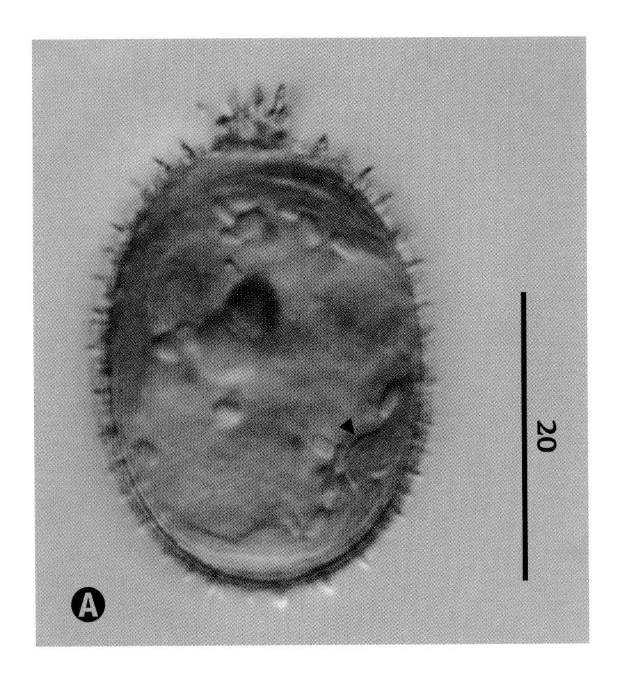

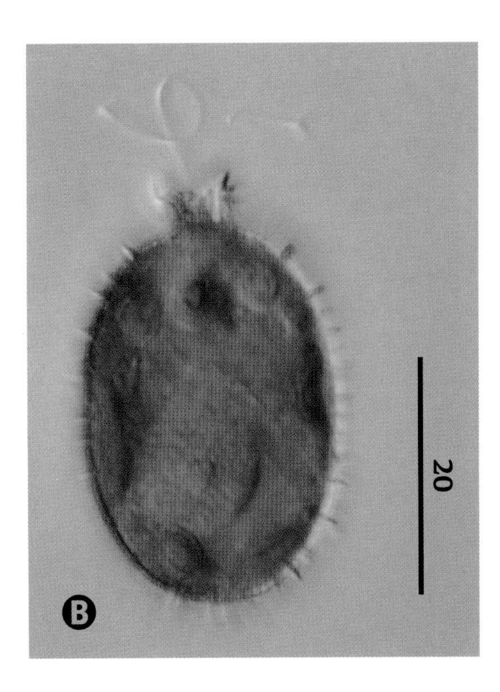

Trachelomonas hispida var. *coronata*

Lemmermann 1913

SIZE: 29–40 µm long × 19–21 µm wide.

Ellipsoid lorica ornamented with short spines and with spiny flagellar collar; red eyespot, small paramylon bodies, and disc-shaped chloroplasts with diplopyrenoids (arrow) (A). Lorica and protoplast with flagellum, red eyespot, chloroplasts, and central nucleus (B).

Trachelomonas hispida var. *duplex*

Deflandre 1926

SIZE: 28–33 µm long × 23–25 µm wide.

Ellipsoidal lorica, exhibiting short spines at both poles, and protoplast with red eyespot (Ⓐ). Cell with disc-shaped chloroplasts and small paramylon grains (Ⓑ, Ⓓ). Lorica surface is densely granulated (Ⓒ).

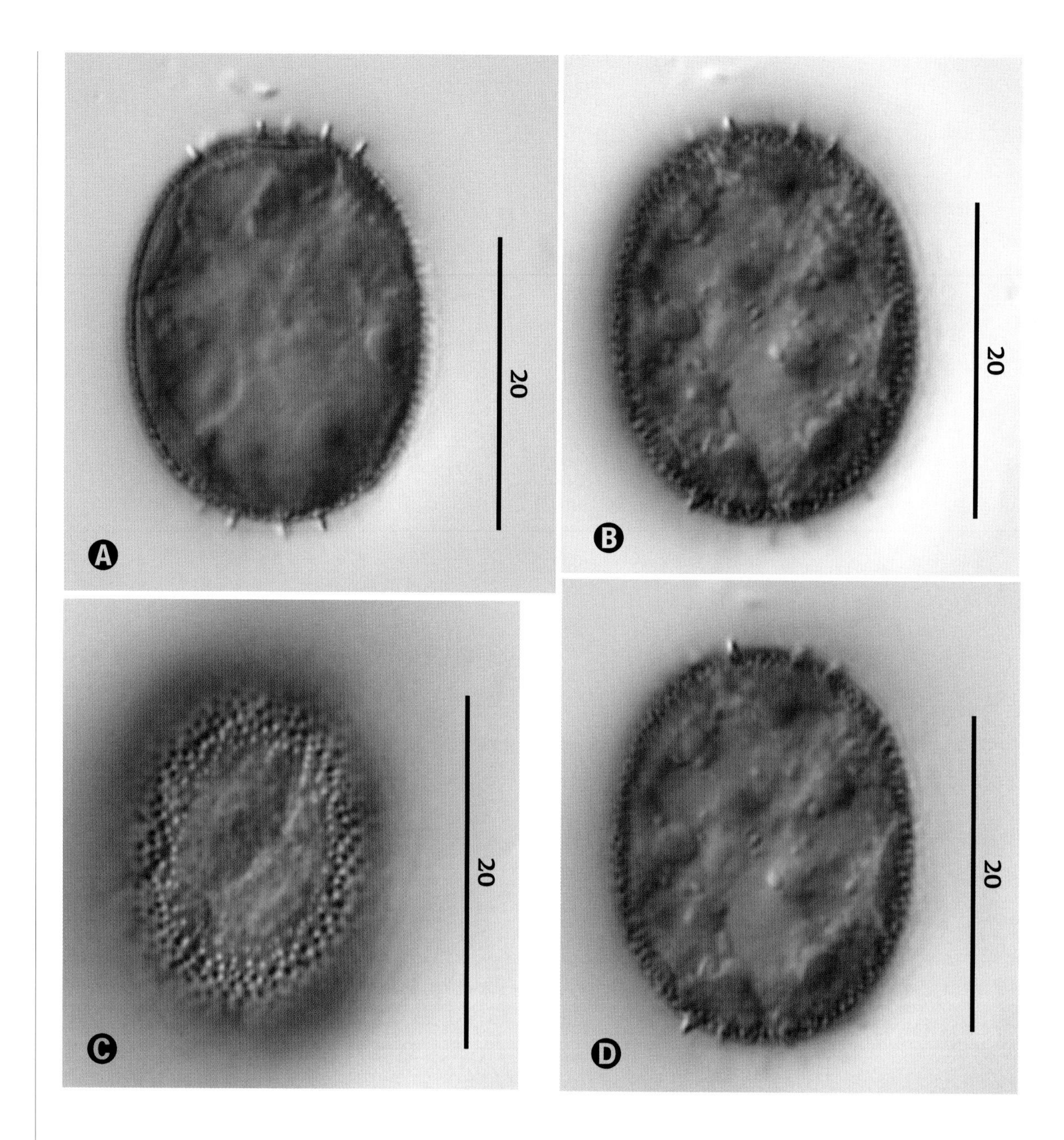

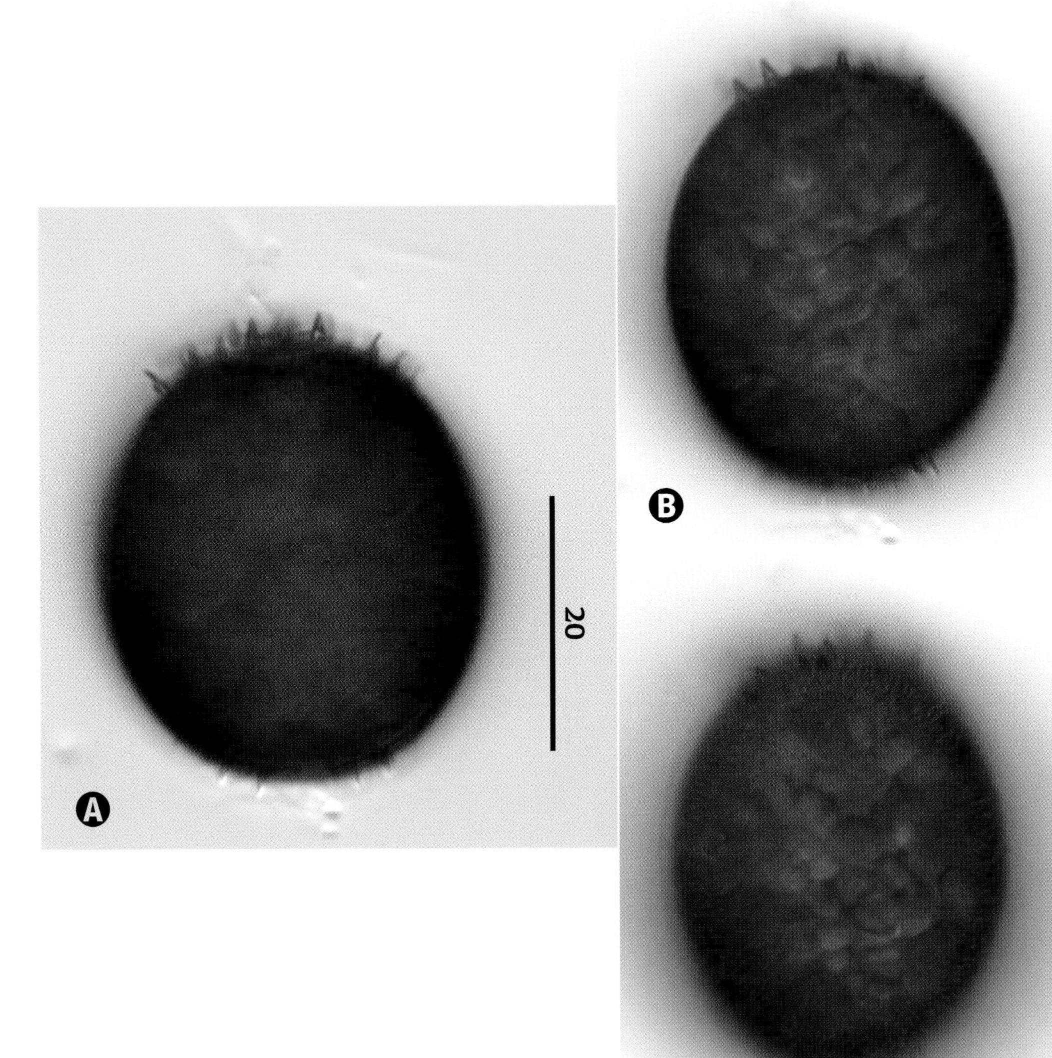

Trachelomonas kellogii

Skvortzov emend. Deflandre 1926

SIZE: 34–39 µm long × 31–35 µm wide.

Ovoid lorica with spines at both anterior and posterior poles; protoplast with flagellum (A). Details of lorica and cell with disc-shaped chloroplasts and ovoid paramylon bodies (B, C).

Trachelomonas lefèvrei

Deflandre 1926

SIZE: 27–31 µm long × 22–24 µm wide.

Cylindrical to ellipsoidal loricas with collar; cells with red eyespots; plate-like chloroplasts with haplopyrenoids (Ⓐ, Ⓑ—arrow, Ⓒ).

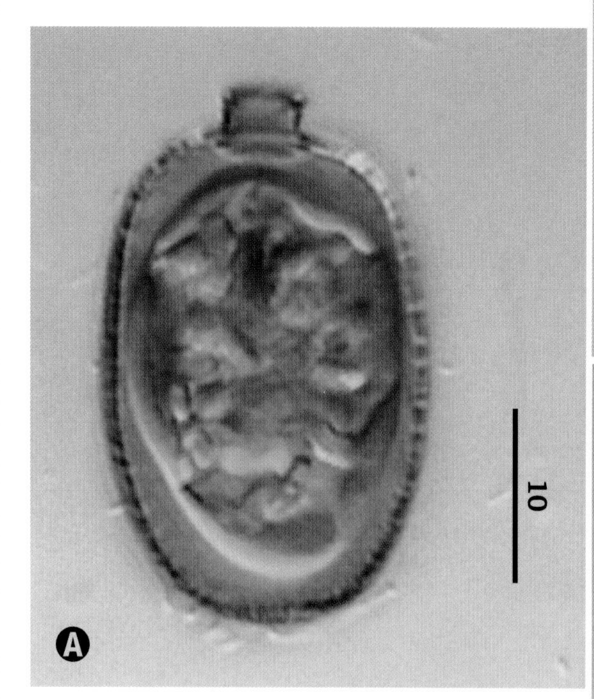

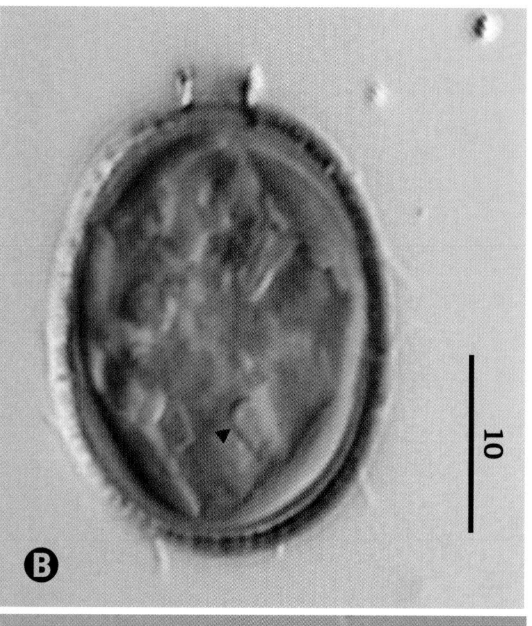

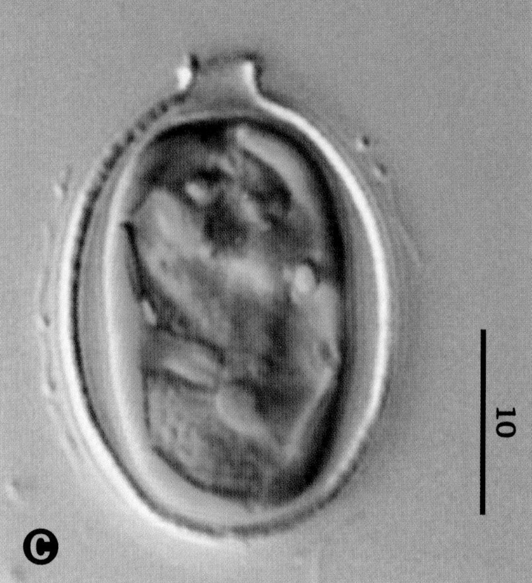

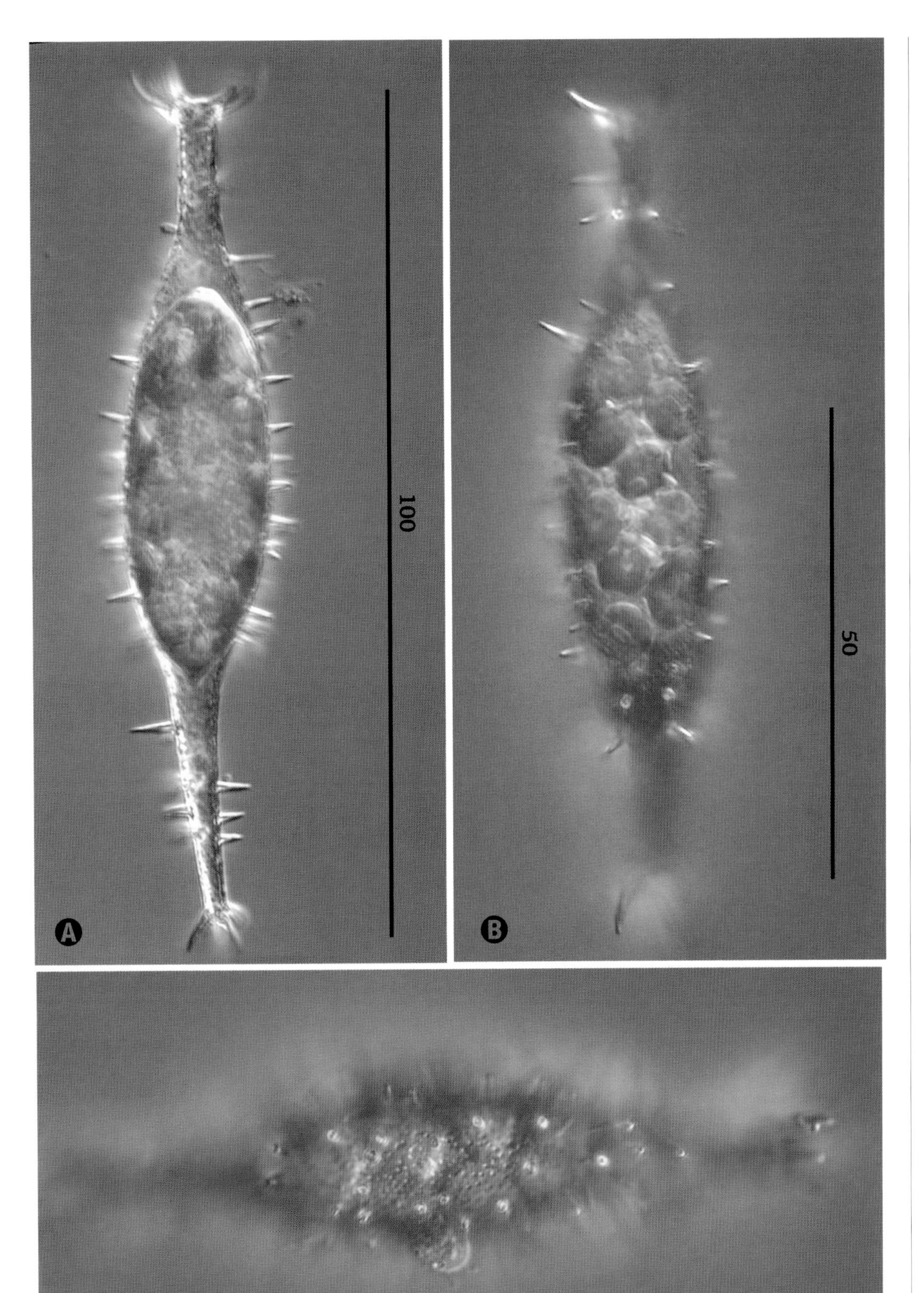

Trachelomonas magdaleniana

Deflandre 1926

SIZE: 81–106 µm long × 16–20 µm wide.

Elongate, spindle-shaped lorica with long spines; cell showing eyespot (A). Numerous parietal disc-shaped chloroplasts visible (B). Lorica surface is finely punctate (C).

Trachelomonas obtusa

Palmer 1925

SIZE: 27–33 µm long × 15–16 µm wide.

Cell showing flagellum, red eyespot, and large disc-shaped chloroplasts; cylindrical lorica with conical end and ornamented with short spines (A). Detail of lorica surface (B, C) and chloroplasts (B).

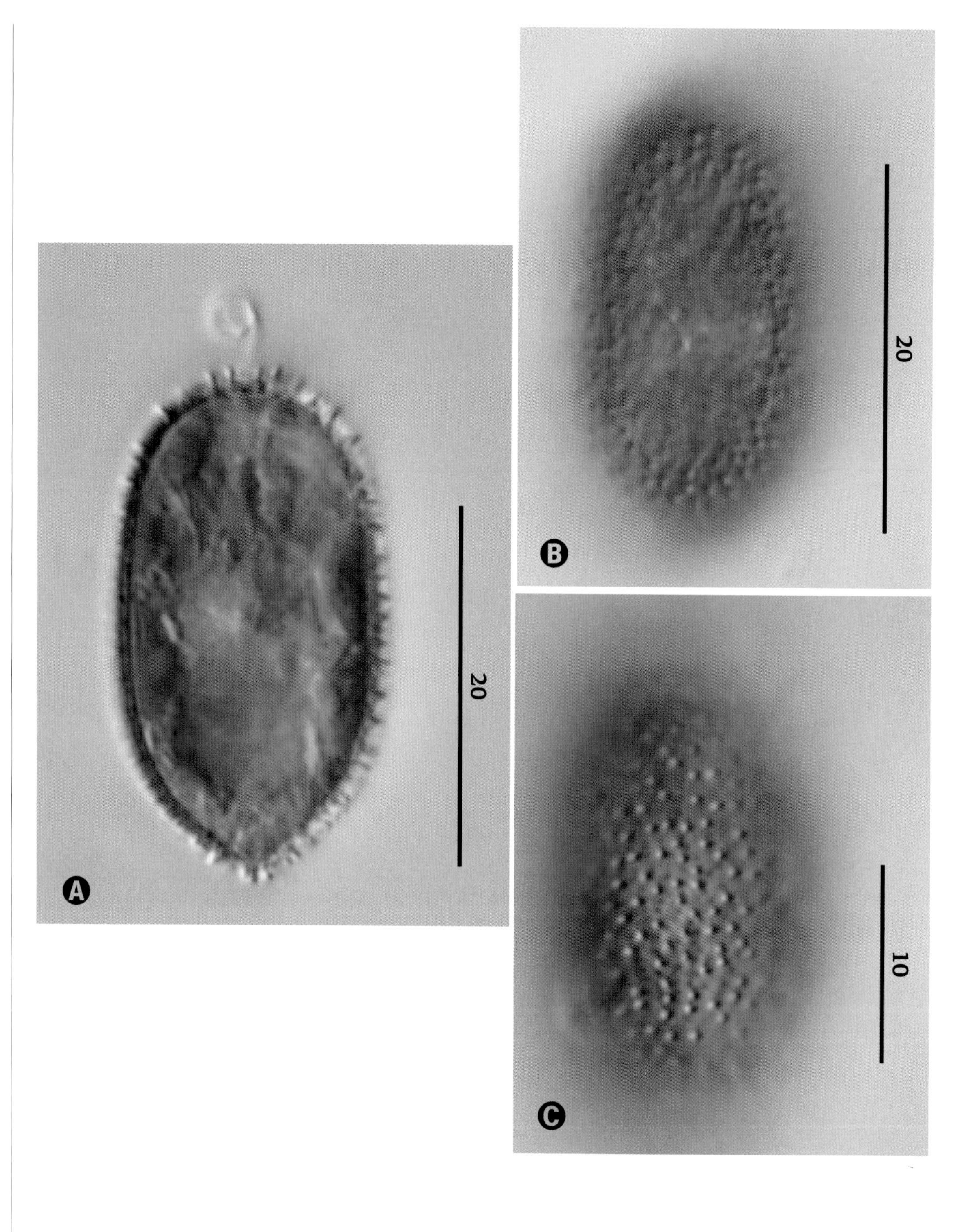

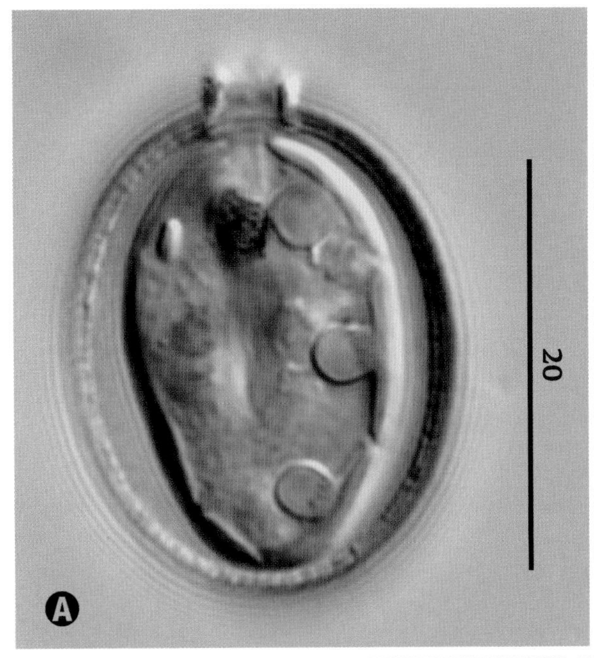

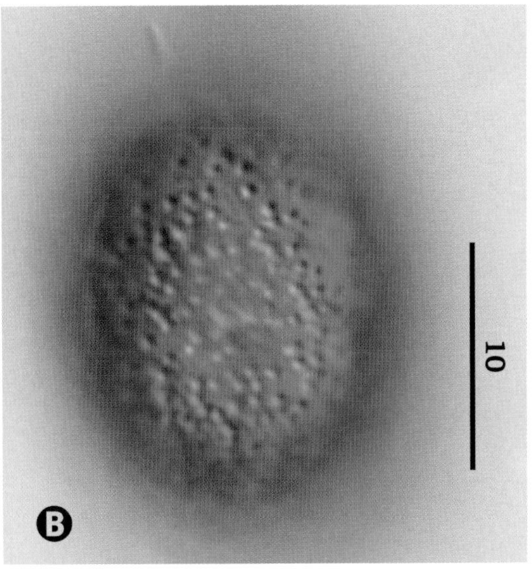

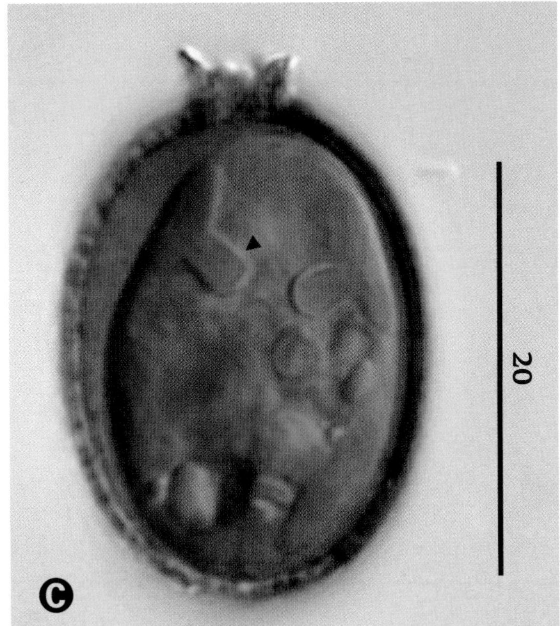

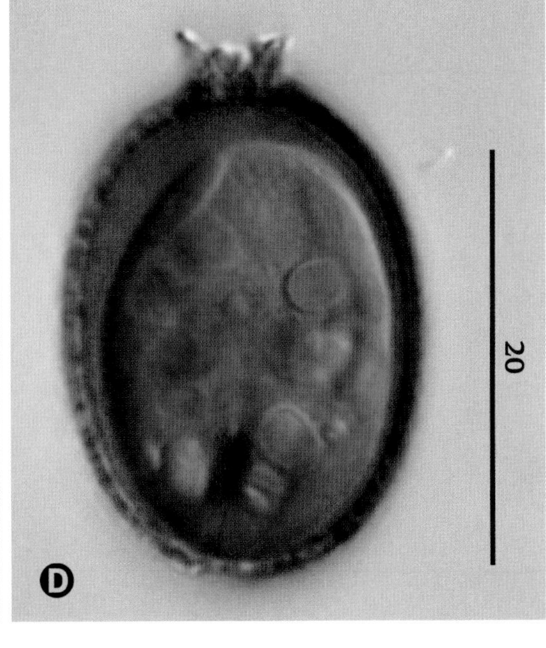

Trachelomonas planctonica

Swirenko 1914

SIZE: 19–30 µm long × 17–22 µm wide.

Ellipsoidal lorica with short collar; protoplast showing one flagellar base in reservoir near eyespot, disc-shaped chloroplasts with haplopyrenoids, and one short paramylon rod (A). Detail of porous lorica surface (B). Lorica and cell in two different focal views showing dentate collar; rotated cell (due to metaboly) with red eyespot, chloroplasts with haplopyrenoids (arrow) (C), and paramylon bodies (C, D).

Trachelomonas raciborskii

Woloszyńska 1912

SIZE: 28–40 µm long × 25–31 µm wide.

Ellipsoid lorica with longer spines located at both anterior and posterior ends and shorter spines over the rest of lorica surface; cell shows red eyespot and disc-shaped chloroplasts, each with a diplopyrenoid (Ⓐ, Ⓑ). Chloroplasts with diplopyrenoids (arrow) and small ovoid paramylon grains (Ⓒ). Lorica surface ornamented in the median region with very short spines (Ⓓ).

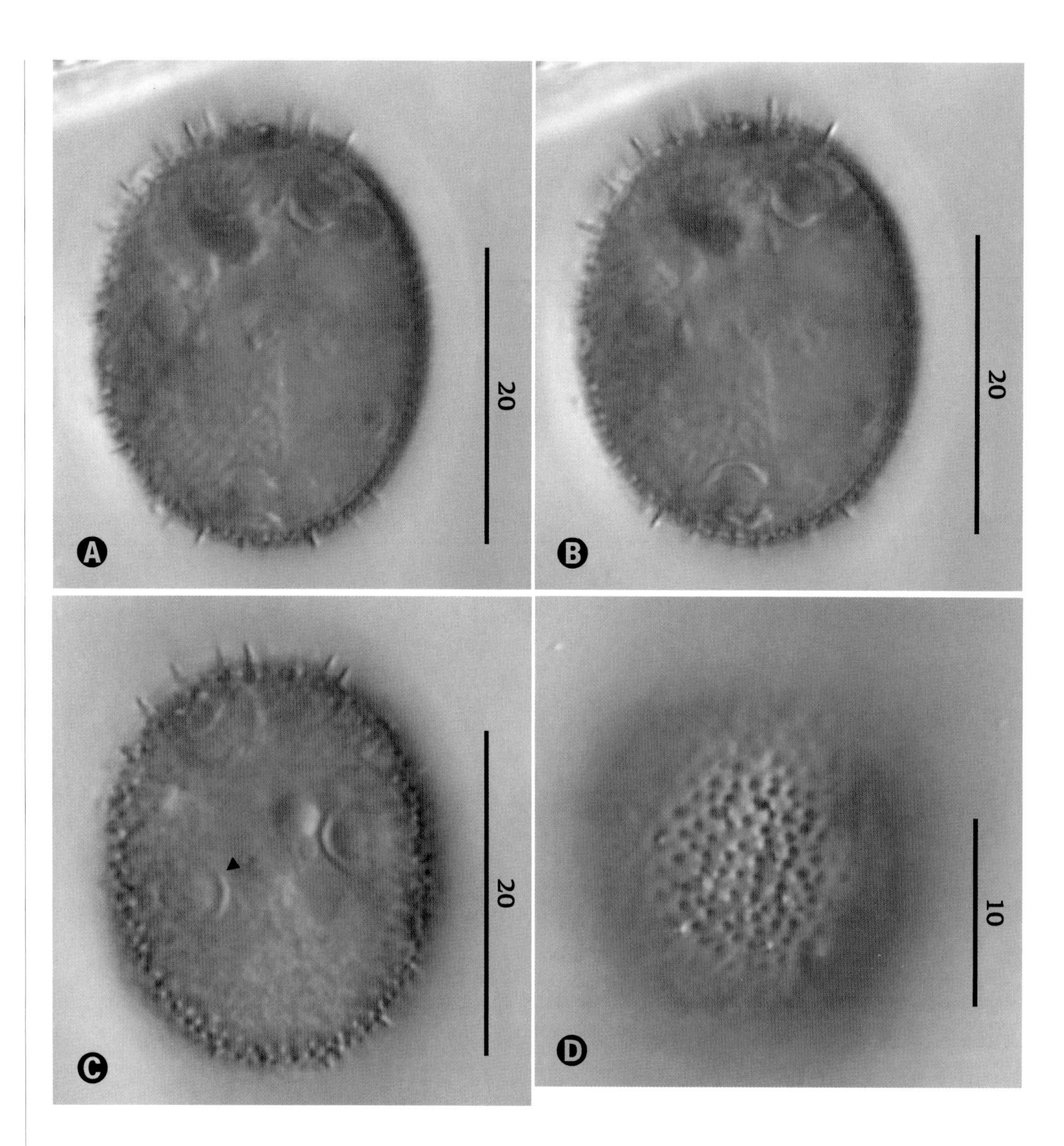

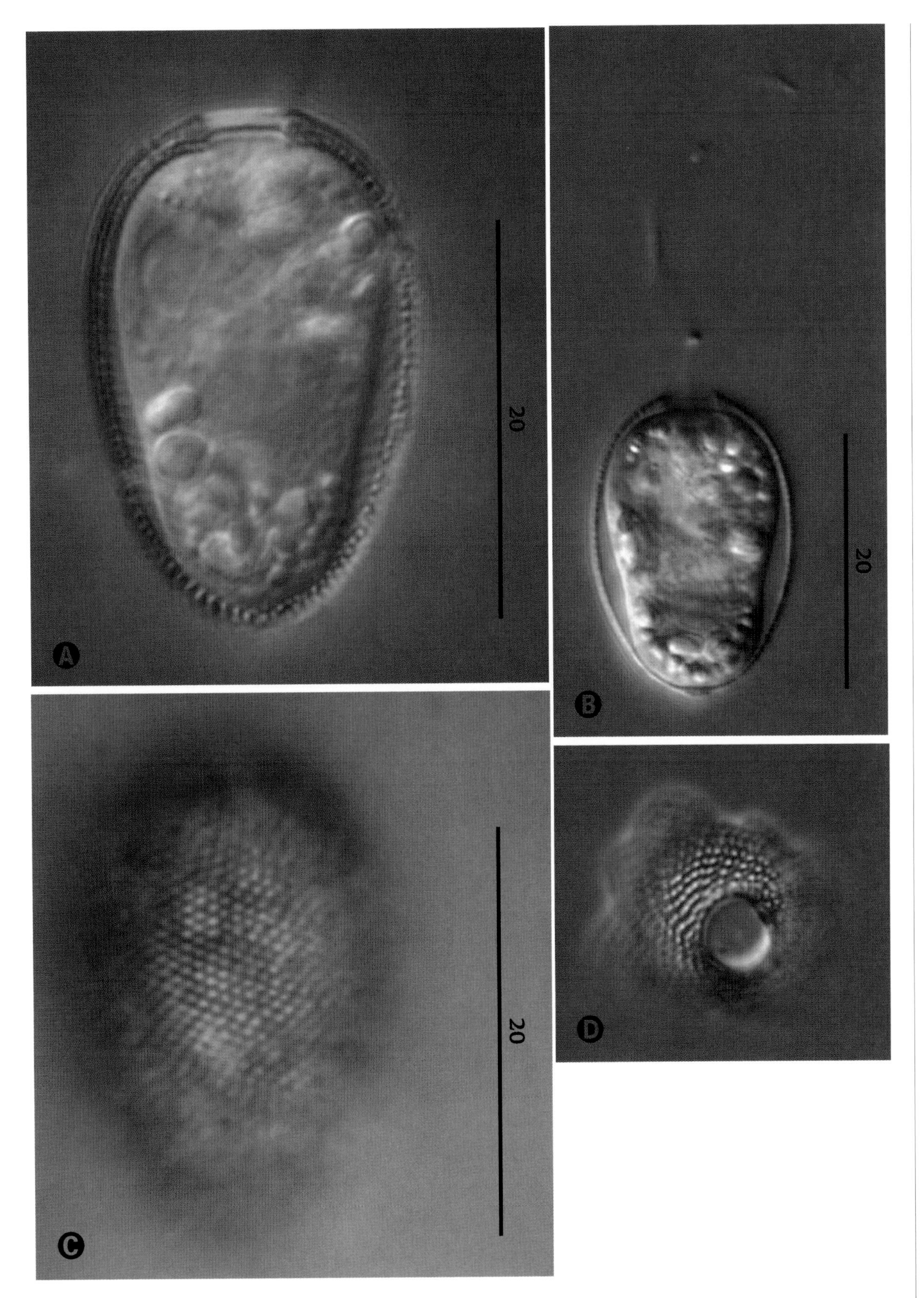

Trachelomonas reticulata

Klebs 1883

SIZE: 23–31 μm long × 17–19 μm wide.

Obovoid lorica; colorless protoplast with red eyespot and paramylon bodies (A). Swimming cell exhibiting a long flagellum (twice body length) (B). Lorica showing spirally disposed scrobiculae (C). Apical view of flagellar pore (D).

Trachelomonas robusta

Swirenko emend. Deflandre 1926

SIZE: 20–30 µm long × 17–24 µm wide.

Subglobose lorica with short conical spines; cell with red eyespot, disc-shaped chloroplasts with diplopyrenoids, and paramylon bodies (A, B). Chloroplasts (C) and lorica surface with fine punctae between spines (D). Cell showing a diplopyrenoid (arrow) and conical short spines on lorica (E).

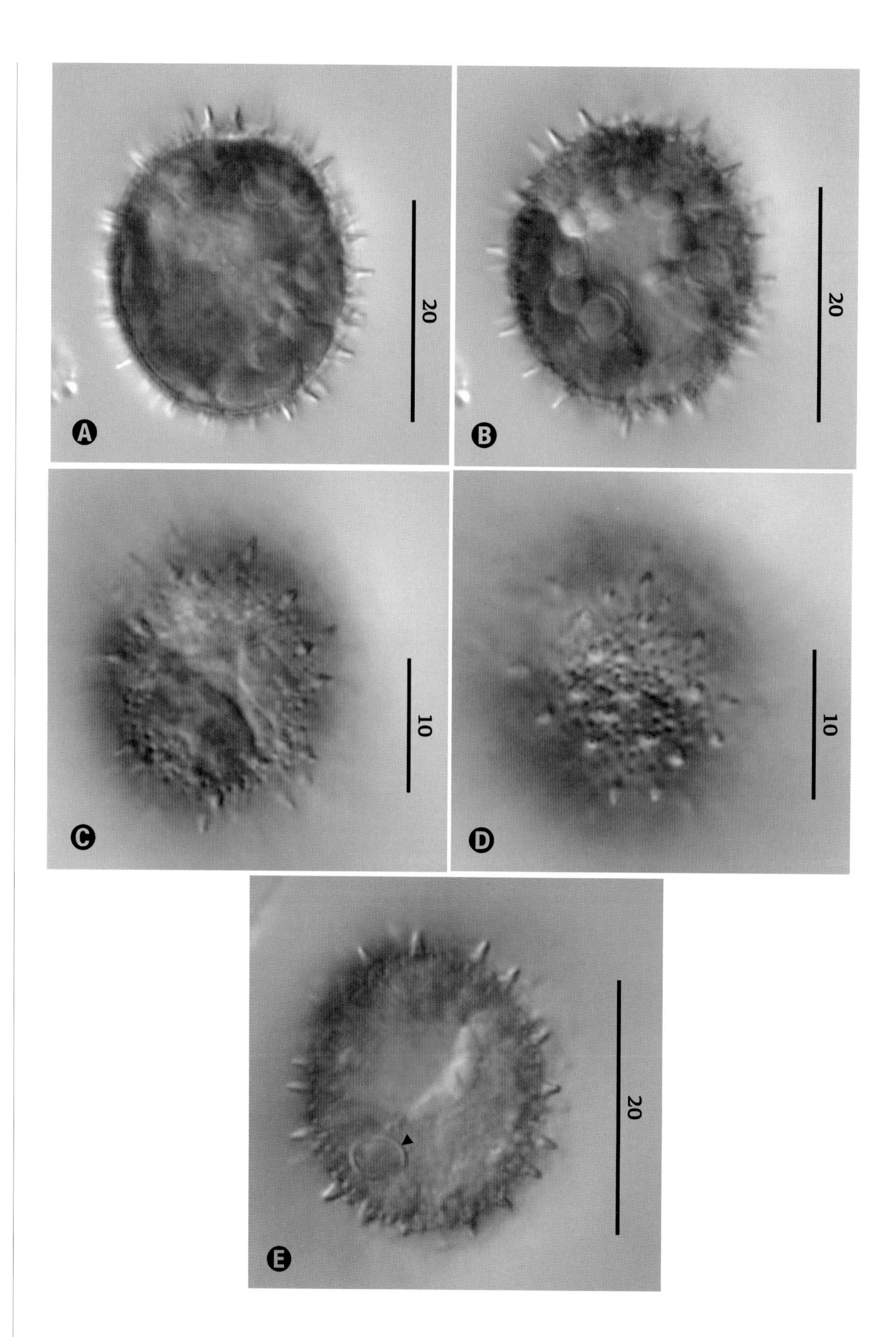

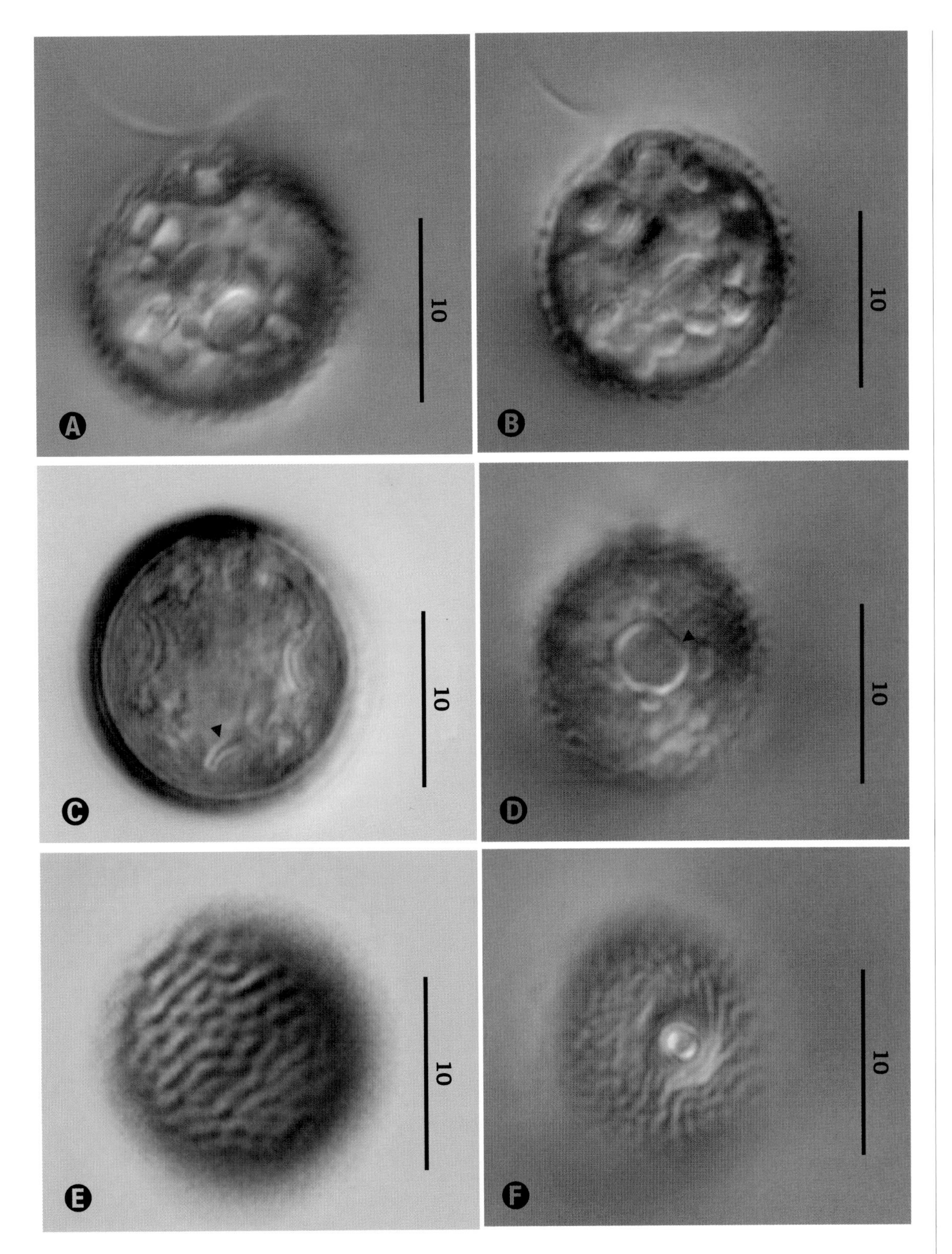

Trachelomonas rugulosa

Stein emend. Deflandre 1926

SIZE: 14–23 µm in diameter.

Round lorica showing flagellar pore with annular thickening and cell with long flagellum, several paramylon bodies, and a haplopyrenoid in face view (A). Same cell in a different focal plane exhibiting red eyespot (B). Protoplast with three haplopyrenoids (arrow) (C). Chloroplast with large haplopyrenoid in face view (arrow) (D). Lorica surface with ridges (E). Detail of flagellar pore in face view and ridged lorica surface (F).

Trachelomonas rugulosa fa. *steinii*

Deflandre 1927

SIZE: 15–23 µm in diameter.

Spherical lorica with annular thickening; cell with flagellum, red eyespot, and plate-like chloroplasts with haplopyrenoids (arrow) (**A**, **B**). Lorica surface with fine ridges (**C**).

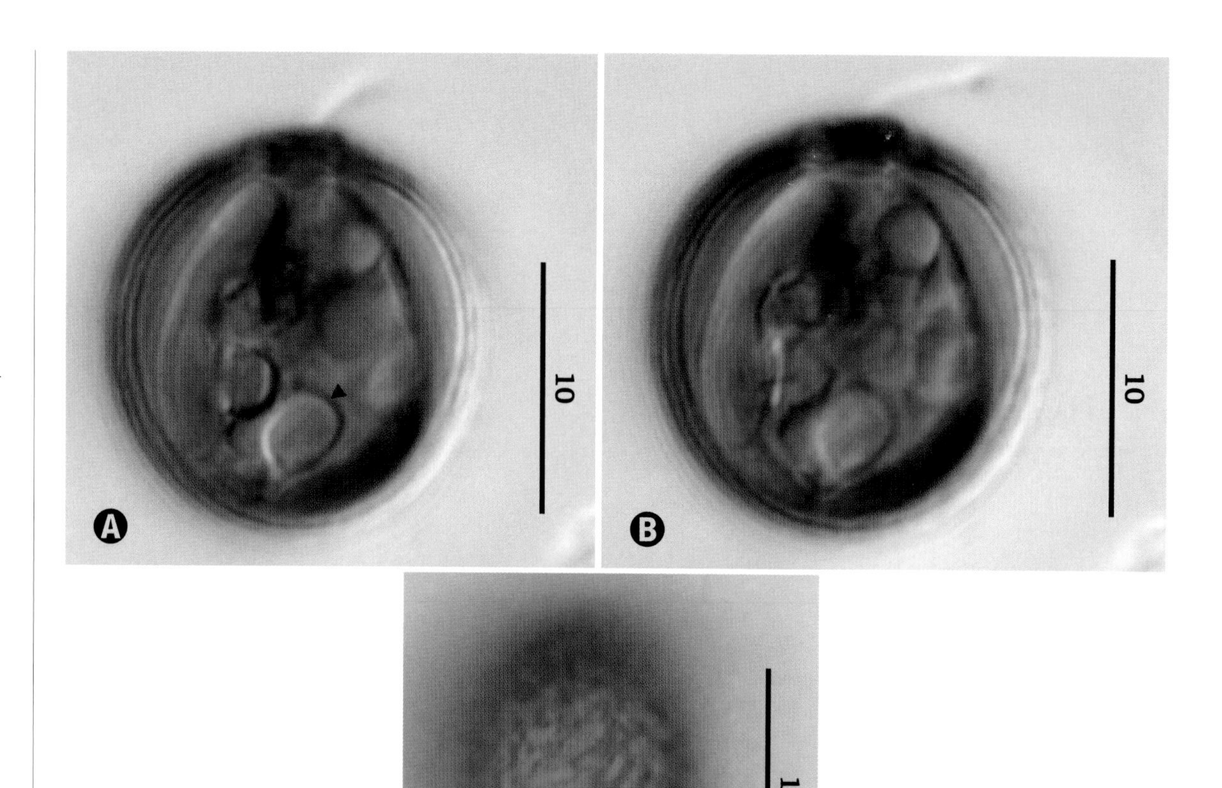

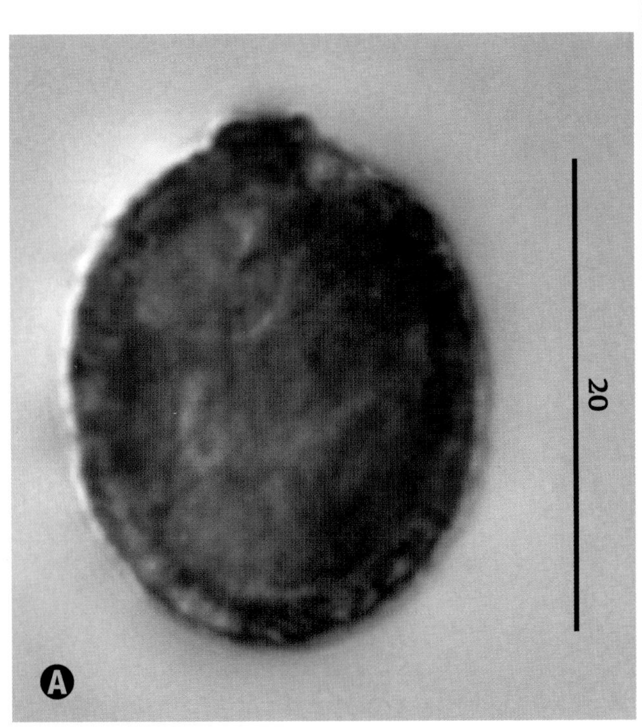

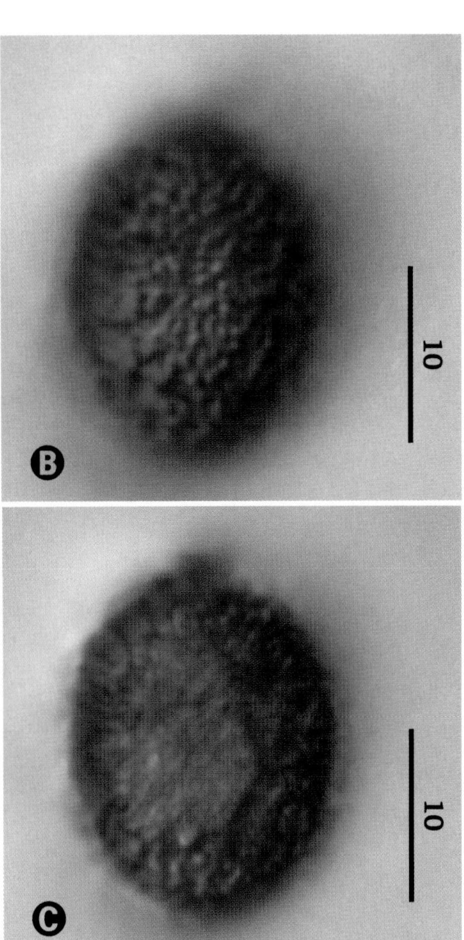

Trachelomonas scabra

Playfair 1915

SIZE: 20–31 µm long × 15–21 µm wide.

Ellipsoidal, reddish brown lorica with short collar; protoplast shows red eyespot (A). Detail of scabrous lorica surface, densely punctate (B), and protoplast showing plate-like chloroplasts (C).

Trachelomonas similis

Stokes 1890

SIZE: 22–40 µm long × 14–23 µm wide.

Ellipsoidal lorica with bent and irregularly dentate collar; protoplast with flagellum, red eyespot, and several disc-shaped chloroplasts (A). Cell showing parietally arranged chloroplasts with haplopyrenoids (arrow) (B). Lorica surface is punctate; the fine spirally arranged pellicle strips are just visible (C). Chloroplasts with haplopyrenoids in face view (arrow) (D). Cell with numerous paramylon bodies (E). Detail of lorica surface and chloroplasts (F).

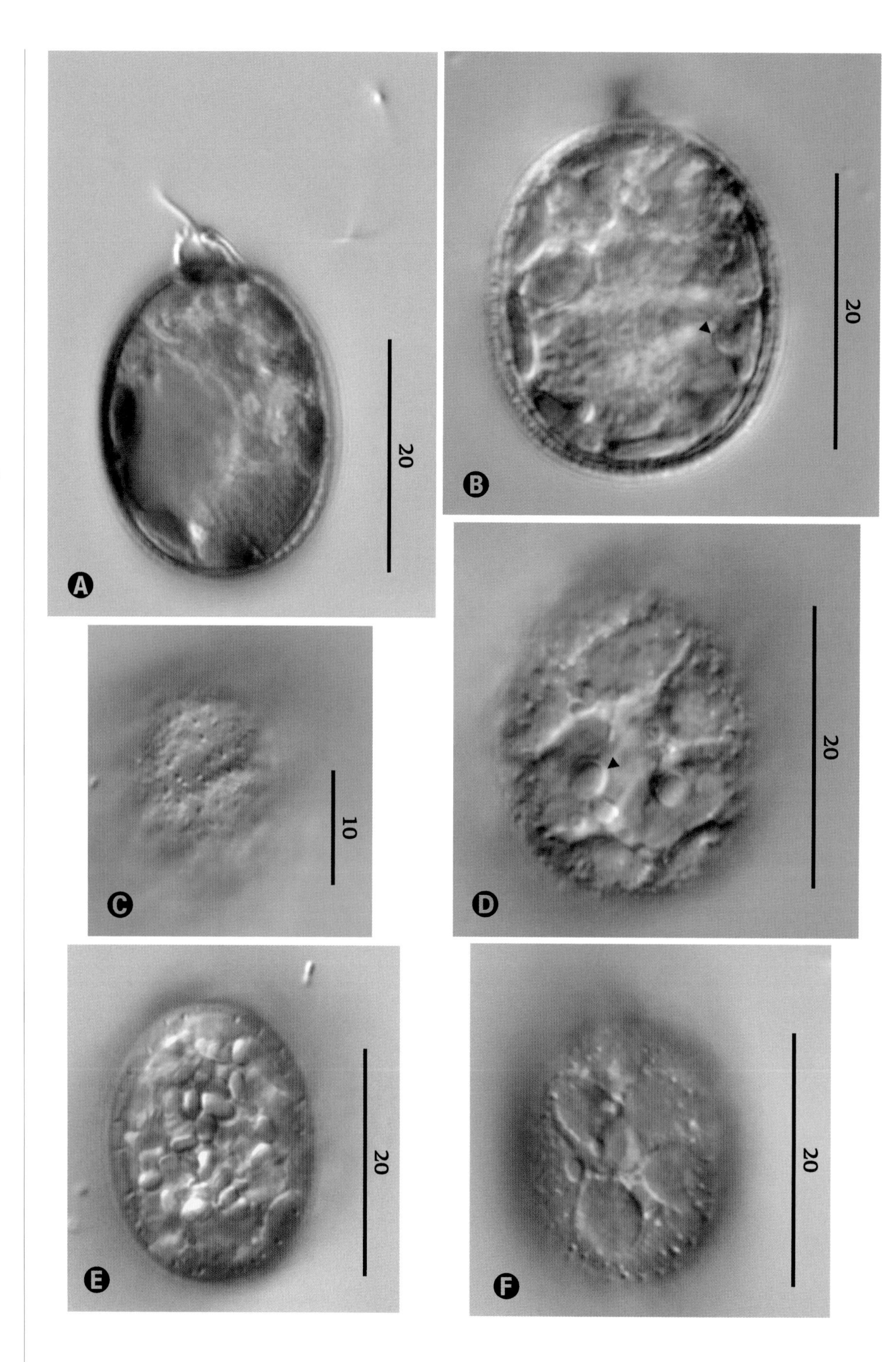

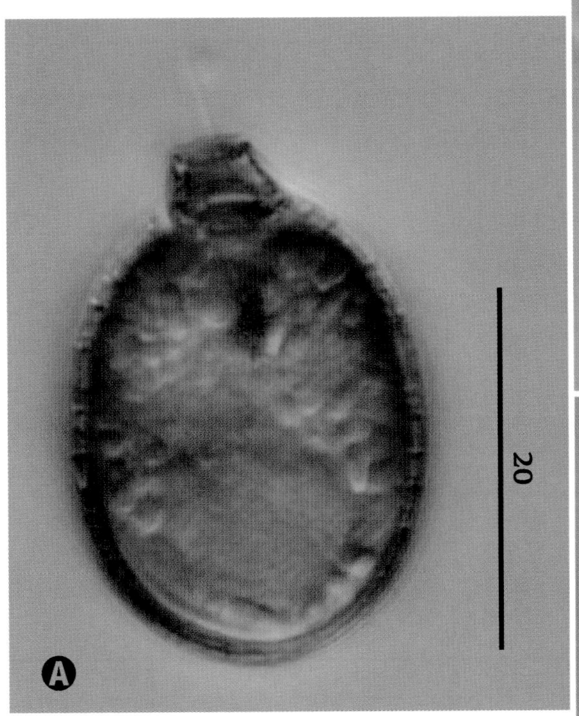

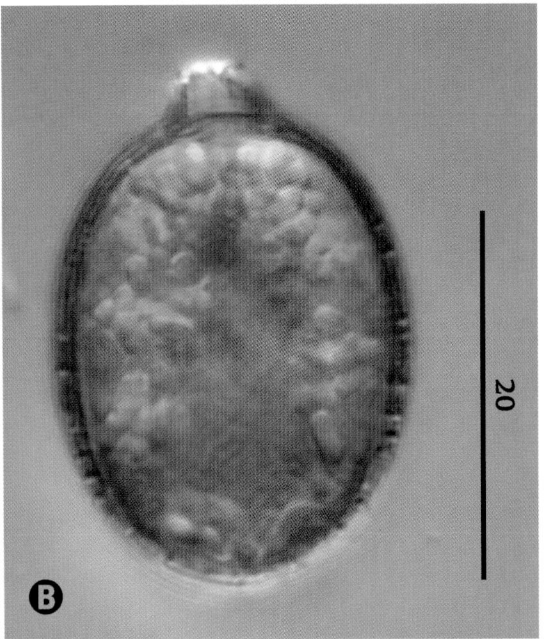

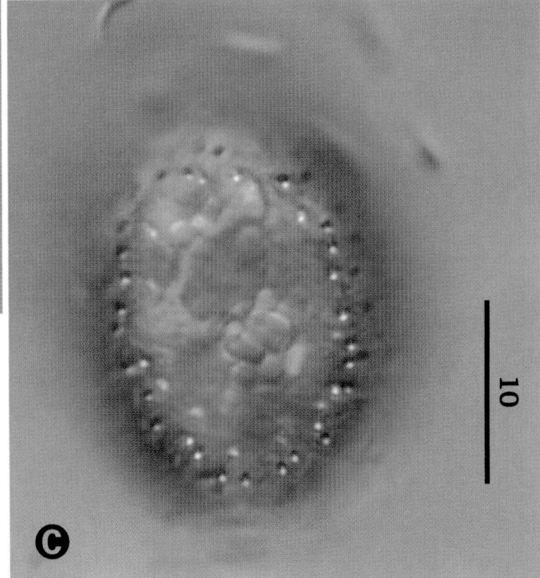

Trachelomonas similis var. *spinosa*

Huber-Pestalozzi 1955

SIZE: 26–36 µm long × 22–25 µm wide.

Ellipsoidal lorica with bent collar; cell showing flagellum (A), red eyespot, chloroplasts, and small paramylon bodies (A, B). Lorica surface bearing short spines; chloroplasts also visible (C).

Trachelomonas stokesiana

Palmer 1905

SIZE: 11–18 μm long × 9–16 μm wide.

Globose lorica with annular thickening; cell showing red eyespot and small paramylon grains (**A**). Cell showing three haplopyrenoids (arrow) (**B**). Lorica surface with longitudinal ridges (**C**).

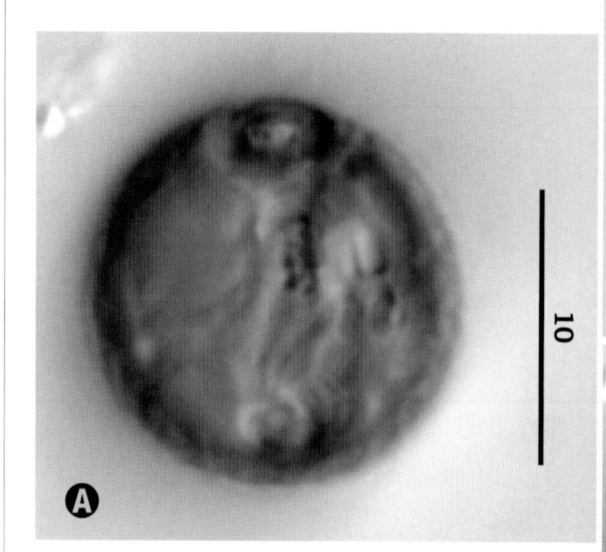

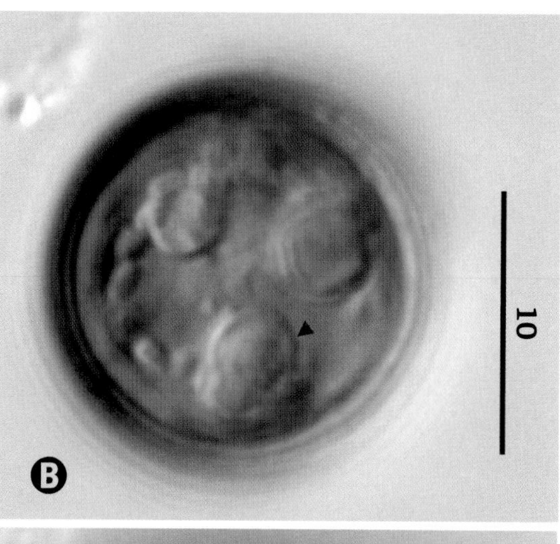

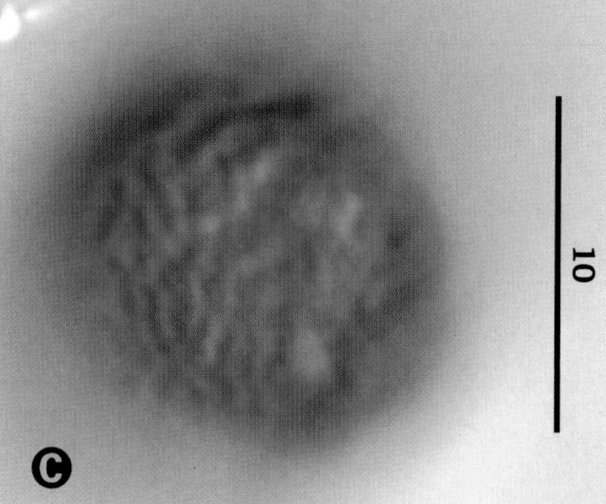

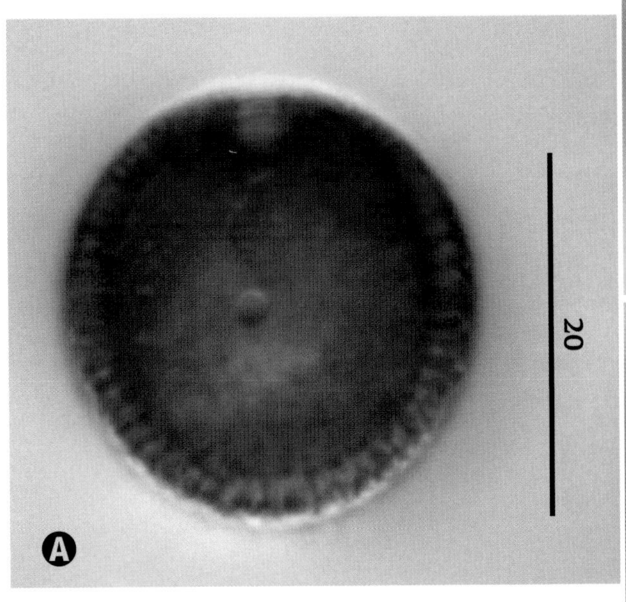

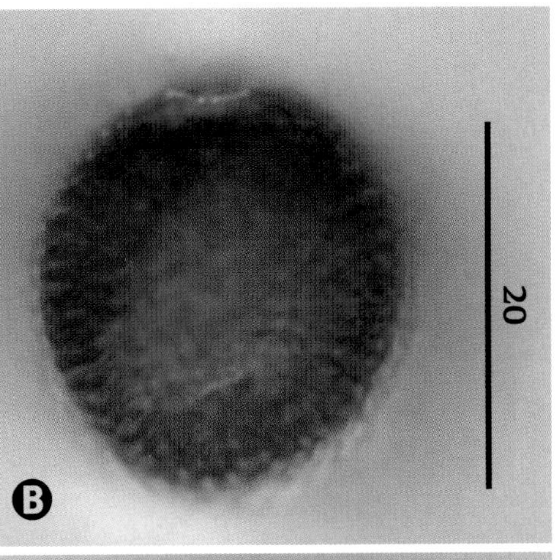

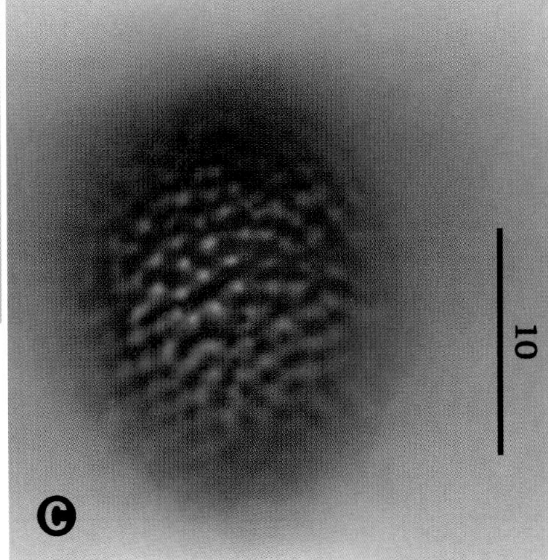

Trachelomonas subverrucosa

Deflandre 1927

SIZE: 23 μm long × 17 μm wide.

Ovoid, thick lorica with flagellar opening and cell showing red eyespot (A). Two different focal planes of lorica showing dense, verrucose ornamentation (B, C).

Trachelomonas superba

Swirenko emend. Deflandre 1926

SIZE: 38–55 µm long × 30–39 µm wide.

Ellipsoid lorica ornamented with thick spines, longer toward posterior pole; flagellar collar deeply crenulated; protoplast filled with numerous paramylon grains, disc-shaped chloroplasts barely visible (A, B). Lorica surface showing longer spines at posterior (C). Detail of densely punctate lorica surface (D).

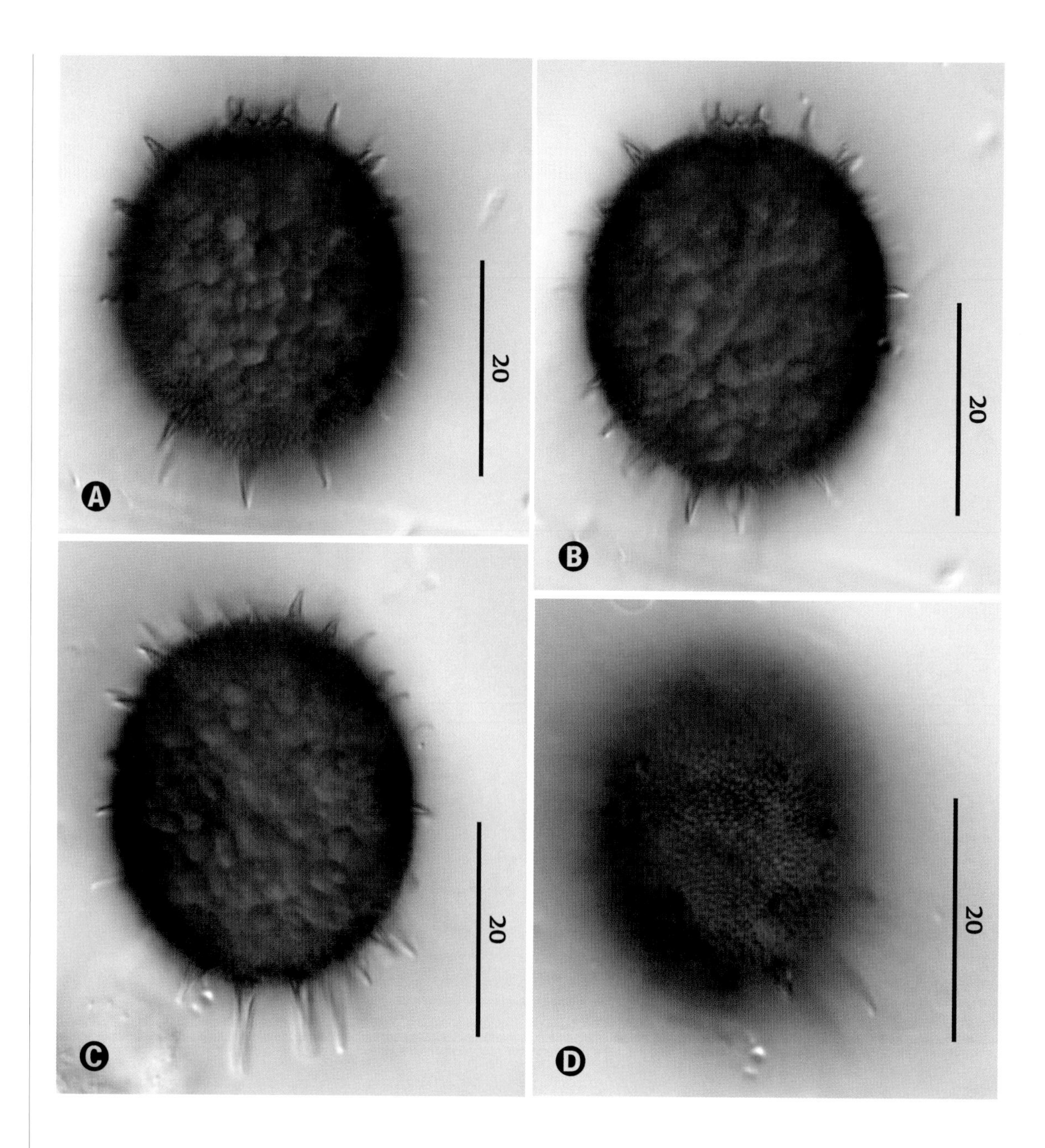

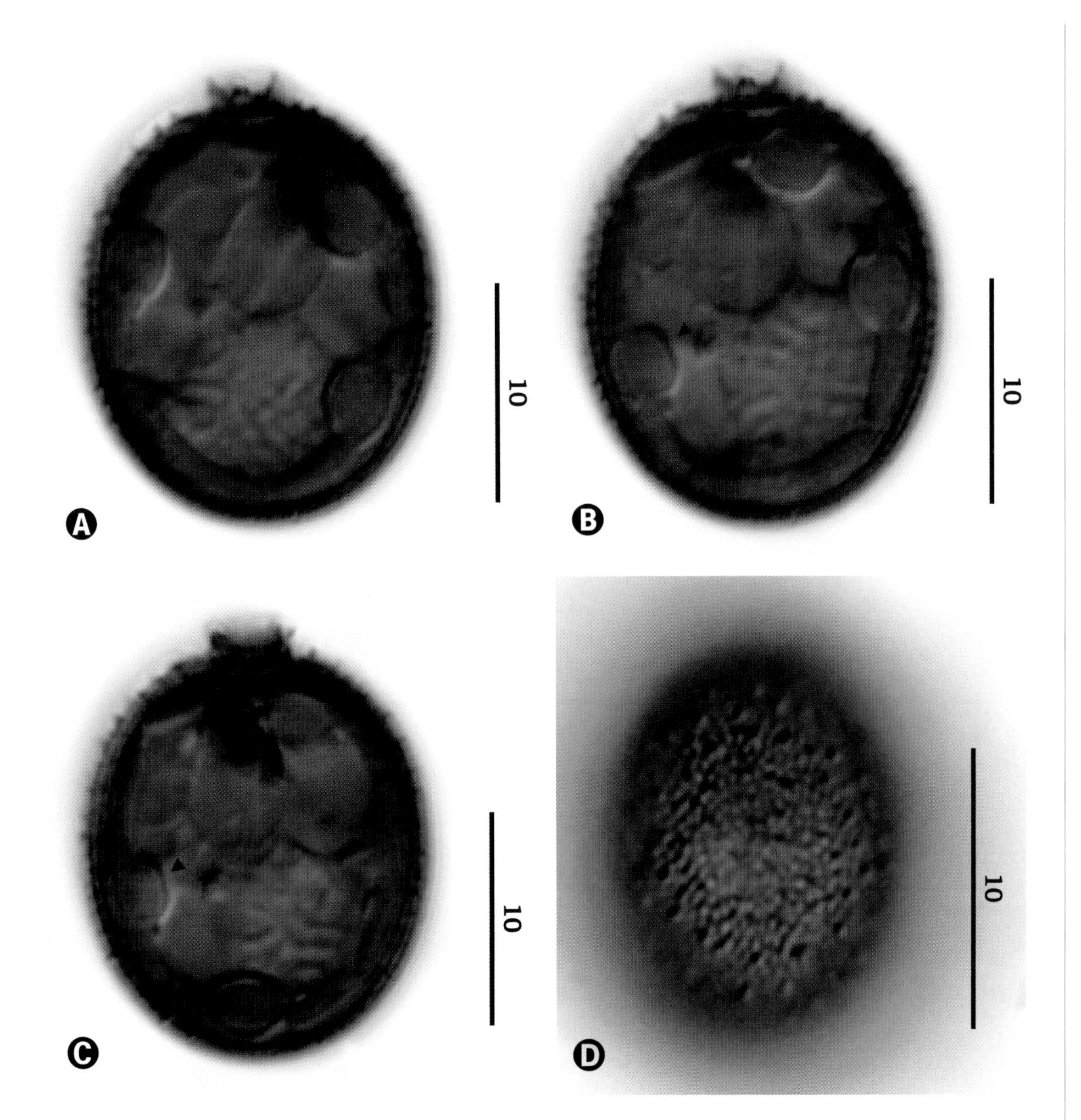

Trachelomonas sydneyensis var. *minima*

Playfair 1915

SIZE: 22–28 µm long × 15–24 µm wide.

Subspherical lorica with short dentate collar and spines; cell with red eyespot and plate-like chloroplasts with diplopyrenoids (A). Same cell in different focal plane revealing more diplopyrenoids (arrows) and eyespot (B, C). Lorica surface is densely punctate (D). Flagellum is up to twice body length.

Trachelomonas urnigera

Skuja 1932

SIZE: 28–40 µm long × 18–23 µm wide.

Urn-like lorica with long collar; cell shows flagellum, red eyespot, small paramylon grains, and disc-shaped chloroplasts with haplopyrenoids (arrow) (Ⓐ). Cell with long flagellum, chloroplasts, and small paramylon grains (Ⓑ). Lorica is finely granulated; cell showing chloroplasts and paramylon (Ⓒ).

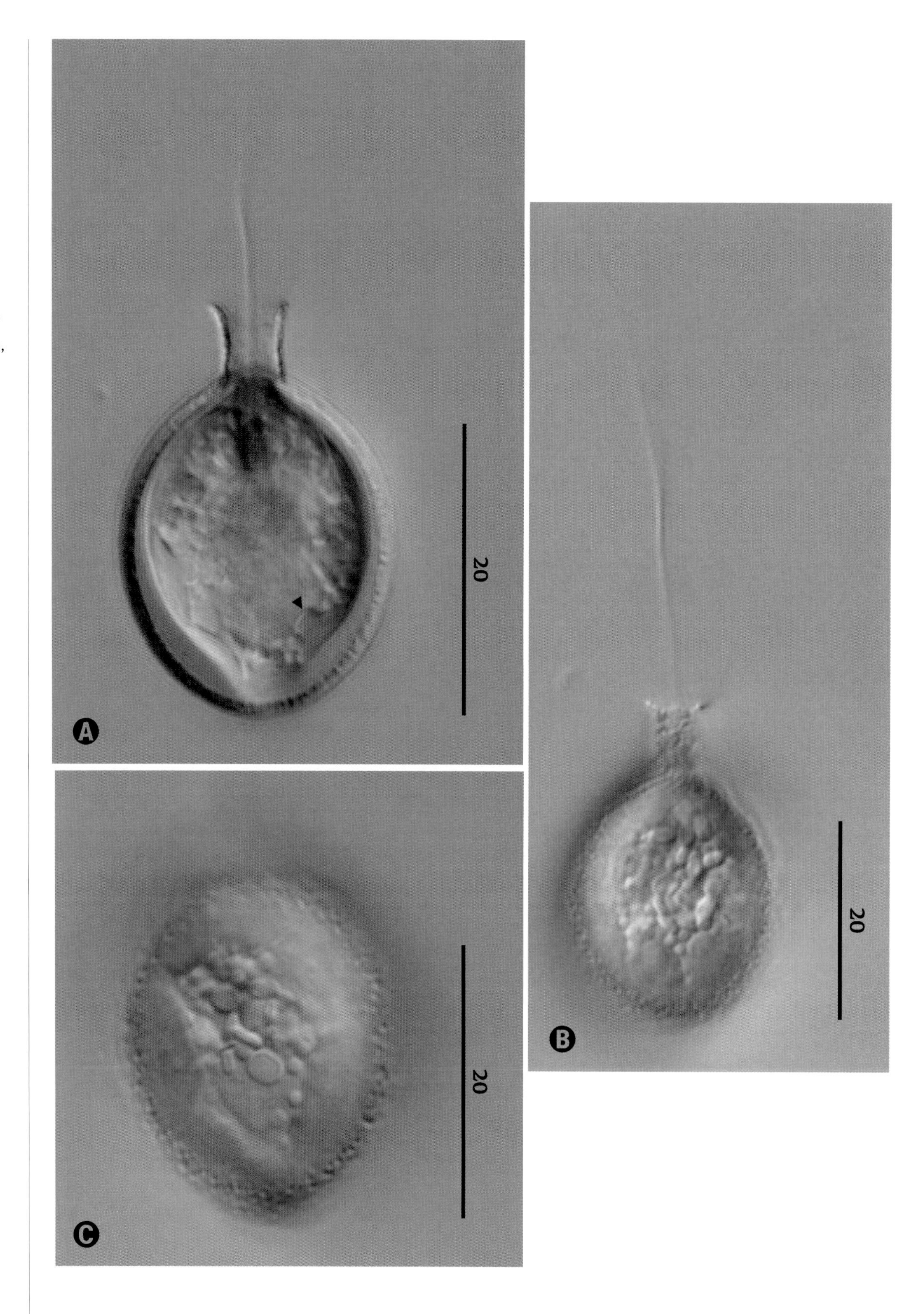

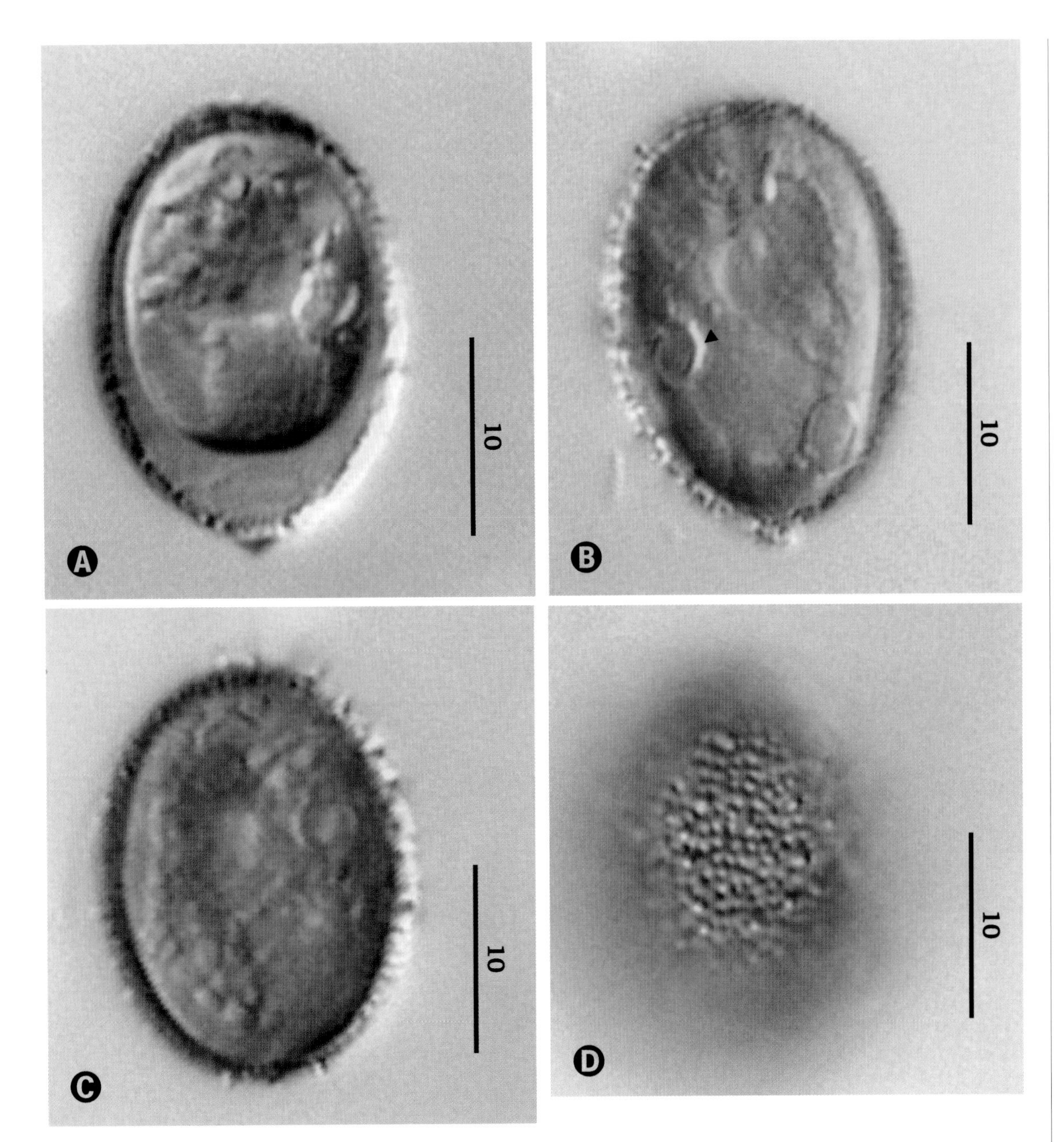

Trachelomonas variabilis

Singh 1956

SIZE: 21–26 µm long × 16–18 µm wide.

Ovoid lorica with protoplast in metaboly (Ⓐ). Lorica and cell showing eyespot, plate-like chloroplasts with diplopyrenoids (arrow), and small paramylon grains (Ⓑ, Ⓒ). Lorica with short spines (Ⓒ, Ⓓ). Flagellum is two to three times body length.

Trachelomonas varians

Deflandre 1926

SIZE: 22–27 µm long × 19–23 µm wide.

Subspherical lorica, flagellar pore with annular thickening, and cylindrical collar projecting inward (A, D). Detail of cell with red eyespot and nucleus (B). Lorica surface is smooth (C).

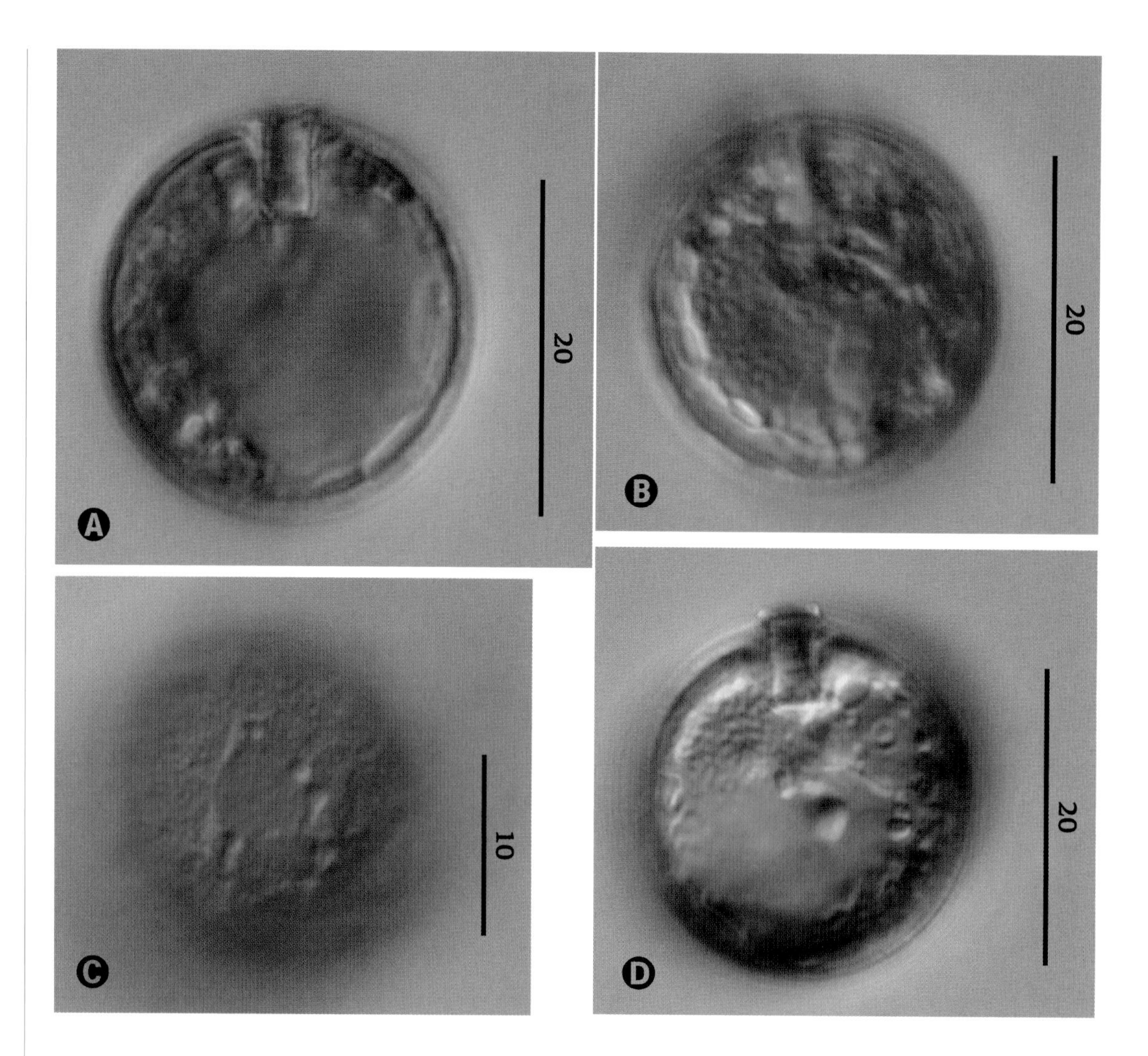

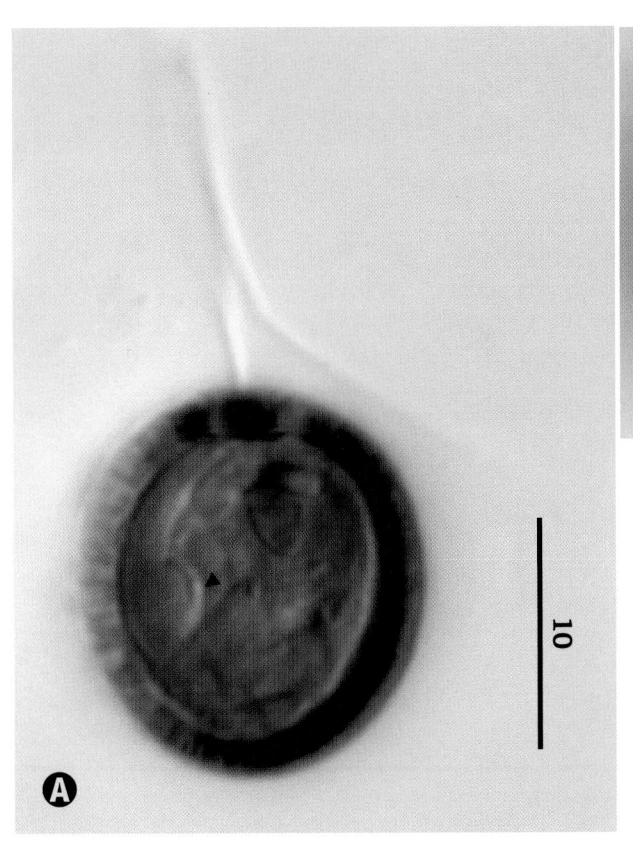

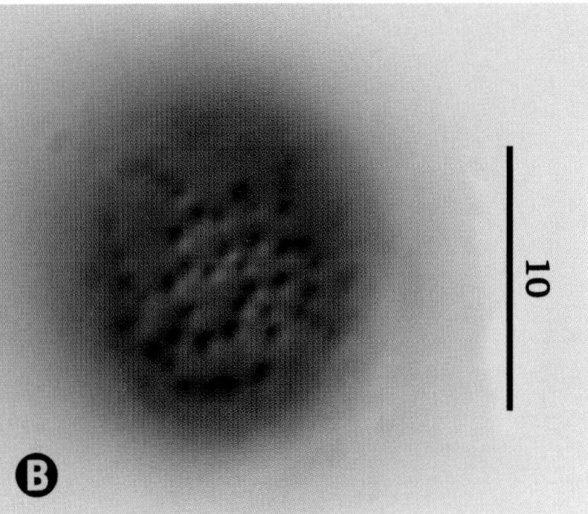

Trachelomonas verrucosa

Stokes 1887

SIZE: 16–24 µm long × 14–22 µm wide.

Subspherical to spherical lorica with annular thickening; cell with long flagellum, red eyespot, and plate-like chloroplasts with haplopyrenoids (arrow) (**A**). Lorica surface is irregularly granulated (**B**).

Trachelomonas verrucosa fa. *irregularis*

Deflandre 1926

SIZE: 12–15 µm long × 11–14 µm wide.

Subspherical to spherical lorica with annular thickening; cell with red eyespot and large plate-like chloroplasts (Ⓐ). Lorica surface with fine, irregular granules (Ⓑ).

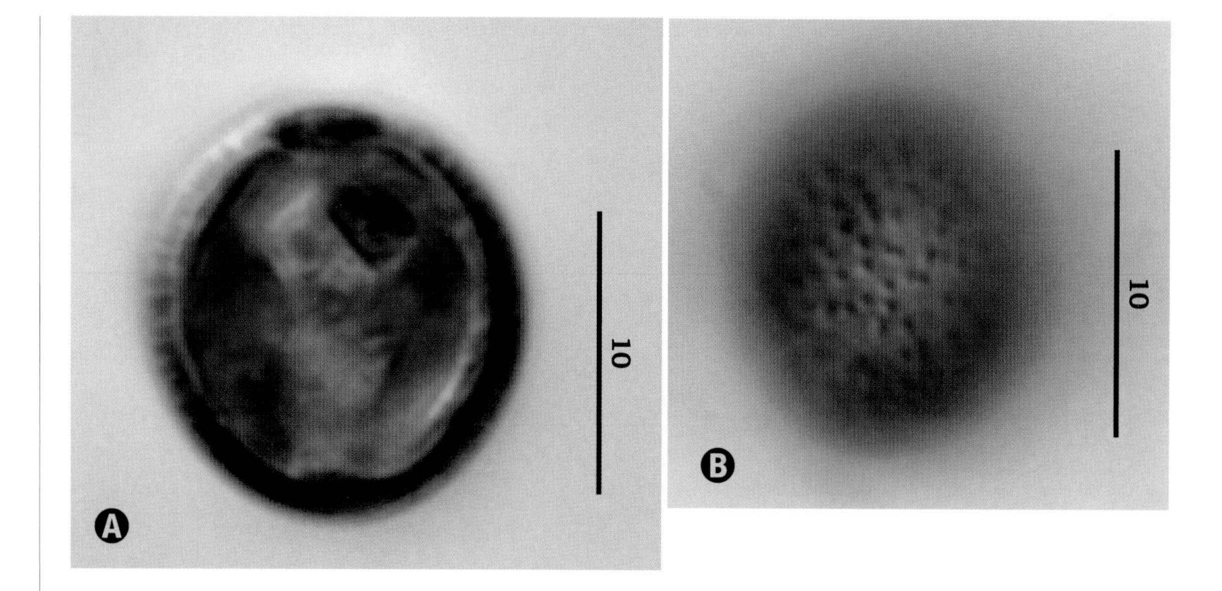

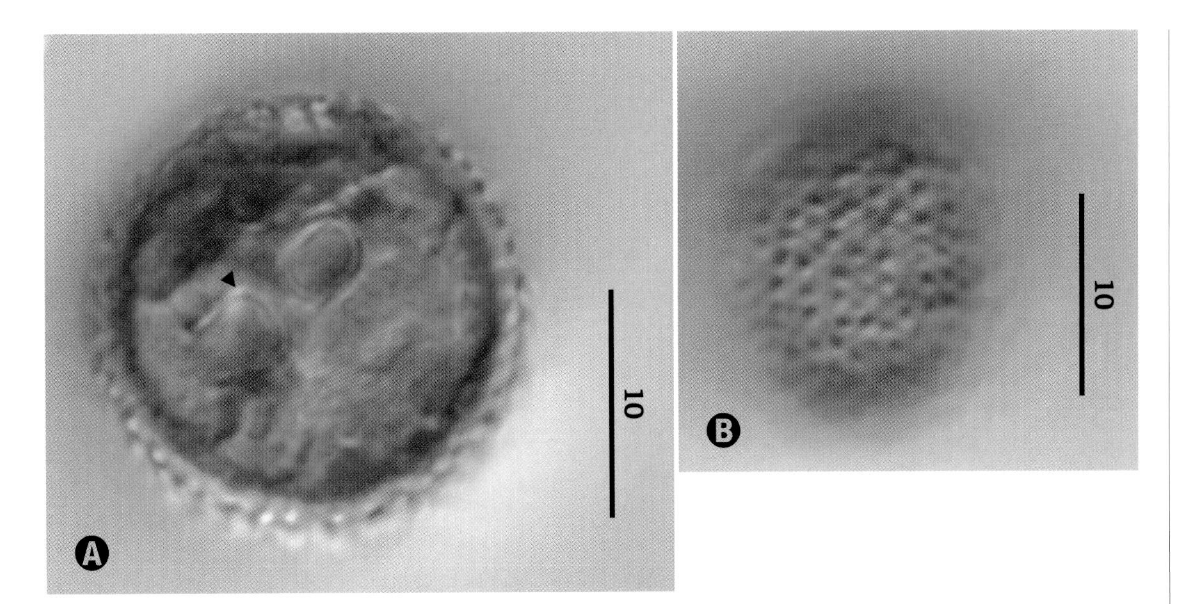

Trachelomonas verrucosa var. *granulosa*

(Playfair) Conrad 1932

SIZE: 11–19 µm in diameter.

Globose lorica with granulated surface; plate-like chloroplasts with haplopyrenoids (arrow) (A). Lorica surface with pearl-like granules (B).

Trachelomonas volvocina

Ehrenberg 1838

SIZE: 6–32 µm in diameter.

Globose lorica; cell showing red eyespot and two haplopyrenoids (A). Cell with eyespot, one of the two plate-like chloroplasts with haplopyrenoid (arrow) (B). Lorica with annular thickening and protoplast with small paramylon grains (C). Lorica surface is smooth (D). Flagellum is three times body length.

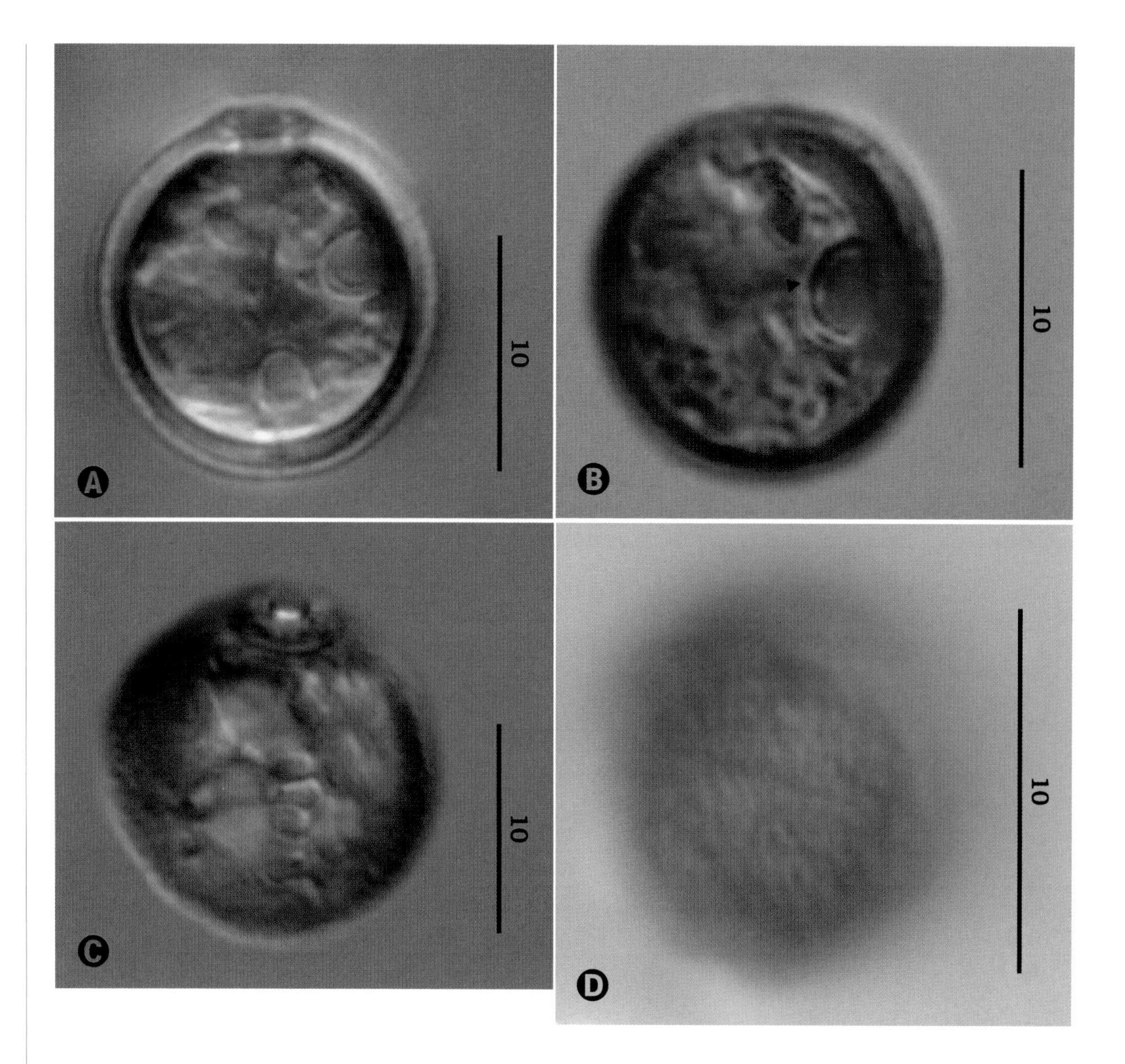

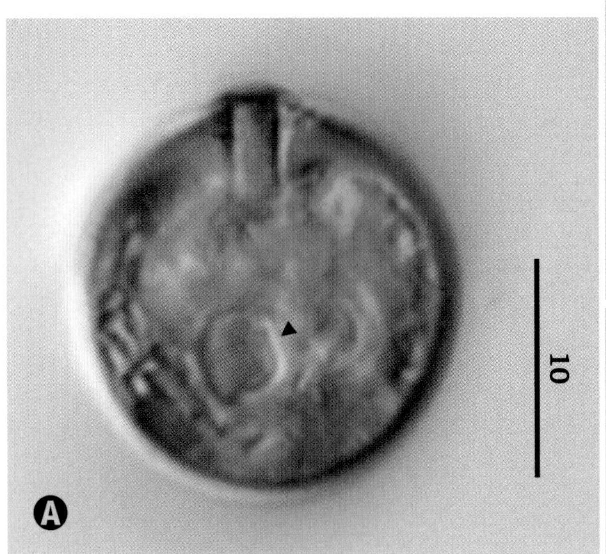

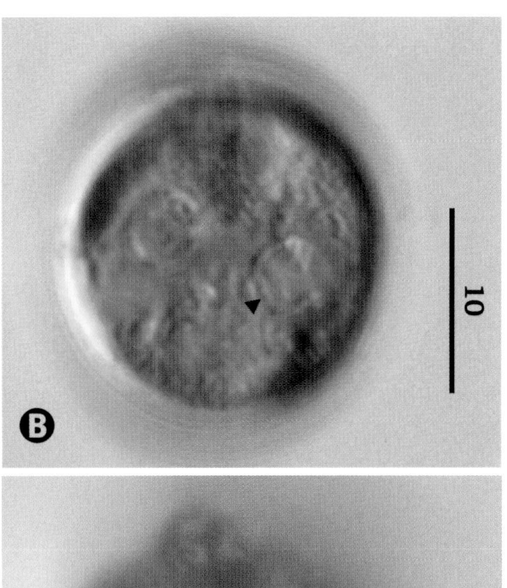

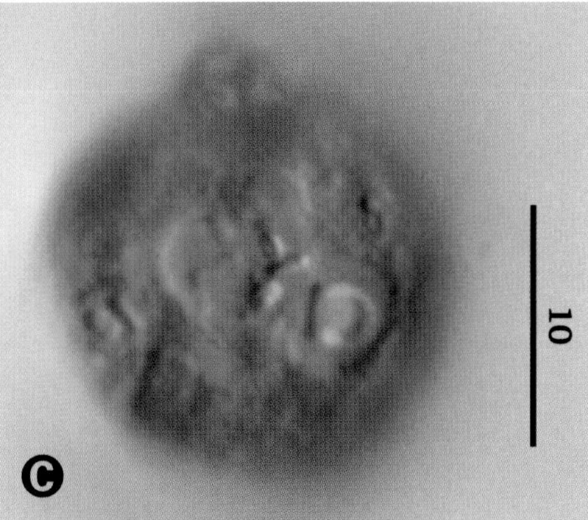

Trachelomonas volvocina var. *cervicula*

(Stokes) Lemmermann 1913

SIZE: 7–32 µm in diameter.

Round lorica with a deep inward projecting tube-like collar; cell with plate-like chloroplasts, haplopyrenoid in frontal view (arrow), and short paramylon rods (A). Cell showing two plate-like chloroplasts, each with a haplopyrenoid (arrow) (B). Lorica surface is smooth (C).

Trachelomonas volvocina var. *derephora*

Conrad 1926

SIZE: 12–21 μm in diameter.

Globose lorica with short, stout collar and protoplast, showing a long flagellum and one of the two haplopyrenoids (arrow) (A). Cell with numerous paramylon grains (B). Lorica shows smooth surface (C).

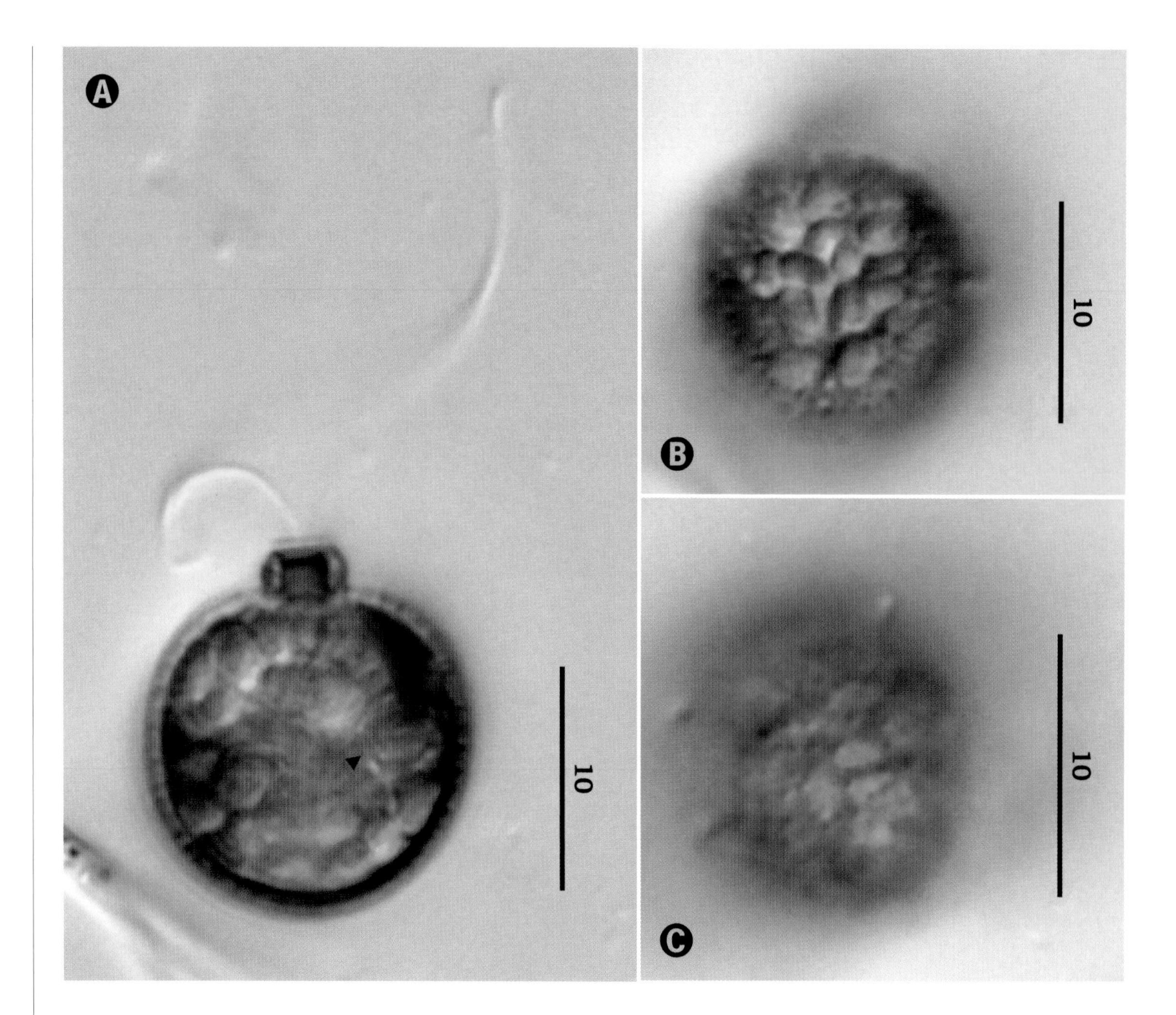

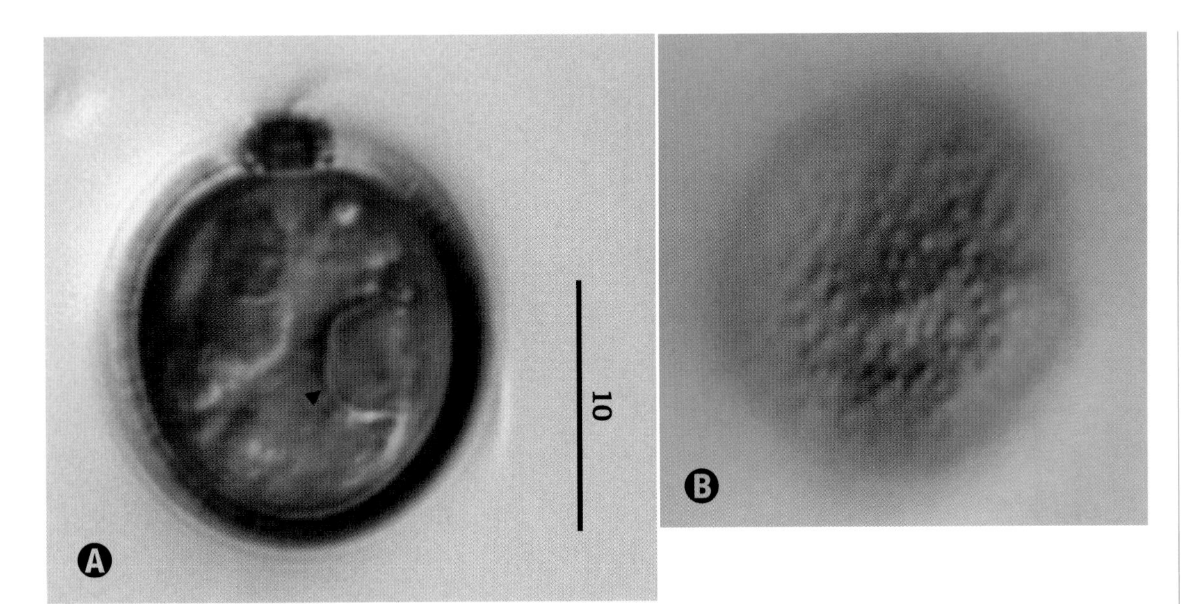

Trachelomonas volvocina var. *punctata*

Playfair 1915

SIZE: 14–21 µm long × 13–20 µm wide.

Lorica with short collar; cell with flagellum, red eyespot, and two chloroplasts, each with a haplopyrenoid (arrow) (A). Lorica surface is densely punctate (B).

Trachelomonas volvocinopsis
Swirenko 1914

SIZE: 14–26 μm in diameter.

Globose lorica; cell with eyespot and disc-shaped chloroplasts (A). Detail of smooth lorica surface and chloroplasts lacking pyrenoids (B). Flagellar pore with annular thickening (C). Lorica and cell with nucleus (D). Flagellum is three to four times body length.

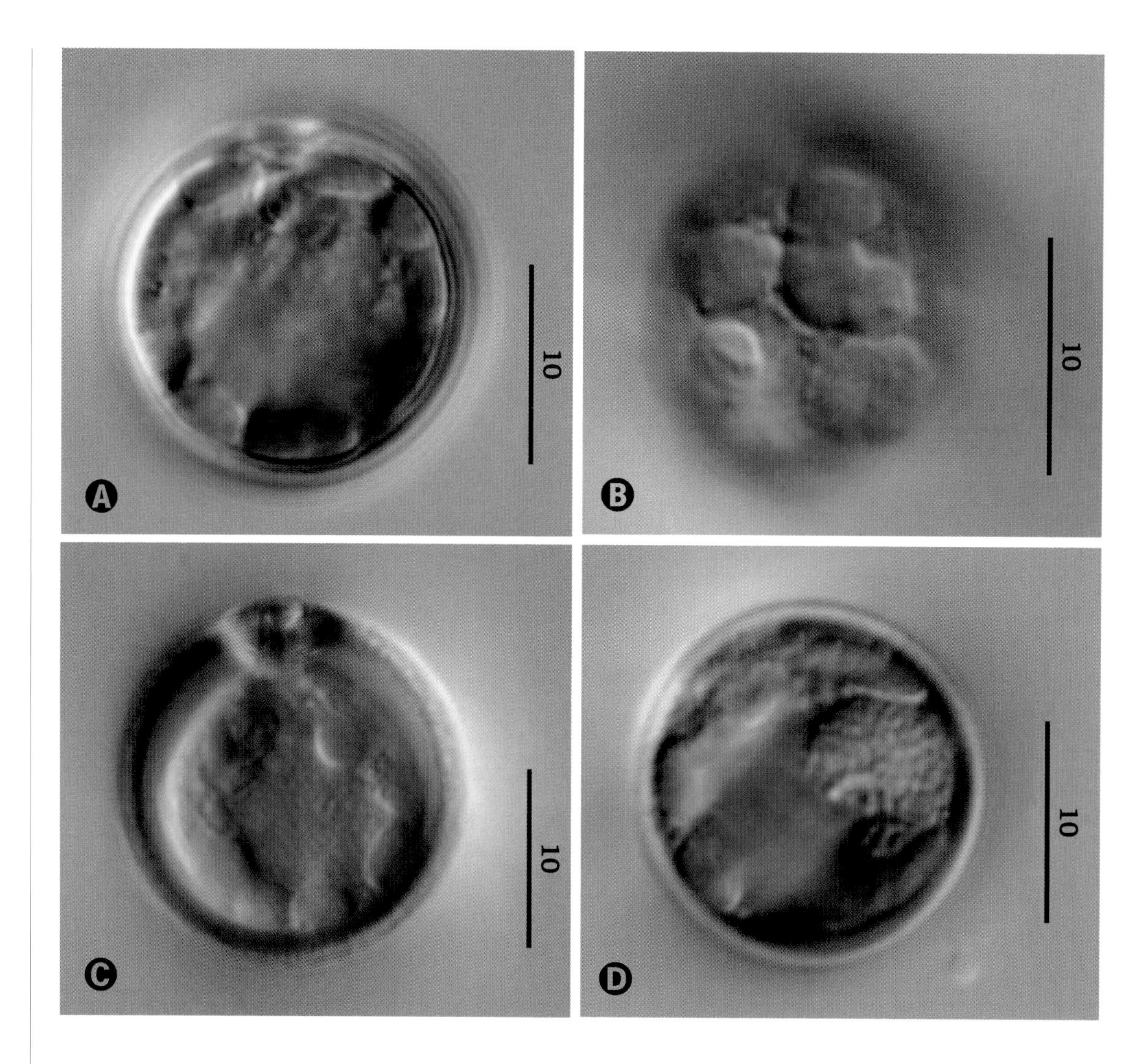

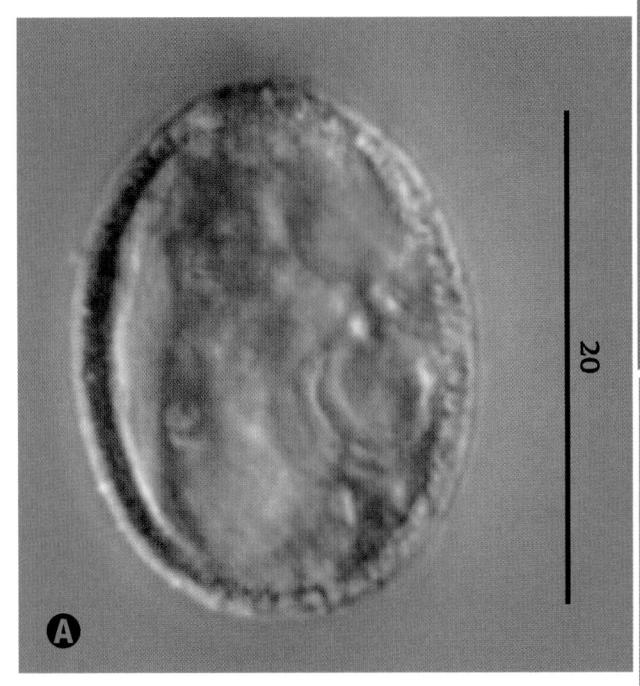

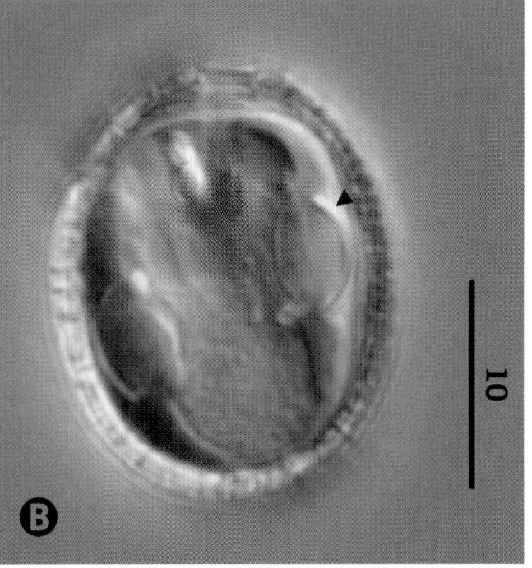

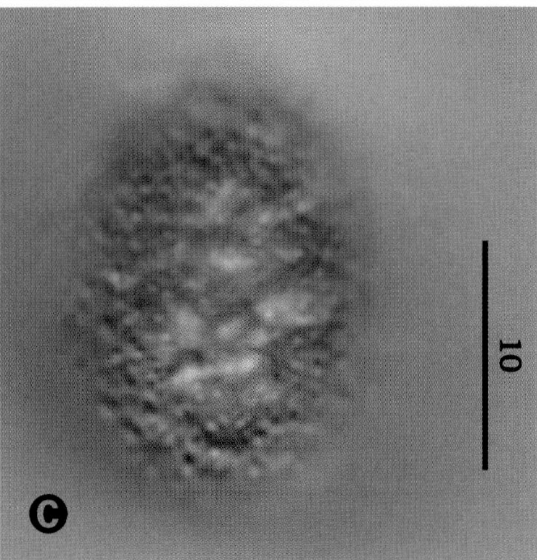

Trachelomonas zorensis

Lefèvre 1933

SIZE: 19–22 μm long × 14–16 μm wide.

Ellipsoid lorica and protoplast with red eyespot; large plate-like chloroplast visible on left and diplopyrenoid on right (A). Lorica showing the flagellar opening at the anterior end and cell with two large plate-like chloroplasts with diplopyrenoids (arrow) and a paramylon body near eyespot (B). Detail of lorica surface showing ornamentation (C). Flagellum is twice body length.

GLOSSARY

accessory pigments light-absorbing pigments, found in photosynthetic organisms, that work in conjunction with chlorophyll a (e.g., chlorophyll b, c, carotenes)

acute ending with a sharp point

adenylyl cyclase an enzyme that converts ATP to 3′, 5′ cyclic AMP and pyrophosphate

annular thickening a ring-shaped thickening around the flagellar pore in the lorica

apical pore the aperture through which the flagellum emerges from the lorica

autotrophic capable of synthesizing complex organic substances from simple inorganic substances

axenic culture a population of individuals of one strain or species free from other strains or species

axial situated around or along an axis

basal body the basal section of the flagellum which anchors the flagellum in the cell

beta carotene one form of the orange photosynthetic accessory pigment

brackish salty; containing a mixture of sea water and fresh water

canal a tube-like area connecting the anterior invagination or reservoir to the surrounding medium in euglenophytes

carotene an orange photosynthetic accessory pigment important for photosynthesis

cauda a tail-like extension at the posterior end of many euglenophytes

chlorophyll green pigments found in photosynthetic organisms, such as plants, algae, and cyanobacteria

chloroplast (chromatophore) the organelle in which photosynthesis takes place

chromosome a structure in all living cells that consists of DNA and various proteins and that carries the genes determining heredity

citriform lemon-shaped

clone a cell, group of cells, or organism that is produced asexually from and is genetically identical to a single ancestor

contractile vacuole an organelle located adjacent to the reservoir and involved in fluid regulation

crenulated having a margin with very small rounded teeth

cross-striations short horizontal strips that connect the longitudinal pellicle striations

cyst a cell protected from unfavorable environmental conditions (drought, cold, heat, nutrient deficiency) by a thick wall

cytokinesis division of the cytoplasm to form two daughter cells

cytoplasm the cell substance between the cell membrane and the nucleus

dendroid tree-like

dentate edged with tooth-like projections

denticulate having a very finely toothed margin

diadinoxanthin a xanthophyll pigment

diatoxanthin a xanthophyll pigment

diplopyrenoid a pyrenoid surrounded with a sheath of paramylon on two sides (also known as a double-sheathed pyrenoid)

DNA deoxyribonucleic acid

endosome the nucleolus in a euglenoid

endosymbiosis a relationship in which an organism lives inside another organism

euglenoid movement (metaboly or contractile body movements) the peristaltic, flowing movements typical of many euglenoid cells

eukaryotic containing a nucleus and cytoplasm

eutrophic a body of water rich in nutrients that promotes a proliferation of algae

eutrophication over-enrichment of a water body with nutrients, resulting in excessive growth of organisms and depletion of oxygen

eyespot (stigma) a red or orange spot, usually found in unicellular or colonial flagellate algae and involved in light perception

fimbriate having the edge or extremity bordered by filiform processes; fringed

flagellar apparatus the whole complex of flagellar basal bodies, microtubular roots, and their associated structures

flagellar canal a tubular opening at the top of the reservoir through which the flagellum emerges to the exterior

flagellar pore the outer aperture of the flagellar canal

flagellate bearing flagella

flagellum a whip-like organelle used for swimming

globose spherical, globular

glucan a polysaccharide

haematochrome red granules present in some euglenoids; these are responsible for the change of cell color from green to red

haplopyrenoid a pyrenoid bordered by a paramylon sheath on only one side

heterodynamic flagella flagella with independent patterns of beating

hyaline glassy or transparent

interphase the period of the cell cycle during which the nucleus is not undergoing division, typically occurring between mitotic or meiotic divisions

lorica a protective shell containing a naked (non-walled) cell

keel a longitudinal ridge

metaboly the ability of the cell to change its body shape through peristaltic movements

mitochondrion organelle that produces energy for the cell through cellular respiration

mitosis nuclear division

mixotrophy utilizing more than one form of nutrition

monad a solitary flagellate cell

monophyly relating to a taxonomic group that contains all the descendants of a single common ancestor

morphology form or shape of a cell or a structure

motility the power of movement conferred on an organism by some specific structure or cell product, e.g., movement by flagellar activity

mucilage bodies vesicles containing mucilage, which lie beneath the plasmalemma and are often discharged upon stimulation

mucocyst (muciferous body) single membrane-bound, mucous-containing organelles located beneath the pellicle; in euglenoids, they are typically spherical or spindle-shaped

naked cell a cell not covered by a wall or other firm material external to the cell membrane

naked pyrenoid a pyrenoid not bordered by paramylon sheaths

neoxanthin a xanthophyll pigment

neuston a community of organisms living at the water-atmosphere interface

neutral red a dye used for staining in histology; in euglenoids it is used to stain mucocysts

nucleolus the small, spherical granular body located in the nucleus of a eukaryotic cell where ribosomal RNA (ribonucleic acid) is synthesized and processed; composed largely of protein and RNA

nucleus the organelle in the cytoplasm of eukaryotic cells that contains nearly all the cell's DNA and controls its metabolism, growth, and reproduction

oblong elongated, with the ends broadly rounded

obovoid inversely egg-shaped

obpyriform inversely pear-shaped

oligotrophic having low primary productivity; pertaining to waters having low levels of nutrients

ontogeny the life history of an organism; the process by which the adult organism develops from the very youngest stage

osmotrophic capable of absorbing organic nutrients directly from the external medium

ovate egg-shaped, broader basally, two-dimensional term

ovoid egg-shaped, broader basally, three-dimensional term

palmella stage (palmelloid stage) cells embedded in an amorphous gelatinous matrix

papilla a protuberance born by a cell wall or envelope

paramylon the β-1, 3-linked glucose reserve of euglenophytes (= paramylum)

paraphyletic relating to a taxonomic group that includes some but not all of the descendants of a common ancestor

paraxial on either side of the axis

parietal chloroplast a chloroplast that lies against the wall of the cell

pellicle the outer surface of euglenoids consisting of the plasma membrane and a proteinaceous layer subtended by microtubules and endoplasmic reticulum

pellicle striations (strips) proteinaceous strips lying beneath the plasma membrane

phagotrophic holozoic, endocytic, phagocytic, or engulfing food particles in nutrition

phagotrophy a form of nutrition involving the engulfment of particles of food

photoautotrophy a form of nutrition involving the formation of organic compounds using the energy from light harvested by photosynthetic pigments

photoreceptor apparatus the entire light recognition complex consisting of the stigma (eyespot) and photoreceptor in flagellate algae cells; the photoreceptor contains a light-sensitive pigment and is situated in a flagellar swelling (Euglenophyta, Heterokontophyta)

phototaxis movement in response to light;

toward light (positive phototaxis) or away from light (negative phototaxis)

phylogeny the evolutionary origin and relatedness of organisms

phytoplankton microscopic free-living algae suspended or swimming freely in the water column

plankton a community of organisms drifting or swimming in a pelagic zone

plasma membrane (plasmalemma) cell membrane

polyphyletic refers to a group in which the last common ancestor is not a member

protist a unicellular, eukaroytic organism belonging to the former taxonomic kingdom Protista

protoplast the contents of a cell including the cytoplasm and plasma membrane

pyrenoid a structure lying in the chloroplast, which is the center for carbon fixation

pyriform pear-shaped

reservoir the base of the flask-like invagination of the plasma membrane in euglenophytes

rhodopsin refers to groups of proteins that represent the molecular basis for a variety of light-sensing systems, such as phototaxis in flagellates and eyesight in animals

rostrate beak-like

RuBisCO an enzyme used in the Calvin cycle to catalyze the first major step of carbon fixation

saprotrophic obtains nourishment from dead or decaying organic matter (saprotroph, saprotrophy)

scabrous having a rough surface

scrobiculate furrowed or pitted

serrate notched on the edge like a saw

spatulate shaped like a spatula; rounded more or less like a spoon

subapical located below or near the apex

subglobose not quite globose

subspherical not quite spherical

systematics the study and classification of organisms with the goal of reconstructing their evolutionary histories and relationships

taxon the general term for any systematic category, e.g., species, genus, family, etc.

taxonomy the classification of organisms in an ordered system

telophase the final stage of mitosis or meiosis during which the chromosomes of daughter cells are grouped in new nuclei

thylakoids the photosynthetic membranes of the chloroplasts

truncate terminating abruptly as if having an end or point cut off

unialgal culture a culture containing one species or strain of alga

urceolate shaped like an urn

vacuole a large fluid-filled cavity within the cell surrounded by a membrane

verrucose a surface bearing wart-like projections

water bloom a massive growth of phytoplankton, visible to the naked eye, in which the water becomes noticeably colored due to the high concentration of algal cells present (often red or green)

xanthophylls yellowish pigments belonging to the carotenoid group

REFERENCES

Allorge, P., and M. Lefèvre. 1925. Algues de Sologne. Fascicule spéciale de la session extraordinaire tenue dans la Sologne en Juillet. *Bull Soc Bot France* 72:122–50.

Blochmann, F. 1895. Die microscopische Pflanzen- und Thierwelt des Süsswassers. Theil II. Abteilung I: Protozoa. Hamburg: Lucas, Gräfe and Sillem, 50–57.

Bourrelly, P., and M. Chadefaud. 1948. Muséum National d'Histoire Naturelle [Paris], Laboratoire de Cryptogamie, catalogues des collections vivantes, herbiers et documents. I. Algothèque, 9.

Braslavska-Spektorova, E.P. 1937. Pro novij vid bezdzhgutikovikh evglen. *Zhurn Inst Bot Acad Sci.* R.S.S. d'Ukraine 11(19):91–99.

Brosnan, S., P.J. Brown, M.A. Farmer, and R.E. Triemer. 2005. Morphological separation of the euglenoid genera *Trachelomonas* and *Strombomonas* based on lorica development and posterior strip reduction. *J Phycol* 41:590–605.

Busse, I., and A. Preisfeld. 2003. Systematics of primary osmotrophic euglenids: a molecular approach to the phylogeny of *Distigma* and *Astasia* (Euglenozoa), part 2. *Int J Syst Evol Microbiol* 53:617–24.

Bütschli, O. 1883–1887. Mastigophora in Bronn's Klassen u. Ordnungen des Thierreichs. Leipzig: Winter's Verlag, :617–1097.

Carter, H.J. 1856. Notes on the freshwater Infusoria on the Island of Bombay. I. Organization. *Ann Mag Nat Hist* Series 2, 18:115–32, 221–48.

Cavalier-Smith, T. 1993. Kingdom Protozoa and its 18 phyla. *Morph Rev* 57:953–94.

———. 1998. A revised six-kingdom system of life. *Biol Rev Camb Philos Soc* 73:203–66.

———. 1999. Principles of protein and lipid targeting in secondary symbiogenesis: euglenoid, dinoflagellate, and sporozoan plastid origins and the eukaryotic family

tree. *J Euk Microbiol* 47:347–66.
Chadefaud, M. 1937. Reserches sur l'anatomie comparée des Eugléniens. *Le Botaniste* 28:85–185.
———. 1944. Une Euglène à sillon prévestibulaire ventral. *Bull Soc Bot France* 91:115–17.
Chu, S.P. 1946. Contributions to our knowledge of the genus *Euglena. Sinensia* 17:75–134.
Ciugulea, I., M.A. Nudelman, S. Brosnan, and R.E. Triemer. 2008. Phylogeny of the euglenoid loricate genera *Trachelomonas* and *Strombomonas* (Euglenophyta) inferred from nuclear SSU and LSU rDNA. *J Phycol* 44:406–18.
Conforti, V., P.L. Walne, and J.R. Dunlap. 1994. Comparative ultrastructure and elemental composition of the envelopes of *Trachelomonas* and *Strombomonas* (Euglenophyta). *Acta Protozool* 33:71–8.
Conrad, W. 1926. Reserches sur les flagellates de nos eaux saumâtres I. *Arch Protistenkd* 55:63–100.
———. 1934. Matériaux pour une monographie du genre *Lepocinclis* Perty. *Arch Protistenkd* 82:203–49.
———. 1938. Flagellates des îles de la Sonde (Euglénacées). *Bull Mus Roy Hist Nat Belg* 14:1–20.
Da Cunha, A.M. 1913. Contribuiçao para o conhecimento da fauna de Protozoarios do Brasil. *Mem Inst Oswaldo Cruz* 5:101–22.
———. 1914. Contribuição para o conhecimento da fauna de Protozoarios do Brazil. II. *Mem Inst Oswaldo Cruz* 6:169–79.
Dangeard, P.A. 1901. Recherches sur les Eugléniens. *Le Botaniste* 8:97–360.
Deflandre, G. 1926. Monographie du genre *Trachelomonas* Ehr. France: Imprimerie André Lesot, Nemours, 162 pp.
———. 1927. Remarques sur la systématique du genre *Trachelomonas.* Ehr. *Bull Soc Bot France* 74:285–7, 657–65.
———. 1930. *Strombomonas,* nouveau genre d'Euglénacées (*Trachelomonas* EHR. pro parte). *Arch Protistenkd* 69:551–614.
———. 1932. Contributions à la connaissance des Flagellés libres 1. *Ann Protistol* 3:219–39.
Drezepolski, R. 1925. Przyczynek do znajomości polskich Euglenin [Supplément à la connaissance des Eugléniens de la Pologne]. *Kosmos* 50:173–257.
Dujardin, F. 1841. Histoire naturelle des Zoophytes: Infusoires. Paris: Libr. Encycl. de Roret, 684 pp.
Dunlap, J.R., P.L. Walne, and P.A. Kivic. 1986. Cytological and taxonomic studies of the Euglenales. II. Comparative microarchitecture and cytochemistry of envelopes of *Strombomonas* and *Trachelomonas. Br Phycol J* 21:399–405.
Ehrenberg, C.G. 1830. Neue Beobachtungen über blutartige Erscheinungen in Ägypten, Arabien und Sibirien, nebst einer Übersicht und Kritik der früher bekannten. *Pogg Ann Physik Chem* 94:477–514.
———. 1831. Über die Entwicklung und Lebensdauer der Infusionsthiere; nebst ferneren Beiträgen zu einer Vergleichung ihrer organischen systeme. *Physik Abh Königl Akad Wiss Berlin*:1–154.
———. 1832 [1830]. Beiträge zur Kenntniss der Organisation der Infusorien und ihrer geographischen Verbreitung, besonders in Sibirien. *Physik Abh Königl Akad Wiss Berlin*:1–88.
———. 1835 [1833]. Dritter Beitrag zur Enkenntniss grosser Organisation in der Richtung des Kleinsten Raumes. *Physik Abh Königl Akad Wiss Berlin*:145–336.
———. 1838. Die Infusionsthierchen als vollkommene Organismen. Leipzig: Verlag von Leopold Voss, 547 pp.
Evangelista, V., V. Passarelli, L. Barsanti, and P. Gualtieri. 2003. Fluorescence behavior of *Euglena* photoreceptor. *Photochem and Photobiol* 78(1):93–97.

Gibbs, S. 1978. The chloroplasts of *Euglena* may have evolved from symbiotic green algae. *Can J Bot* 56:2883–89.

———. 1981. The chloroplasts of some algal groups may have evolved from endosymbiotic eukaryotic algae. In Origins and evolution of eukaryotic intracellular organelles, ed. J. Fredrick. *Ann NY Acad Sci* 361:193–218.

Gockel, G., and W. Hachtel. 2000. Complete gene map of the plastid genome of the nonphotosynthetic euglenoid flagellate *Astasia longa. Protist* 151:347–51.

Gojdics, M. 1934. The cell morphology and division of *Euglena deses* Ehrenberg. *Trans Amer Microsc Soc* 53:299–310.

———. 1953. The genus *Euglena.* Madison: University of Wisconsin Press, 268 pp.

Hardy, A.D. 1911. On the occurrence of a red *Euglena* near Melbourne. *Victoria Naturalist* 27:215–20.

Hollande, A. 1942. Étude cytologique et biologique de quelques F1agellés libres. *Arch Zool Exp Gen* 83:1–268.

Hort, G. 1957–1958. The plagues of Egypt. Zeitschrift für die alttestamentliche Wissenschaft 69:84–103; 70:48–59.

Hortobagyi, T. 1943. Beiträge zur Kenntniss der im Boglárer Seston, Psammon und Lasion lebenden Algen des Balaton-Sees. *Arb Ungar Biol Forschungsinst* 15:75–127.

Huber-Pestalozzi, G. 1955. Das Phytoplankton des Süßwassers, Systematik und Biologie, 4. Teil: Euglenophyceen. In *Die Binnengewässer. Band 16, 4. Teil,* ed. A. Thienemann. Stuttgart: Schweitzerbart'sche Verlagsbuchhandlung, 606 pp.

Hübner, K. 1886. Euglenaceenflora von Stralsund. Program d. Stralsund, Germany: Realgymnasiums, Stralsund. K. Regierungsbuchdruckerei, 20 pp.

Iseki, M., S. Matsunaga, A. Murakami, K. Ohno, K. Shiga, K. Yoshida, M. Sugai, T. Takahashi, T. Hori, and M. Watanabe. 2002. A blue-light-activated adenylyl cyclase mediates photoavoidance in *Euglena gracilis. Nature* 415:1047–51.

Jahn, T.L., and M. Shawhan. 1942. *Phacus quinque-marginatus* sp. nov. (Protozoa, Mastigophora, Euglenoidina). *Trans Amer Microsc Soc* 61:30–32.

Johnson, L.P. 1944. *Euglena* of Iowa. *Trans Amer Microsc Soc* 63:97–135.

Karnkowska-Ishikawa, A., R. Milanowski, J. Kwiatowski, and B. Zakryś. 2010. Taxonomy of the *Phacus oscillans* (Euglenaceae) and its close relatives—balancing morphological and molecular features. *J Phycol* 46.

Kiss, J.Z., A.C. Vasconcelos, and R.E. Triemer. 1987. Structure of the euglenoid storage carbohydrate, paramylon. *Amer J Bot* 74:877–82.

Klebs, G. 1883. Über die Organisation einiger Flagellatengruppen und ihre Beziehungen zu Algen und Infusorien. *Unters Bot Inst Tübingen* 1:233–61.

Korshikov, A.A. 1941. On some new and little known flagellates. *Arch Protistenkd* 95:21–44.

Kosmala, S., M. Bereza, R. Milanowski, J. Kwiatowski, and B. Zakryś. 2007. Morphological and molecular examination of relationships and epitype establishment of *Phacus pleuronectes, Phacus orbicularis* and *Phacus hamelii. J Phycol* 43:1071–82.

Kosmala, S., A. Karnkowska, R. Milanowski, J. Kwiatowski, and B. Zakryś. 2005. Phylogenetic and taxonomic position of *Lepocinclis fusca* comb. nov. (=*Euglena fusca*) (Euglenaceae): Morphological and molecular justification. *J Phycol* 41:1258–67.

Kosmala, S., R. Milanowski, K. Brzóska, M. Pękala, J. Kwiatowski, and B. Zakryś. 2007. Phylogeny and systematics of the genus *Monomorphina* (Euglenaceae) based on morphological and molecular data. *J Phycol* 43:171–85.

Kusber, W.-H. 1998. A study on *Phacus smulkowskianus* (Euglenophyceae)—a rarely reported taxon found in waters of the Botanic Garden Berlin-Dahlem. *Willdenowia* 28:239–47.

Leander, B.S., and M.A. Farmer. 2000a. Comparative morphology of the euglenid pellicle. I. Patterns of strips and pores. *J Eukaryot Microbiol* 47:469–79.

———. 2000b. Epibiotic bacteria and a novel pattern of strip reduction on the pellicle of *Euglena helicoideus* (Bernard) Lemmermann. *Europ J Protistol* 36:405–13.

———. 2001a. Comparative morphology of the euglenid pellicle. II. Diversity of strip substructure. *J Eukaryot Microbiol* 48:202–17.

———. 2001b. Evolution of *Phacus* (Euglenophyceae) as inferred from pellicle morphology and SSU rDNA. *J Phycol* 37:143–59.

Leedale, G.F. 1967. Euglenoid flagellates. Englewood Cliffs, NJ: Prentice-Hall, 242 pp.

———. 1978. Phylogenetic criteria in euglenoid flagellates. *BioSystems* 10:183–87

Lefèvre, M. 1933. Contribution à la connaissance des flagelles d'Indochine. *Ann Cryptogam Exotique* 6:258–64.

Lemmermann, E. 1901. Beiträge zur Kenntniss der Planktonalgen, XII: Notizen über einige Schwebealgen. *Ber Deutsch Bot Ges* 19:85–95.

———. 1905. Brandenburgische A1gen III. *Neue Forschungsber Biol Station Ploen* 12:145–68.

———. 1910. Kryptogamenflora der Mark Brandenburg und angrenzender Gebiete (herausgegeben von dem Bot. Ver. der Prov. Brandenburg). Algen I (Schizophyceen, Flagellaten, Peridineen). Eugleninen, vol. 3. Leipzig: Borntraeger, 484–562.

———. 1913, Euglenineae. In Die Süsswasserflora Deutschlands Osterreichs und der Schweiz, ed. A. Pascher. Jena: Verlag von Gustav Fischer, 115–74.

Linton, E.W., D. Hittner, C. Lewandowski, T. Auld, and R.E. Triemer. 1999. A molecular study of euglenoid phylogeny using small subunit rDNA. *J Euk Microbiol* 46:217–23.

Linton, E.W., A. Karnkowska-Ishikawa, J.-I. Kim, W. Shin, M. Bennett, J. Kwiatowski, B. Zakryś, and R.E. Triemer. 2010. Reconstructing euglenoid evolutionary relationships using three genes: Nuclear SSU and LSU, and chloroplast 16S rDNA sequences and the description of *Euglenaria* gen. nov. (Euglenophyta). *Protist*, 2010.

Linton, E.W., M.A. Nudelman, V. Conforti, and R.E. Triemer. 2000. A molecular analysis of the euglenophytes using SSU rDNA. *J Phycol* 36:740–46.

Mainx, F. 1926. Einige neue Vertreter der Gattung *Euglena* Ehr. *Arch Protistenkd* 54:150–60.

———. 1927. Beiträge zur Morphologie und Physiologie der Eugleninen. I. Teil. Morphologische Beobachtungen, Methoden und Erfolge der Reinkultur. II. Teil. Untersuchungen über die Ernährungs- und Reizphysiologie. *Arch Protistenkd* 60:305–414.

Marchessault, R.H., and Y. Deslandes. 1979. Fine structure of (1→3)-β-D-glucans: Curdlan and paramylon. *Carbohidr Res* 75:231–42.

Marin, B., A. Palm, M. Klingberg, and M. Melkonian. 2003. Phylogeny and taxonomic revision of plastid-containing euglenophytes based on SSU rDNA sequence comparison and synapomorphic signatures in the SSU rDNA secondary structure. *Protist* 154:99–145.

Matvienko, O.M. 1938. Materiali do vivtchennia vodorostii. *Trudy N D Inst Bot Kharkov* 3:29–70.

McFadden, G.I. 2001. Primary and secondary endosymbiosis and the origin of plastids. *J Phycol* 37:951–59.

Mereschkowsky, K.S. 1877. Etjudy nad prostejsimi zivotnymi severa Rossii. Trudy S-Peterburgsk. *Obshch Estestvoisp* 8:1–299.

Montegut-Felkner, A., and R.E. Triemer. 1997.

Phylogenetic relationships of selected euglenoid genera based on morphological and molecular data. *J Phycol* 33:512–19.

Moroff, T. 1904. Beitrag zur Kenntnis einiger Flagellaten. *Arch Protistenkd* 3:96–103.

Müller, O.F. 1786. Animalcula infusoria fluviatilia et marina Havniae et Lipsiae. Copenhagen, 367 pp.

Müllner, A.N., D.G. Angeler, R. Samuel, E.W. Linton, and R.E. Triemer. 2001. Phylogenetic analysis of phagotrophic, phototrophic and osmotrophic euglenoids by using the nuclear 18S rDNA sequence. *Int J Syst Evol Microbiol* 51:783–91.

Nudelman, M.A., P.I. Leonardi, V. Conforti, and R.E. Triemer. 2006. Fine Structure and Taxonomy of *Monomorphina aenigmatica* comb. nov. (Euglenophyta). *J Phycol* 42:194–202.

Nygaard, G. 1949. Hydrobiological studies on some Danish ponds and lakes II. *Det Kong Dansk Vid Selsk Biol Skr* 7:1–293.

Osorio-Tafall, B.F. 1942. Estudios sobre el plancton de Mexico II. El género *Trachelomonas* Ehrenberg con descripción de nuevas especies. *Ciencia* 3:249–54.

Palmer, C.M. 1980. Algae and water pollution. UK: Castle House, 123 pp.

Palmer, T.C. 1905. Delaware Valley forms of *Trachelomonas. Proc Acad Nat Sci Philad* 57:665–75.

———. 1925. *Trachelomonas* new or notable species and varieties. *Proc Acad Nat Sci Philad* 77:15–22.

Perty, M. 1852. Zur Kenntnis kleinster Lebensformen nach Bau, Funktionen, Systematik, mit Spezialverzeichniss der in der Schweiz beobachteten Arten. Jent und Reinert (Bern), i–viii, 1–228.

Playfair, G.I. 1915. The genus *Trachelomonas. Proc Linn Soc N S Wales* 40:1–41.

———. 1921. Australian freshwater flagellates. *Proc Linn Soc N S Wales* 46:99–146.

Pochmann, A. 1942. Synopsis der Gattung *Phacus. Arch Protistenkd* 95:81–252.

Popova, T.G. 1955. Evglenovyje vodorosli. Opredelitel presnovodnych vodoroslej SSSR. 7. *Sov Nauka*, 282 pp.

Preisfeld, A., S. Berger, I. Busse, S. Liller, and H.G. Ruppel. 2000. Phylogenetic analyses of various euglenoid taxa (Euglenozoa) based on 18S rDNA sequence data. *J Phycol* 36:220–26.

Preisfeld, A., I. Busse, M. Klingberg, S. Talke, and H.G. Ruppel. 2001. Phylogenetic position and inter-relationships of the osmotrophic euglenids based on SSU rDNA data, with emphasis on the Rhabdomonadales (Euglenozoa). *Int J Syst Evol Microbiol* 51:751–58.

Pringsheim, E.G. 1956. Contributions toward a monograph of the genus *Euglena. Nova Acta Leopoldina* 18:1–168.

Provasoli, L. 1958. Nutrition and ecology of protozoa and algae. *Ann Rev Microbiol* 12:279–308.

Provasoli, L., and I.J. Pintner. 1953. Ecological implications of *in vitro* nutritional requirements of algal flagellates. *Ann NY Acad Sci* 56:839–51.

Rosowski, J.R. 2003. Photosynthetic euglenoids. In Freshwater algae of North America: Ecology and classification, ed. J.D. Wehr and R.G. Sheath. Academic Press, 383–422.

Schmarda, S.K. 1846. Kleine Beitraege zur Naturgeschichte der Infusorien. Wien: Verlag der Carl Haas'schen Buchhandlung.

Schmitz, F. 1884. Beiträge zur Kenntnis der Chromatophoren. *Pringsh Jb Wiss Bot* 15:1–177.

Senn, G. 1900. Euglenineae. In Engler-Prantl. Leipzig: Die natürlichen Pflanzenfamilien, 1:173–85.

Shin, W., S. Brosnan, and R.E. Triemer. 2002. Are cytoplasmic pockets (MTR/pocket) present in all photosynthetic euglenoid genera? *J Phycol* 38:790–99.

Shin, W., and R.E. Triemer. 2004. Phylogenetic analysis of the genus *Euglena* (Euglenophyceae) with particular reference to the type species *Euglena viridis*. *J Phycol* 40:226–35.

Singh, K.P. 1956. Studies in the genus *Trachelomonas*. I. Description of six organisms in cultivation. *Amer J Bot* 43:258–66.

Sittenfeld, A., M. Mora, J.M. Ortega, F. Albertazzi, A. Cordero, M. Roncel, E. Sánchez, M. Vargas, M. Fernández, J. Weckesser, and A. Serrrano. 2002. Characterization of a photosynthetic *Euglena* strain isolated from an acidic hot mud pool of a volcanic area of Costa Rica. *FEMS Microbiol Ecol* 42:151–61.

Skuja, H. 1932. Beitrag zur Algenflora Lettlands I. *Acta Horti Bot Univ Latv* 7:25–86.

———. 1948. Taxonomie des Phytoplanktons einiger Seen in Uppland, Schweden. *Symbolae. Bot Upsaliensis* 9:1–399.

Skvortzov, B.W. 1925. Über neue und wenig bekannte Formen der Euglenaceengattung *Trachelomonas* Ehr. *Ber Deutsch Bot Ges* 43:306–15.

———. 1928. Die Euglenaceengattung *Phacus* Dujardin. Eine systematische Übersicht. *Ber Deutsch Bot Ges* 46:105–25.

Sogin, M.L., and J.H. Gunderson. 1987. Structural diversity of eukaryotic small subunit ribosomal RNAs: Evolutionary implications. *Ann NY Acad Sci* 573:125–39.

Stein, F.R. 1878. Der Organismus der Infusionsthiere. III. Abt. Der Organismus der Flagellaten. 1. Germany: Verlag von Wilhelm Engelmann Leipzig, 154 pp.

Stokes, A.C. 1885. Notices on fresh-water infusoria. IV. *Amer Monthly Microscopical J* 6:183–90.

———. 1887. Notices of new freshwater Infusoria. *Proc Amer Philos Soc Philad* 24:244–55.

———. 1890. Notices of new freshwater Infusoria. *Proc Amer Philos Soc Philad* 28:74–80

Swirenko, D.O. 1914. Zur Kenntnis der russischen Algenflora l. Die Euglenaceengattung *Trachelomonas*. *Arch Hydrob Planktonk* 9:630–47.

———. 1915. Matérial pour servir à l'étude des algues de la Russie. Études systématique et géographique des Euglénacées. *Trav Inst Bot Univ Kharkoff* 26:1–84.

Tell, G., and V. Conforti. 1984. Ultrastructura de la loriga de cuatro especies de *Strombomonas* Defl. (Euglenophyta) en MEB. Nova Hedwigia 40:123–31.

Triemer, R.E., and M.A. Farmer. 1991a. An ultrastructural comparison of the mitotic apparatus, feeding apparatus, flagellar apparatus and cytoskeleton in euglenoids and kinetoplastids. *Protoplasma* 164:91–104.

———. 1991b. Ultrastructural organization of the heterotrophic euglenids and its evolutionary implications. In The biology of free-living heterotrophic flagellates, ed. D.J. Patterson and J. Larsen. New York: Oxford University Press, 185–204.

———. 2007. A decade of euglenoid molecular phylogenetics. In Unravelling the algae: The past, present and future of algal systematics, ed. J. Brodie and J. Lewis. Systematics Association Series, CRC Press, 315–30.

Triemer, R.E., E. Linton, W. Shin, A. Nudelman, A. Monfils, and M. Bennett. 2006. Phylogeny of the Euglenales based upon combined SSU and LSU rDNA sequence comparisions and description of *Discoplastis* gen. nov. (Euglenophyta). *J Phycol* 42:731–40.

Van Leeuwenhoek, A. 1674. The collected letters of Antoni van Leeuwenhoek. The complete works of van Leeuwenhoek, issued and annotated under the auspices of the Leeuwenhoek-Commission of the Royal Netherlands Academy of Arts and Sciences. Amsterdam, 1939–present.

Walne, P.L., V. Passarelli, P. Lenzi, L. Barsanti, and P. Gualtieri. 1998. Rhodopsin: A photopigment for phototaxis in *Euglena gracilis. Crit Rev Plant Sci* 17:559–74.

Wehrle, E. 1939. Zur Kenntnis der Algen im Naturschutzgebiet Weingartener Moor bei Karlsruhe a. Rh. Beitr. naturkundl Forsch Südwestdeutschl Band IV (Heft 1):1–84.

Woloszyńska, J. 1912. Das Phytoplankton einiger javanischer Seen mit Berucksichtigung des Sawa-Planktons. *Bull Int Acad Sci Cracovie* 1:649–709.

Wołowski, K., and F. Hindák. 2005. Atlas of Euglenophytes. VEDA, 136 pp.

Zakryś, B. 1986. Contribution to the monograph of Polish members of the genus *Euglena* Ehrenberg 1830. *Nova Hedwigia* 42(2–4):491–540.

Zakryś, B., and P.L. Walne. 1994. Floristic, taxonomic and phytogeographic studies of green Euglenophyta from the Southeastern United States, with emphasis on new and rare species. *Algological Studies* 72:71–114.

———. 1998. Comparative ultrastructure of chloroplasts in the subgenus *Euglena* (Euglenophyta): Taxonomic significance. *Cryptogam Algol.* 19(1–2):3–18.

Zimba, P.V., M. Rowan, and R.E. Triemer. 2004. Identification of euglenoid algae that produce ichthyotoxin(s). *J Fish Dis* 27:115–17.

INDEX

Page numbers in italics refer to figures.

ABOUT THE AUTHORS

Ionel Ciugulea is a Visiting Research Associate in the Department of Zoology at Michigan State University.

Richard E. Triemer is a Professor and Chair of the Department of Plant Biology at Michigan State University.